MW01633732

SPACE PHYSIOLOGY

SPACE PHYSIOLOGY

Jay C. Buckey, Jr., M.D.

UNIVERSITY PRESS

2006

OXFORD
UNIVERSITY PRESS

Oxford University Press, Inc., publishes works that further
Oxford University's objective of excellence
in research, scholarship, and education.

Oxford   New York
Auckland   Cape Town   Dar es Salaam   Hong Kong   Karachi
Kuala Lumpur   Madrid   Melbourne   Mexico City   Nairobi
New Delhi   Shanghai   Taipei   Toronto

With offices in
Argentina   Austria   Brazil   Chile   Czech Republic   France   Greece
Guatemala   Hungary   Italy   Japan   Poland   Portugal   Singapore
South Korea   Switzerland   Thailand   Turkey   Ukraine   Vietnam

Published by Oxford University Press, Inc.
198 Madison Avenue, New York, New York 10016

www.oup.com

Library of Congress Cataloging-in-Publication Data
Buckey, Jay C.
Space physiology / Jay C. Buckey, Jr.
p. cm.
Includes bibliographical references.
ISBN-13 978-0-19-513725-5
ISBN 0-19-513725-6
1. Space flight—Physiological effect—Handbooks, manuals, etc.
2. Space medicine—Handbooks, manuals, etc.
[DNLM: 1. Aerospace Medicine—Handbooks. WD 701 B922s 2006]
I. Title.
RC1150.B83 2006
616.9'80214—dc22   2005013758

To Sarah

Foreword

Human space exploration to destinations beyond the moon was first envisioned during the Apollo era. Over the course of a decade the world followed the transition from suborbital missions to witnessing humans walking on the surface of the moon. The rate of technological advancement to enable the lunar missions exceeded anything previously seen in history and would be difficult to achieve even with today's resources. While the dream of sending humans to Mars after the Apollo program was not fulfilled, the experience of the Shuttle, Skylab, NASA-MIR, and International Space Station programs was critical in following the roadmap to Mars first proposed so many years ago. The initial steps on the path to Mars have been from robotic explorers. These planetary rovers have played a critical role in exploring the surface of Mars, helping to provide insight into the central question of the origins, complexity, and possible diversity of life within our solar system. Ultimately, though, joint human–robotic missions will be required to answer one of the major scientific questions of this millennium: Does life exist elsewhere in outer space?

As government agencies have acquired more experience and technical capability to support long-duration spaceflight in low earth orbit, there has been the exciting emergence of privately funded spacecraft with the potential to make low earth orbit accessible to members of the public. This capability marks an important transition in human spaceflight. With increased accessibility to low earth orbit, government agencies may now shift their focus to extending the capability for human space exploration. Developing the technical capability to send humans farther into space and support them on longer duration missions will provide numerous challenges that will be the ultimate test of the technical capabilities of the world's space-faring nations. The recent NASA vision for space exploration has clearly focused attention on the path to the Moon and Mars. It will be exciting to follow the development of the Crew Exploration Vehicle, the next generation planetary spacesuit, and habitats required to support these missions.

Developing the technical capabilities for planetary exploration is only one part of the equation for human missions. Understanding the long-term physiological adaptation of humans living for months in microgravitational and partial gravitational environments will be critical in minimizing the health consequences of these missions and safely returning the exploration crews to Earth. The transition from the force of gravity on Earth to 0 G while traveling to a planetary destination will be followed by a period of living with a different planetary gravitational force (in the case of Mars, approximately one-third of the Earth's gravity). This process will be reversed when the crew returns to Earth. These transitions will cause significant adaptive changes in a number of physiological systems that may make it more difficult for the crew to function normally after landing on the planetary surface or back on the Earth.

This book provides an excellent overview of the significant biomedical issues associated with sending humans to explore other planets. The chapters discuss all of the relevant physiological changes associated with astronauts adapting to living in 0 G, and recommendations have been made to minimize the deleterious aspects of these changes and optimize the capability of astronauts to quickly transition to performing work after arriving at their destination. Although these recommendations may change as new knowledge and technical capabilities emerge in the future, they focus our current efforts to provide the appropriate support for crews on long-duration missions.

The *Spirit* rover on the surface of Mars took the first picture of the Earth as seen from another planet. The image of the Earth appears as a pale, likely blue, dot on the Martian horizon, making one wonder about the profound behavioral challenges associated with human planetary exploration. The psychosocial issues of long-duration flight are an integral component of this text, and the critical behavioral issues are discussed in detail.

As we plan for human space exploration beyond low earth orbit, my dream can be summarized in the simple statement: "by the fiftieth." In other words, it is a dream to have humans walking again on the surface of the Moon, and possibly on the surface of Mars, by the fiftieth anniversary of the *Apollo 11* lunar landing in 2019. Let us all strive to make this dream reality.

Dave Williams
NASA Johnson Space Center
Houston, Texas

Preface

The possibility of sending people to Mars no longer resides solely in the realm of science fiction. Advances in propulsion, power systems, and materials make a Mars mission an extremely challenging, but potentially achievable, goal. One critical part of a Mars mission, or of any long-duration spaceflight, is keeping the crew healthy and safe. This book outlines the issues that must be addressed to make long-duration space travel successful and provides suggestions on the approaches to use.

Mars has long held a special fascination. At the turn of the twentieth century, many thought Mars harbored an old and dying civilization. They speculated that the Martians had resorted to building canals to irrigate their planet and stay alive. This notion, based on astronomer Percival Lowell's belief that he observed canals on Mars, was eventually thoroughly discredited. But there may have been a grain of truth in it. Mars may once have harbored life, although of a very simple kind. Knowledge about where life can exist on Earth has greatly expanded our willingness to believe that life may exist elsewhere in the universe. On Earth, life has been found in rocks a kilometer below the surface, in hot springs, in salt flats, and in the cooling systems of nuclear power plants. The fact that life can arise and survive in so many hostile environments makes it possible to consider that life may exist, or have once existed, on Mars. This makes Mars exploration compelling. Finding life on another planet could show that life on Earth is not unique and that life may exist in many places throughout the universe.

Travel to Mars, however, takes a long time. Although there are many possible mission designs and scenarios, the most common ones project a total trip time of 2.5–3 years. With current technology, the Mars trip needs to take place when the orbits of Earth and Mars are in the proper arrangement to allow for a low energy transfer between orbits. At these times, getting to Mars takes approximately 6 months. To make the return voyage, however, the Mars crew would need to wait until Mars and Earth were back in the right alignment to make a low energy return trip. This would mean a stay on the Martian surface of around a year and a half. The trip back would require approximately another 6 months. Overall, a mission to Mars would be more like the 3-year voyages of Ferdinand Magellan and Captain Cook than the 2-week missions to the Moon during the Apollo program.

A mission of this variety and length presents a series of medical and physiological obstacles. On the trips to and from Mars, the crew would have to deal with the effects of weightlessness on the human body, such as bone loss and muscle atrophy. The crew would need to do spacewalks during the journey, and once on Mars, they would need to work outside in spacesuits. This could expose the crew to the risk of decompression sickness and would require that they be in good physical condition. A variety of different medical and surgical problems could arise at any point. The isolated crew would need superior psychological functioning and support to maintain a high level of individual and team performance. A variety of other problems, such as changes in the balance, cardiovascular, and immunological systems also would need to be understood and addressed.

This book is designed to enable space crews to live and work effectively in space and to provide guidance for the wide community of scientists, physicians, and engineers who support space crews. The objective of this book is to provide a practical handbook and reference to enable flight surgeons, astronauts, and their support teams to make informed decisions about medical care and physiological maintenance. The 12 chapters cover the main medical and physiological issues of spaceflight. Within each chapter, the book provides relevant background material on the area, followed by a summary of space flight experience. The chapters examine both the physiological effects of spaceflight and their clinical treatments. The chapters close with a series of recommendations based on current knowledge. The overall focus of this book is on practical problems and their solutions, and it is not designed to be a comprehensive review of all the physiology in a particular area.

One feature of the book is that each chapter provides specific recommendations. This approach has drawbacks. Sometimes recommendations will be based on incomplete knowledge, so they may suggest that more is known about a particular topic than is actually the case. Other times, the recommendation may choose a side on a controversial issue where the facts may not be completely clear. Also, recommendations may have to change over time as new knowledge becomes available. Nevertheless, this approach does require that conclusions be drawn from the physiological and medical data currently available. This often can be more useful than a simple presentation of the facts. The reader should be aware, however, that the recommendations represent just one perspective. Others may view the same data and arrive at different conclusions.

A human mission to Mars would be extremely difficult. The technological issues are daunting. Inadequate attention to the physiological and medical aspects could cause it to fail. On the other hand, it is possible. The aim of this book is to help surmount the physiological and medical problems so that a mission to Mars could succeed.

Jay C. Buckey, Jr.
Hanover, New Hampshire

Acknowledgments

I was fortunate to have advice and assistance from several knowledgeable and helpful people in preparing this book. Discussions with my crewmate Jim Pawelczyk shaped the book at the outset of the project. A review by my crewmate Dave Williams provided a needed and valuable check at the end.

Jacque Havelka and the entire staff at the Life Sciences Data Archive were extremely helpful in gathering data for the book. Elizabeth Barry, David Goldfarb, and Heinz Valtin gave me key information on calcium metabolism and acid-base balance. My knowledge of the effects of radiation on the body was expanded by several people including Paul Todd, John Weaver, Harold Swartz, and Marc Kenton. In the psychosocial area I was fortunate to have feedback from Al Holland and Jim Carter. Brad Arrick guided me to important new findings in the area of cancer genetics.

I would like to thank William Scavone from Kestrel Illustrations for his ability to distill key points into simple illustrations. Virginia Wedell was instrumental in finding and summarizing information about the psychosocial stresses of space flight. Joan Austin deserves great thanks for all her work in keeping the piles of references and documents needed for book organized.

My editor, Jeffrey House, deserves thanks for his patience, faith in the project, and insightful suggestions throughout. His encouragement was greatly appreciated and kept the book moving forward. Carrie Pedersen's, Matthew Winfield's, and Karla Pace's careful guidance helped me through the publication process.

The whole project would not have been possible without the loving encouragement I received from my family. I would like to thank Jay, Alexandra, and Jessica for their patience with a father who spent many hours in front of a laptop. But most of all I would like to thank my wife Sarah for her unfailing support. She provided essential encouragement and advice throughout. I cannot thank her enough.

Contents

SPACE PHYSIOLOGY

1

Bone Loss: Managing Calcium and Bone Loss in Space

Introduction

November 11, 1982: aboard the *Salyut 7* space station, Valentin Lebedev finds his crewmate Anatoli Berezovoy writhing in pain. The pain is on the left side of his abdomen. Both the crew and mission control are concerned, but fortunately the symptoms resolve on their own, and the mission moves forward [1]. In retrospect, the leading possible cause of the pain was a kidney stone. Although the details are not known, one explanation is that unloading of the skeleton led to calcium leaving the bones. This calcium entered the urine, increasing the risk of kidney stone formation during the flight. Berezovoy's symptoms started when the stone, which had been growing in the kidney, broke loose and traveled down a ureter. The case highlights one of the major physiological changes facing crews in weightlessness: bone and calcium loss.

Low light levels, high ambient CO_2 concentrations, and minimal skeletal loading—all known consequences of long-duration spaceflight—can have a profound effect on the skeleton. Within a few days of entering weightlessness, urinary calcium excretion increases by 60–70%. Data from the Skylab program in the early 1970s showed that approximately 0.3% of total body calcium is lost per month while in space [2, 3]. This loss is not distributed equally throughout the skeleton. Data compiled from the Mir program show that the hip may lose greater than 1.5% of bone mass per month [4, 5]. The upper extremities show minimal or no bone loss, and bone mass in the skull may actually increase. All the data to date collected in space have been done in the setting of an active exercise countermeasure program.

Although bone is lost at a rapid rate, recovery is slow. Recent data from the Mir program in one individual showed that while 12% of bone was lost during 4.5 months in space, recovery of 6% took 1 year [6]. Follow up of the Skylab crew members 5 years after their 1- to 3-month flights suggested that not all the bone lost on the mission had been recovered [7]. In patients who recover completely or partially after spinal cord injury (where bone is lost in a similar way to spaceflight), bone is still not recovered completely after 1 year [8]. The quality of the recovered bone in these instances is not known.

These data indicate that bone loss and the accompanying risk of kidney stone formation present a significant problem for long-duration space missions and must be adequately monitored and controlled. This chapter reviews the basic physiology of bone, presents the effects of weightlessness on bone, and discusses of some potential countermeasures. The chapter concludes with recommendations based on current knowledge.

Calcium and Bone Physiology Relevant to Spaceflight

Bone serves two important functions. One is to provide the rigid structure needed for movement and activity. The other is to serve as a reservoir for calcium within the body. Three basic cell types (the osteoblast, osteoclast, and osteocyte) function within a complex regulatory network to maintain bone integrity, respond to changing skeletal loading conditions, and provide a stable level of blood calcium.

Certain aspects of calcium metabolism are particularly relevant to spaceflight. The changes that occur in response to alterations in dietary intake, lighting, ambient CO_2 concentrations, and loading are especially important for understanding the interventions needed for preserving bone during a long duration spaceflight.

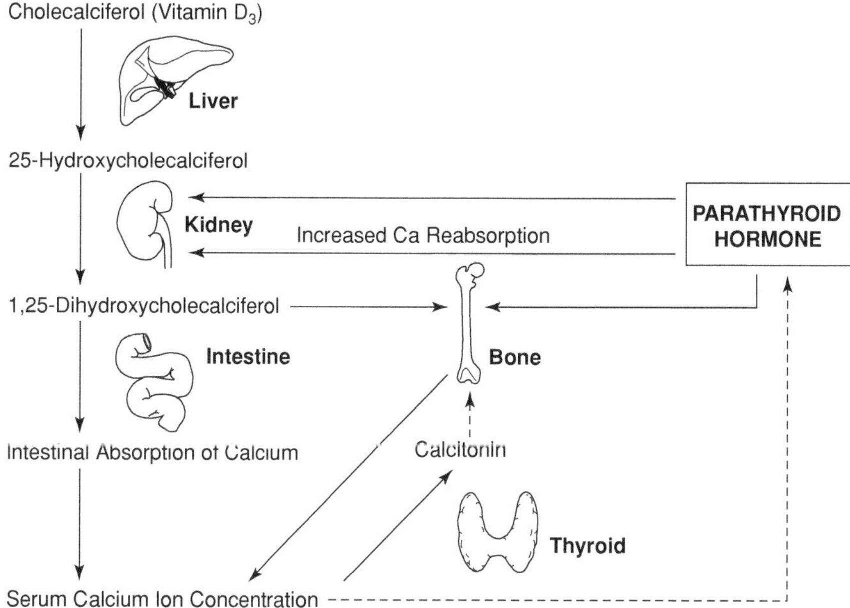

Figure 1-1. A schematic diagram of the overall regulation of calcium homeostasis by the major calcium regulatory hormones. Parathyroid hormone stimulates 1,25-dihydroxyvitamin D3 production, increases renal calcium absorption, and promotes bone resorption. These actions are suppressed when the calcium level increases. Figure reprinted from Breslau [122], with permission of Oxford University Press.

Response to a Decrease in Dietary Calcium

The recommended daily intake of calcium is 1000 mg a day [9]. If calcium intake drops below 400 mg/day, calcium deficiency results [10]. With calcium deficiency, serum calcium levels can fall and cause parathyroid hormone levels to increase. Parathyroid hormone in turn works at a variety of sites to increase calcium absorption, reduce calcium excretion, and restore blood calcium levels to the range needed for proper physiological functioning (figure 1-1).

One important action of parathyroid hormone is to increase bone resorption. Parathyroid hormone works at the osteoblast, which in turn signals the osteoclasts to increase bone resorption. Parathyroid hormone also increases the conversion of 25-hydroxyvitamin D_3 to 1,25-dihydroxyvitamin D_3. 1,25-dihydroxyvitamin D_3 in turn acts on both the intestine and the kidney. In the intestine, 1,25-dihydroxyvitamin D_3 increases dietary calcium absorption, and in the kidney calcium excretion is reduced. These changes are outlined in figure 1-1. Through a combination of these effects blood calcium levels are well maintained—but at the expense of the skeleton. A persistent low calcium intake will lead to bone loss, which underscores the importance of adequate calcium in the diet.

Response to Low Light Levels

Vitamin D plays an important role in the mineralization of the skeleton [11]. Pre-vitamin D_3 is formed in the skin by the action of ultraviolet light on 7-dehydrocholesterol. Pre-vitamin D_3 is then converted to vitamin D_3. Vitamin D_3 and vitamin D_2 (the usual form of vitamin D in dietary supplements) are jointly referred to as vitamin D. Vitamin D is further metabolized to 25-hydroxyvitamin D in the liver, and 1,25- dihydroxyvitamin D in the kidney. 1,25-dihydroxyvitamin D is the most active form of vitamin D, and it stimulates intestinal calcium absorption and bone calcium mobilization.

Without sunlight or ultraviolet light exposure (such as in spacecraft or submarines), vitamin D deficiency will occur if dietary intake is inadequate. Vitamin D deficiency can lead to poor mineralization of bone (osteomalacia), diminished calcium absorption in the intestine, decreased serum ionized calcium levels, and an increase in parathyroid hormone. The combination of these effects will weaken bone.

Response to High Ambient Carbon Dioxide Levels

Carbon dioxide levels on Earth are 0.03% of the atmosphere, but on a space station or in a submarine CO_2 levels can rise to 0.7–1%. The increased carbon dioxide levels affect acid-base balance and can have secondary effects on bone. Bone plays a role in neutralizing excess acid. The carbonates and phosphates in bone serve as buffers for acid. At the cellular level, acidosis has been shown to increase bone resorption [12]. Providing base in the form of potassium citrate can decrease bone resorption in postmenopausal women [13]. Presumably the mechanism is that the citrate helps to neutralize endogenous acid, eliminating the need for skeletal buffering.

Chronic exposure to high ambient CO_2 levels produces a compensated respiratory acidosis and contributes to the acid load that the body must neutralize. The magnitude of this effect is not clear. Drummer et al. [14] examined calcium balance in four subjects exposed to 0.7% and 1.2% carbon dioxide atmospheres. At the higher carbon dioxide level, markers of bone resorption were increased, consistent with the idea that bone calcium is being used to help neutralize the acid load. A study done on submarines, however, showed no increase in urinary hydroxyproline (a marker of bone resorption) while the crew was exposed to carbon dioxide levels of 0.8–1.0% [15]. In-vitro studies show that a respiratory acidosis has a much less profound effect on bone than does metabolic acidosis [16]. Overall, the net effect of high ambient CO_2 levels on bone is not clearly established and seems significantly less important than the effects of a metabolic acidosis. Nevertheless, the increased acid load in respiratory acidosis may decrease urinary pH and decrease urinary citrate. Both of these factors would increase the risk of kidney stones because a high urinary pH and high urinary citrate concentration help to prevent stone formation [17].

Bone Remodeling and Genetic Factors

Bone is constantly remodeled. Osteoclasts remove bone, creating a resorption area that is subsequently filled in by osteoblasts. The constant bone turnover and continual remodeling is thought to be essential for maintaining the strength and integrity of bone. Without it, the continual impacts on the skeleton could lead to the accumulation of microfractures throughout the bone, progressively producing mechanically weaker bones. Different physical and dietary factors can tip the balance in this complex interplay and enhance either the formation or loss of bone.

The baseline level of bone remodeling and the changes in remodeling that occur in response to various stimuli (such as spaceflight) can be profoundly affected by genetics. For example, Boyden et al. [18] identified a group of people with a high bone mass who all shared a mutation in the *LRP5* gene. Judex et al. [19] showed that in rats the bone response to various anabolic and catabolic stimuli is strongly influenced by the genome.

Factors that favor bone formation

Mechanical loading.   How bone senses and responds to loading is not fully understood. According to Frost [20], bone has a "mechanostat" that senses strain and maintains bone mass at an appropriate level to keep the strain within range. In this concept, when the strain within a bone exceeds a setpoint, modeling of the bone is initiated to reduce the strain back to that setpoint.

Even though the mechanism for sensing strain may not be firmly established, bone clearly changes in response to the loads placed upon it. Cross-sectional studies in athletes show that, on average, weight lifters have a greater bone mass than swimmers [21]. Gymnasts have a high bone mass, which may be attributable to the significant impact loads that their skeletons experience [22, 23]. The playing arm of tennis players has a higher bone mass than the nonplaying arm [24]. In addition, there is evidence that peak loads, rather than the frequency of loading, is important [23]. This suggests that short periods of high-impact loading may be as effective as frequent low-level loading for maintaining bone.

Gravity clearly has a role in skeletal loading. Heavier people have a higher bone mass than individuals of normal weight [25, 26]. Patients with spinal cord injuries lose considerable bone mass in the lower extremities, but can maintain bone mass in the lumbar spine, which still experiences gravitational loading in a wheelchair [27, 28]. Part of the loading on the hip and lower extremities during walking or running is the ground reaction force, generated by the gravitational acceleration of the body to the ground during locomotion. These gravitationally induced forces are key components of the loading on the hip and lower extremity [29].

Static loading and ground reaction forces are not the only loading forces lost when gravity is no longer present. Muscle contractions also play a major role in generating forces on the bone in 1 *G* [29, 30], and muscular effort is needed to work against gravity both supine and upright. During rehabilitation after hip replacement, very high hip pressures can be generated with muscular movements while supine [31], indicating that the upright posture is not necessary for creating substantial hip pressures. The absence of gravity could affect loading on the skeleton both through a loss of ground reaction forces and also due to a marked reduction in forces needed to move the weightless limbs.

Hormonal factors.   Sex steroids play an important role in bone biology. Low testosterone levels lead to osteoporosis in men, and replacing testosterone can help to restore bone mass. Also, increasing testosterone levels in men who are not androgen deficient can increase bone mass further, suggesting that testosterone can be used to increase bone formation [32]. Estrogen deficiency is the critical factor in the increased bone turnover and bone loss produced after menopause.

Growth hormone has been shown to increase bone mass [33]. When given to young men, growth hormone increased bone mass, but the results with older individuals have been disappointing [34]. Insulinlike growth factor-1 (IGF-1) has also shown promise as an agent that can increase bone mass [35]. IGF-1 was given to rats that flew for 10 days on the Space Shuttle and increased bone formation in the humerus. Whether the same effect would be seen in humans, however, is not known.

Parathyroid hormone has complex effects on the skeleton. A chronic increase in parathyroid hormone levels, such as occurs with secondary hyperparathyroidism from reduced calcium intake or vitamin D deficiency, will lead to increased bone resorption and a reduction in bone mass. Parathyroid hormone, however, can also have anabolic effects on bone. Cyclic administration of parathyroid hormone can increase bone mass. [36, 37]. Studies of postmenopausal osteoporosis show that the cyclic administration of parathyroid hormone can reduce fractures and improve bone mass [38, 39]. Animal studies have shown an increased risk of osteosarcoma during high-dose parthyroid hormone treatment.

Calcitonin is used to treat postmenopausal osteoporosis. It has not proven to be effective at preventing bone loss in immobilization in either animals [40] or humans [41].

Dietary factors.   Calcium and vitamin D are two critical dietary factors for bone formation that have been discussed above. Other dietary factors can also be important. Phytoestrogens are a family of plant compounds that have a variety of estrogenic and antiestrogenic properties. Isoflavones are one class of phytoestrogens that may be effective in preserving bone mass [42]. In one study, postmenopausal women who consumed a diet high in soy protein (which is a good source of isoflavones) showed a significant increase in lumbar spine bone density [43]. Ipriflavone, a synthetic isoflavone, has also shown effectiveness against postmenopausal osteoporosis [44]. Whether isoflavones would be effective in the bone loss caused by immobilization or weightlessness is unknown. The physiology of postmenopausal bone loss differs substantially from that of immobilization or weightlessness.

Vitamin K is another nutritional factor that can affect bone health [45]. Available evidence suggests that vitamin K is important in osteocalcin metabolism, and osteocalcin seems to play an important role in bone formation. Vitamin K_2 may also affect bone through the osteoprotegerin/osteoprotegerin ligand system (discussed later in this chapter). In rat studies, vitamin K_2 prevented bone resorption after ovariectomy, reduced the increase in bone turnover after orchiectomy, ameliorated the increase in bone resorption after sciatic neurectomy, and prevented the decrease in bone formation after glucocorticoid treatment. Clinically, vitamin K sustains lumbar bone mineral density (BMD) and prevents fractures in patients with age-related osteoporosis, although data are limited. There is some evidence that taking vitamin K in space may help to reduce bone turnover [46, 47].

Physical factors.   Bone is sensitive to electric fields. The application of low frequency, low intensity electric fields to marrow culture inhibited the recruitment of osteoclasts [48]. Applied electric fields are used in clinical medicine to assist with fracture healing.

Low magnitude (0.25 G, producing < 5 microstrain), high frequency (30–90 Hz) mechanical signals (vibration) have been shown to inhibit bone loss and increase bone formation [49, 50]. Vibration has been shown to be effective in increasing bone mass in a rat model of postmenopausal osteoporosis [49]. Rubin et al. [50] mechanically stimulated the hindlimbs of adult sheep every day for a year with 20-minute bursts of very low magnitude, high frequency vibration. This increased trabecular bone density in the proximal femur by 34% compared to controls [50]. Human studies have also shown promise [51, 52].

Factors that favor bone resorption

Immobilization.   Reduced bone loading occurs on Earth though prolonged bed rest, immobilization, or paralysis. Animal models of immobilization show reduced bone formation, increased resorption, and loss of bone down to a plateau [53]. In some animal studies up to 60% of trabecular bone mass is lost before bone mass stabilizes at a lower level. Studies of patients with spinal cord injuries show that approximately 30–50% of lower extremity bone mass can be lost before reaching a plateau [28, 54]. The plateau is reached approximately 16 months after spinal cord injury.

Ischemia-reperfusion.   Bone perfusion can have a marked effect on bone turnover. Stress fractures can result from overuse of bone. Although it might be expected that bone loading would increase bone mass, in some settings excessive bone loading can create transient bone ischemia. The subsequent increase in blood flow to the ischemic area once the activity stops can markedly increase bone turnover, with a net loss of bone. The area weakens rather than strengthens due to overuse, and stress fractures can result [55].

Hormones.   Chronically increased parathyroid hormone levels increase bone resorption and are a cause of osteoporosis. Low calcium intake and vitamin D levels can increase parathyroid hormone levels. As discussed previously, however, parathyroid hormone can also be anabolic when given intermittently. Excess thyroid hormone is a cause of osteoporosis, but is unlikely to be a factor in spaceflight. Similarly, glucocorticoid excess markedly reduces bone formation and produces osteoporosis but is unlikely to occur in space.

Dietary factors.   A high sodium intake increases urinary sodium excretion, which in turn increases urinary calcium excretion. The increased calcium excretion could aggravate ongoing bone loss, and there is some evidence that a high sodium intake can reduce bone mass in the hip [56]. Typically, sodium intake has been high on Space Shuttle flights (> 4 g/day) [57].

A diet high in protein provides an acid load that may increase the need for skeletal buffering and increase bone loss [13].

Cellular control of bone mass

The ability of bone to sense and respond to changing loading conditions is regulated by a complex series of cytokines and receptors on the osteoblast, osteocyte, and osteoclast. Understanding the action of these factors is critical for designing a rational countermeasure program.

Sensing mechanical loads.   One critical ability for modifying bone mass to accommodate different loading conditions is the ability to sense and respond to mechanical strain. How the bone converts a change in strain into a biological signal is not understood. Whether it is mechanical strain, piezoelectricity, or intraosseous fluid pressures that form the main stimulus to bone is not known. This has some relevance for countermeasure design because exercise countermeasures can increase stress and strain on bone but cannot reproduce the hydrostatic gradients that existed in the bone on Earth.

Although the exact stimulus may not be known, recent evidence suggests that the response to increased strain may work at the cellular level through osteoprotegerin, which can markedly reduce the activity and activation of osteoclasts [58] (see below).

Regulating bone resorption.   The discovery of the cytokine osteoprotegerin (also known as osteoclastogenesis inhibitory factor) and its receptor osteoprotegerin ligand (also known as osteoclast differentiation factor or RANK ligand) has led to a better understanding of how bone resorption is controlled at the cellular level [59, 60]. Fig-

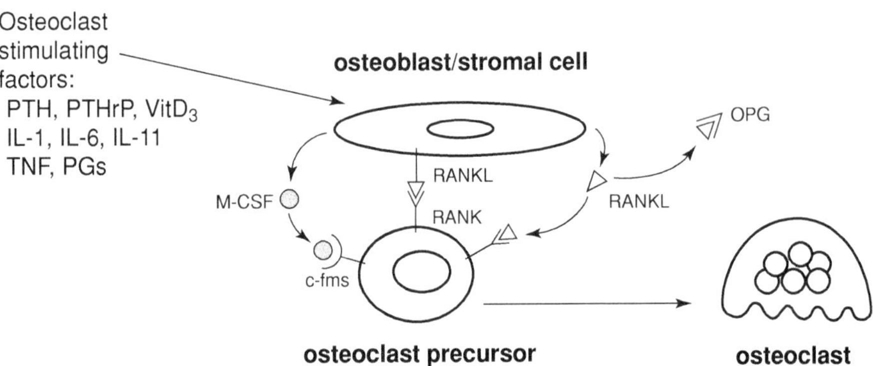

Figure 1-2. Schematic representation of the function of the osteoprotegerin/osteoprotegerin ligand system. Osteoprotegerin ligand (also known as RANKL as shown in the graph) stimulates the production of mature osteoclasts from osteoclast precursors. Osteoprotegerin (OPG) also binds to osteoprotegerin ligand (RANKL) and prevents osteoclast development. Transgenic mice that overexpress osteoprotegerin develop osteopetrosis because they do not form osteoclasts. Many of the factors that stimulate osteoclast formation seem to work through the osteoprotegerin ligand system. Figure reprinted from Boyce et al. [123], with permission of *Laboratory Investigation.*

ure 1-2 shows how this system of bone regulatory cytokines works. Osteoprotegerin ligand is the essential cytokine for stimulating the production of osteoclasts and for normal osteoclast differentiation and activation [61]. The effects of this ligand are counterbalanced by osteoprotegerin, which competes with osteoprotegerin ligand for a receptor site on the osteoclast. If osteoprotegerin is given exogenously, this inhibits osteoclast differentiation and markedly reduces bone resorption. Most of the calciotropic cytokines, peptides, and steroid hormones have been shown to regulate the expression of these two factors. These factors seem to serve as the final common effector system to regulate the formation of osteoclasts [61].

Space-related Bone Loss

The most comprehensive study of calcium metabolism in space occurred during the three manned Skylab missions [2]. The three-man crews spent 29 (*Skylab 2*), 59 (*Skylab 3*) or 84 (*Skylab 4*) days in space. Before, during, and after the flight, the crew members all took part in a metabolic balance study on the effects of weightlessness on calcium metabolism. Aspects of calcium metabolism have also been studied on various other spaceflights.

Markers of Bone Resorption and Formation

Calcium excretion rises promptly in space and remains elevated for several months, as shown in Figure 1-3. Whether calcium excretion stays elevated throughout weightlessness exposure or eventually returns to normal is not known. Limited data from a Russian long-duration flight showed no increase in calcium excretion after 218 days in space for 2 people [62]. It is not known whether this indicates that calcium excretion returns to normal eventually in space or that the countermeasure program was effective in minimizing bone loss. Experience from bed-rest studies and from studies on spinal cord injury suggest that the increase in calcium excretion in space should eventually taper off.

Urine samples that had been saved from the Skylab missions were examined recently using newer markers for bone resorption [63]. These data show that *N*-telopeptide, a sensitive marker for bone resorption, was increased throughout the flight. Another study also showed increased bone resorption markers in space. Caillot-Augusseau et al. [64] measured *C*-telopeptide on a 180-day *Mir* flight. *C*-telopeptide remained elevated throughout the flight. Together, these studies demonstrate that the increase in urinary calcium excretion was due to an increase in bone resorption.

Data on bone formation markers are limited. The one subject studied on a 180-day *Mir* flight showed decreased osteocalcin levels, which is a marker for bone formation [64]. Animal studies have shown that bone formation is reduced in weightlessness [65].

Location of Bone Loss

Bone mass and bone mineral content have been measured after spaceflight using a variety of techniques. Figure 1-4 provides a compilation of the data through the Mir

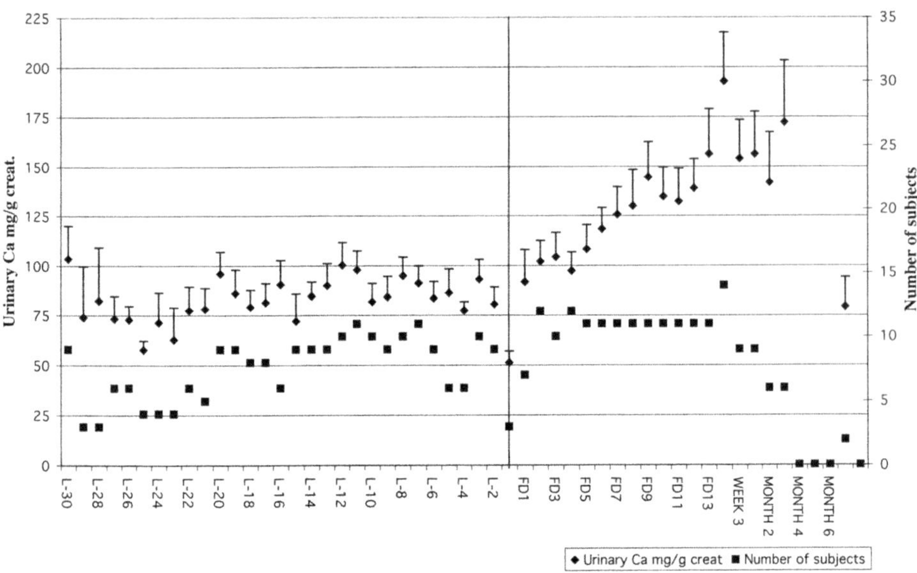

Figure 1-3. Compilation of all available urinary calcium excretion data in space done by the Life Sciences Data Archive at NASA-Johnson Space Center (www.lsda.jsc.nasa.gov). Data include *Gemini*, *Skylab*, *Salyut*, *Mir*, and Shuttle missions. L indicates the number of days before launch (e.g., L-30 is 30 days before launch); FD is flight day. Spaceflight increases calcium excretion markedly within a few days of weightlessness exposure, and calcium excretion stays high throughout the flight. Two data points taken from day 217 of a long-duration *Mir* mission show no increase in urinary calcium excretion. It is not known if this represents individual characteristics of those cosmonauts, a tapering off of a previously elevated urinary calcium level, or an effective countermeasure program.

program on changes in bone mineral after spaceflight. Data from the International Space Station are similar [66]. In weight-bearing areas such as the hip, losses of up to 1.7% per month have been documented in space. The upper extremities are spared, and evidence suggests that the skull may even gain bone mass. Taken together the data suggest the bone loss in space is concentrated in the lower spine and lower extremities, those areas most loaded by gravity on Earth. These data, however, show average results. Some individuals showed substantial reductions, while in others the changes were quite small. These large interindividual differences need to be taken into account by the countermeasure program [67].

Parathyroid Hormone and Vitamin D

Serum calcium was measured during the Skylab program and showed a small increase within the normal range. This increase might be expected to decrease parathyroid hormone, and data collected on parathyroid hormone suggest this is true. Parathyroid hormone levels decreased on the *Spacelab Life Sciences 1* and *2* flights (9 and 14 days, respectively) and on a 180-day *Mir* flight [64]. Reduced parathyroid hormone

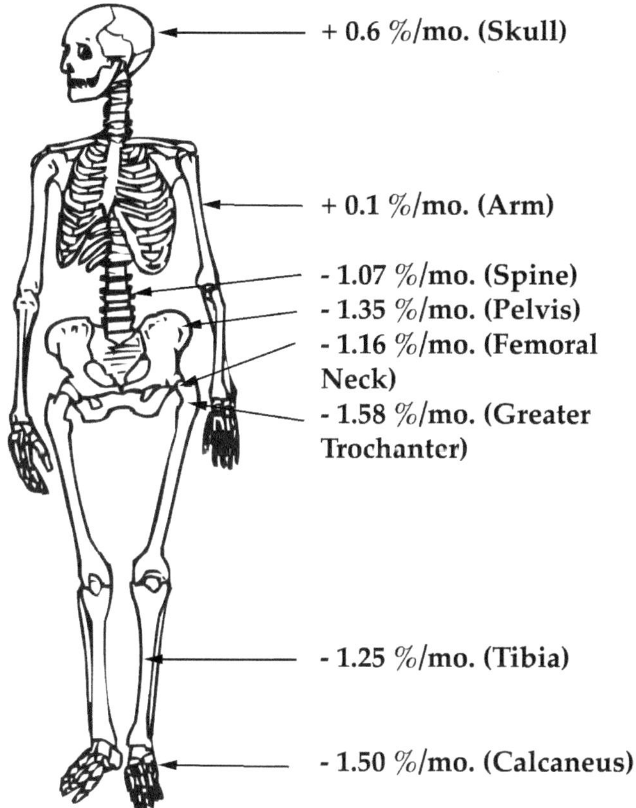

Figure 1-4. Compilation of bone loss rates from data presented in LeBlanc et al. [4] and Oganov and Schneider [83]. Bone loss is concentrated in the lower extremities; the upper extremities are unaffected, and the head may actually gain bone mass.

concentration would also reduce the production of 1,25-dihydroxyvitamin D, with an accompanying reduction in calcium absorption.

Bone Loading during Spaceflight

Astronauts and cosmonauts exercise in space using a treadmill, ergometer, and resistance exercise device. Bungees and expanders are also available. As a result of the coutermeasure program, bone loading is not zero in space, but is instead intermittent. Data collected on ground reaction forces on the foot in weightlessness show that, in general, the impact forces that crew members experience are usually below the level for similar activities on Earth. The treadmill used in space has bungees to keep the crew member in contact with the treadmill surface, but the forces generated are still less than those generated during similar activities on the ground. This indicates that, in general, even when crew members are taking part in the countermeasure program, they are not generating high loads [68–70].

Physiology of Space-related Bone Loss: Summary

The picture that emerges from the existing data on bone loss in space is an increase in bone resorption, a decrease in bone formation, bone loss concentrated in weight-bearing areas, and a suppression of parathyroid hormone. These changes are similar to those that would be seen with immobilization or spinal cord injury. This similarity is encouraging because it indicates that research in these other areas may be applicable to spaceflight and vice versa.

Countermeasures for Space-related Bone Loss

Data from immobilization studies and from patients with spinal cord injuries show that bone is difficult to regain once it is lost. In one study of patients who recovered from their spinal cord injuries, bone was not fully recovered even 1 year after the injury [8]. The data available from spaceflight suggest that this slow recovery holds for astronauts as well [6, 7]. Bone may be recovered, but it takes longer to recover than it does to lose it. Not all the lost bone may be regained.

This principle suggests that, in space, preventing bone loss is a more effective strategy than aggressive postflight rehabilitation. If a significant bone loss can be prevented, this will minimize the amount of postflight rehabilitation needed and will extend the time people can remain weightless.

Calcium and Vitamin D

The low light levels in the spacecraft mean that little vitamin D will be formed. Because parathyroid hormone is suppressed, there will be minimal conversion of the vitamin D created. Low vitamin D levels impair the absorption of calcium. To ensure that vitamin D deficiency or a low calcium diet does not aggravate the loss of calcium from bone, adequate dietary intake of vitamin D and calcium is necessary. Although 400 IU of vitamin D per day is usually adequate to maintain bone health in most individuals, in the absence of sunlight, 600–800 IU daily may be needed to prevent deficiency [71]. It is not known if administering vitamin D as the 1,25-dihydroxyvitamin D_3 (calcitriol) form would be more effective than the vitamin D_2 in dietary supplements.

In at least one bed-rest study, calcium supplementation minimized the negative calcium balance during bed rest. Approximately 1000 mg of elemental calcium (usually given as calcium carbonate, which is approximately 40% elemental calcium) would be necessary to achieve this effect and is currently recommended for space station flights up to 360 days [9]. Calcium citrate might be a superior oral supplement because it has better oral absorption [72].

One concern about calcium (and vitamin D) supplementation in a setting where there is hypercalciuria due to skeletal unloading is that the extra calcium may aggravate the hypercalciuria and increase the tendency to form kidney stones. Several lines of evidence suggest this will not be the case [73]. Despite the use of aggressive oral calcium and vitamin D supplementation to treat postmenopausal osteoporosis, there has not been an increase in kidney stone incidence in patients treated this way. One possible reason for this is that calcium may bind to oxalate in the gut and reduce oxalate absorption. Because oxalate is necessary for the formation of calcium oxalate stones, a reduced risk of kidney stone formation from reduced oxalate absorption may overcome any increase in risk from increased calcium ingestion [74]. To achieve this effect, calcium supplements during spaceflight should be taken with meals.

Exercise to Prevent Bone Loss

Exercise has been tried in bed-rest studies and is already part of existing countermeasure programs. Exercise has not fully prevented bone loss in a variety of bed-rest studies [75], and the existing data on bone loss in the space program were collected on individuals who were already participating in an exercise countermeasure program. The current exercise program for the International Space Station consists of 2.5 hours, 6 days a week with a combination of resistive and aerobic components (table 1-1).

Data from spaceflight show that bone loss occurs mainly in the lower extremities and lower back (figure 1-4). These are also the areas that receive substantial loading during daily activities on Earth (figure 1-5). Thus, any exercise countermeasure program should focus on the lower extremities and lower back to preserve bone mass. Studies that have shown an increase in bone mass with exercise suggest that impact loading may be more important than static loading [23]. Weightlifters and gymnasts show higher bone mass than runners or swimmers [21].

Hip loads

One of the key areas where bone loss needs to be prevented is the hip. Some of the greatest bone losses seen to date in the space program have been in the hips [4]. Walking and running on Earth provide substantial hip loading. Peak joint forces can range from 3–4 times body weight with walking, 5.5 times body weight with jogging, and as high as 8.7 times body weight with stumbling [76]. In weightlessness, these forces do not seem to be generated during the exercise program.

The advent of instrumented hip prostheses has allowed direct measurements of hip joint contact pressures during a variety of activities. These studies provided some

Table 1-1. Daily exercise countermeasures for long-duration Space Station flights

- 2.5 hours per day, 6 days per week, consisting of set-up and stowage (10 min each), cool-down and personal hygiene (20 min each), aerobic/anaerobic exercise (30 min each), and resistive exercise (60 min)
    - Day 1: 60 min—lower limb resistance exercise, 30 min—treadmill
    - Day 2: 60 min—lower limb resistance exercise, 30 min—treadmill
    - Day 3: 60 min—lower/upper limb resistance exercise, 30 min—cycle
    - Day 4: 60 min—lower limb resistance exercise, 30 min—treadmill
    - Day 5: 60 min—lower limb resistance exercise, 30 min—treadmill
    - Day 6: 60 min—lower/upper limb resistance exercise, 30 min—cycle
    - Day 7: Active rest
- May be accelerated or increased before entry—with increased exercise times
- No exercise 24 hours before the periodic fitness evaluation

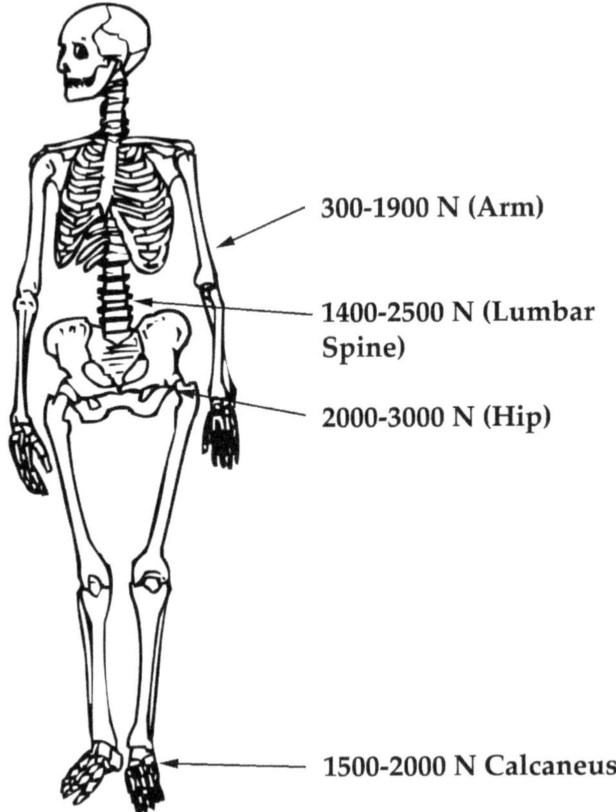

Figure 1-5. Compilation of estimated loads on the lower extremity from data presented in Lindh [78] and Hall [76]. To provide data in Newtons, a body weight of 70 kg was used. The data show that the hip, lower back, and legs are heavily loaded in 1 *G*.

surprising results. Strickland et al. [31] showed that peak pressures in the hip joint during supine isometric abduction (3.78 mPa) were as high as the pressures generated during walking (3.64 mPa) in a patient undergoing rehabilitation. Hodge et al. [77], also using an instrumented hip prosthesis, measured a peak pressure of 7.14 mPa while rising from the seated position, which exceeded the values measured during gait. The conclusion from these studies is that muscular contractions are a significant component of hip loading and can exceed the forces produced by body weight. In a study of spinal-cord-injured patients, one patient who had retained iliopsoas function and could flex the hip maintained hip bone content, which supports the importance of muscle power in maintaining bone mass [28].

These results have implications for countermeasure design for use in space. It may be that treadmill running in weightlessness (using a harness to provide loading) may be an inefficient way to maintain hip bone mass. Although treadmill running would provide frequent repetitive loads and would be effective at maintaining aerobic capacity, it may not provide the highest hip loads. It may be that short periods of high loading using abduction, adduction and squatting exercises would provide high peak hip loading pressures and provide a more effective anabolic stimulus.

Lumbar spine loads

In 1 G, loads on the third lumbar vertebrae are approximately 70 kg in a 70-kg man (686 N or 1 body weight), while standing. During relaxed sitting the loads are 75% higher. Spinal bending, flexing, or rotation also further increase the loads [78]. In weightlessness the spine is unloaded, height is gained, and some spinal muscles atrophy. The intrinsic muscles of the back show significant atrophy even after short duration spaceflights [79, 80].

In a study of patients with spinal cord injury, Biering-Sorensen et al. [27, 28] found significant losses in the lower extremities, but preservation of lumbar spine bone mass. They attributed this to the fact that the patients spent most of their time seated in wheelchairs and so had gravitational loading. As noted above, seated gravitational loads in the spine can be significant. What is not known, however, is the minimum amount of loading necessary for maintaining bone mass in the lumbar spine. Some bed-rest studies have shown that standing for a few hours a day can markedly reduce the calcium loss seen in bed rest [81, 82]. Providing even this duration of gravitational loading, however, is difficult in weightlessness. The relevant question is whether a shorter duration of a higher load can provide the same bone protection as a longer duration of a lower load. Also, the role of muscular loading in maintaining spine bone mass is not clear. Whether spinal extension exercises would help preserve both muscle and bone mass is not known, but this seems like a reasonable approach.

Femoral loads

The neck of the femur and the greater trochanter were considered above in the discussion of hip loads. These areas of the femur show significant losses in bone mass after spaceflight. The femoral shaft mainly consists of cortical bone and on average has shown minimal losses after long-duration spaceflight (-1.6% after 4–6 months in space) [83]. In spinal cord injury, loss of bone mineral content in the shaft does occur, but it takes much longer to develop than in other areas of the lower extremity [28]. Ultimately, however, bone mineral content in the shaft seems to stabilize at approximately 75% of normal values in patients with spinal cord injuries [27]. Which exercise would be most effective to prevent bone loss in the femur is not known.

Tibial loads

In 1 G, the compressive force at the junction between the femur and tibia (tibiofemoral joint) has been reported to be greater than 3 times body weight during walking and up to 4 times body weight during stair climbing [76]. The proximal portion of the tibia, which consists of a mix of trabecular and cortical bone, loses approximately 1.25% of its mass a month in space (figure 1-2). In spinal-cord-injured patients, bone is lost in the proximal tibia until the bone mass reaches 40–50% of normal values [28]. Squatting exercises under load may help prevent or ameliorate this bone loss.

Calcaneal loads

Running in 1 G provides a ground reaction force of 2–3 times body weight to the foot. Ground reaction forces are provided throughout the day in 1 G by walking, sitting, and standing. Such a loading history would be difficult to replicate in space. In spinal-cord-injured patients, calcaneus bone content decreases rapidly after injury [84]. In gymnasts, the calcaneal bone density is increased [85]. The data from gymnasts suggest that peak loading—not just the number of loading cycles—is important in increasing bone mass in the calcaneus. Impact loads will be important in space to maintain calcaneal bone mineral.

Drugs to Prevent Bone Loss

The available evidence on space-related bone loss shows that resorption is increased and formation is decreased, so drugs that either decrease resorption or increase formation could be effective. As stated earlier in the chapter, the key point from spinal cord injury recovery studies is that bone loss should be minimized at the outset of immobilization to prevent a protracted and perhaps only partial recovery. Bone mass is usually adequate at the beginning of a spaceflight, so there is no need to increase bone mass further in space. What is necessary, however, is to prevent the loss of bone that occurs when loading is removed. Several drugs may be useful in either minimizing bone resorption or increasing bone formation. A summary of potential drugs for preventing bone loss in space appears in table 1-2.

Bisphosphonates

Chemically, bisphosphonates (etidronate, alendronate, pamidronate, risedronate, tiludronate, zoledronate) resemble pyrophosphate, and they bind to the hydroxyapatite found in bone matrix. They are inhibitors of osteoclastic bone resorption. Early bisphosphonates, such as etidronate, impaired mineralization if administered daily, but newer bisphosphonates, such as alendronate and zoledronate, are much more potent inhibitors of bone resorption and do not appear to interfere with mineralization. Bisphosphonates also interfere with an intermediate step in cholesterol biosynthesis which is essential for osteoclast function [86].

Bisphosphonates have been shown to be effective in preventing bone loss during immobilization or bed rest [87–90]. When used to prevent postmenopausal osteoporosis, alendronate has shown effects on bone mass that exceed what would be expected from a pure inhibitor of resorption. Once areas that had been resorbed had been filled in, further increases in bone mass would not be expected. Alendronate, however, has been shown to continue to increase bone mass with continued use. The stimulation of nocturnal parathyroid hormone release has been presented as one potential mechanism [91]. Whether this effect would be seen during spaceflight or immobilization is not known. Also, recent evidence suggests that bisphosphonates may possess some anabolic activity by inhibiting osteoblast and osteocyte apoptosis [92].

Although the bisphosphonates are well tolerated, they have some disadvantages [93]. Oral alendronate, like other bisphosphonates, can cause local irritation of the

Table 1-2. Potential drugs for treating bone loss in space.

Drug	Dose	Advantages	Disadvantages	Reference
Alendronate	10–20 mg/day	Works in immobilization	Esophagitis, poor absorption	[90]
Zoledronate	0.5 mg IV, given once every 3 months, 2 mg IV once every 6 months, or 4 mg IV once every 12 months	Very infrequent dosing, can be given yearly	Parenteral administration, long-term effects not known	
Cyclic parathyroid hormone	200 U IV daily	Counteracts the low PTH levels in space	Limited data in immobilization, parenteral administration, rare risk of osteosarcoma	[39, 116]
Raloxifene	60 mg qd	Lowers lipids, may prevent breast/prostate cancer	Limited data in men, thrombophlebitis	
Hydrochlorothiazide	12.5–25 mg/day	Prevents kidney stones	Limited data in immobilization, may cause hypercalcemia	[104]
Potassium citrate	25–30 mEq bid	Prevents kidney stones	Limited data in immobilization	[13]
Vitamin D	600 IU/day	Avoids deficiency which could aggravate bone loss	Excessive dosing may cause hypercalcemia	[117]
Calcium	1000–1200 mg/day	Avoids deficiency which could aggravate bone loss	Excessive dosing in combination with Vitamin D may cause hypercalcemia	[41]
Statins	Various	May increase bone mass	No clinical data, not bone specific	[111]

upper gastrointestinal mucosa [94]. Therefore, it should be taken with 6–8 ounces of water and the patient should remain upright for at least 30 minutes after ingestion and until after their first food of the day to facilitate delivery to the stomach. The requirement to remain "upright" could not be met in space. The absorption of oral alendronate from the gastrointestinal tract is very poor (< 1%) and drastically impaired by food, beverages or other medication [95, 96]. Patients must take the drug after an overnight fast and at least 30 minutes before any other food or beverage. The bioavailability of oral alendronate is similar in men and women [95]. It is not known at present whether weightlessness would further impair absorption. The bisphosphonates pamidronate and zoledronate are given intravenously and would avoid problems with absorption. Zoledronate, for example, can be given as a yearly injection.

Once absorbed, bisphosphonates are not metabolized. They are rapidly cleared from plasma (< 2 hours) by being deposited into bone (20–80%), and the remainder is excreted in the urine. The skeletal half-life is very long, on the order of years. Because bisphosphonates are relatively new drugs, their long-term safety has not yet been established. The effect of long-term suppression of bone resorption in normal individuals in space is not known. In animal studies, however, alendronate has not interfered with fracture repair or normal bone remodeling [97]. Some patients who have been treated with bisphosphonates for long durations have develop osteonecrosis in their jaw after dental extractions [98]. Whether chronic suppression of bone turnover had made the maxillary bone less able to respond to damage is not clear, but this is a concern.

Studies comparing the effect of varying doses of alendronate, from 1 to 40 mg, on bone mineral density in postmenopausal women suggest that 10 mg is the most favorable dose [99–101]. The dose to use in space to prevent bone loss is not established.

Thiazide diuretics

Thiazide diuretics (hydrochlorothiazide, chlorthalidone) are not usually considered drugs for preventing bone loss, but they are commonly used for kidney stone prevention. They markedly reduce urinary calcium excretion, which makes them attractive for reducing stone risk. Several studies have shown, however, that thiazides can also affect bone. Data on thiazide diuretics against bone loss come mainly from epidemiological, retrospective, or cross-sectional studies. A meta-analysis of all thiazide studies showed that thiazide users have a 20% reduction in fracture risk [102]. A longitudinal study showed sizable reductions in bone loss rates (29% for radius, 49% for hip) in men taking hydrochlorothiazide [103]. Preliminary data from a randomized trial of two doses of thiazide in older adults showed that hydrochlorothiazide prevented bone loss [104]. The presumed mechanism of action is that thiazides prevent bone loss through a reduction in calcium excretion. Some evidence exists, however, that thiazides could have a direct effect on osteoblasts [105], which might be anabolic. Hydrochlorothiazide has not been well studied for preventing bone loss during immobilization or spaceflight.

Hydrochlorothiazide is well absorbed by the gastrointestinal tract. Diuresis begins within 2 hours, peaks at 4 hours, and lasts approximately 6–12 hours [106]. Hydrochlorothiazide dosing ranges from 12.5 mg once a day to 50 mg twice a day for hypertension and kidney stone prevention [106–108]. Doses of 12.5 mg and 25 mg once a day have been used in longitudinal studies of osteoporosis [103, 104]. The major side effect of thiazides is hypokalemia, and this, via changes in intracellular pH, leads to increased renal citrate reabsorption and reductions in urinary citrate excretion. The reduced urinary citrate is undesirable for renal stone prevention. Therefore thiazide use is often accompanied by supplementation with potassium citrate, which replaces the potassium and replenishes urinary citrate.

Thiazide diuretics have been used in clinical practice for approximately 40 years. The side effect profile of thiazides is well known [107] and includes hyponatremia, hypokalemia, hyperuricemia, hyperglycemia, and occasional hypercalcemia. Patients with recurrent kidney stones take thiazide chronically, and Yendt and

Cohanim [108] reported on their experience giving thiazides to 376 renal stone patients. Side effects necessitated discontinuation in 10% of subjects. The most common dose used in their population was 50 mg twice a day. Fatigue, loss of energy, and an increased need for sleep were the most common side effects that lead to stopping thiazides. Yendt and Cohanim had six patients who developed hypercalcemia. Two of these had hyperparathyroidism. It is not known if thiazides would cause hypercalcemia in space.

While thiazides have some disadvantages for spaceflight (diuresis, possible hypokalemia, possible hypercalcemia), they are very commonly used on Earth and could possibly be part of an astronaut's treatment for hypertension.

Potassium citrate

Potassium citrate is often given in conjunction with thiazide diuretics for the prevention of kidney stones. The citrate is metabolized to bicarbonate and provides extra base, which alkalizes the urine and helps to neutralize endogenously produced acid.

Although the main use of potassium citrate is in kidney stone prevention, some evidence exists for using it to prevent bone loss. Sebastian et al. [13] administered potassium citrate to postmenopausal women with osteoporosis and demonstrated an improvement in calcium balance, a reduction in bone resorption, and an increased rate of bone formation. They concluded that the ability to neutralize endogenous acid minimized skeletal buffering of acid and helped to prevent bone loss.

Since spaceflight can produce a low-grade chronic respiratory acidosis due to the high carbon dioxide concentrations in most space vehicles, there is some theoretical support for the notion that neutralizing acid in space may help preserve bone. Ground-based studies, however, suggest that the effects of respiratory acidosis on bone are much less than metabolic acidosis [16].

Potassium citrate has not been studied in the context of immobilization-induced bone loss.

Selective estrogen receptor modulators

Estrogen clearly has a beneficial effect on bone and has been shown to be important in both men and women. Estrogen, however, can have significant side effects when given to men or to women who are estrogen replete. Selective estrogen receptor modulators, or SERMs (e.g., tamoxifen, toremifene, raloxifene) comprise a group of compounds that are distinguished from estrogens by their ability to act as either an estrogen agonist or antagonist at the estrogen receptor depending on the target tissue and hormonal milieu. Raloxifene has been shown to increase bone mass in postmenopausal women, without the undesirable estrogen effects on the breast and endometrium [109]. Although raloxifene has been given to men [110], no studies exist demonstrating whether it could be effective in increasing bone mass in men or whether it would be effective in immobilization osteoporosis.

There is no evidence that estrogen plays a role in the bone loss seen in spaceflight, so it is not clear if stimulating bone estrogen receptors would be beneficial. While generally well tolerated, these drugs, like estrogen, carry the risk of thrombophlebitis, which would be a very significant side effect if it were to occur in space.

Statins

Recent evidence suggests that statins (lovastatin, simvastatin, pravastatin, atorvastatin) may have bone effects, in addition to their usual role in lowering serum cholesterol. Studies in animals show that statins can increase bone formation [111], and epidemiological studies show that patients on statins have a lower risk of fracture [112–115]. Although a drug that stimulates bone formation would be attractive for spaceflight use, there are no data on using statins in analogs of spaceflight, such as bed rest.

Parathyroid hormone

Cyclic parathyroid hormone (hPTH 1-34) administration has been shown to increase bone mass in patients with osteoporosis [39]. Parathyroid hormone was also effective in maintaining bone mass during immobilization in rats [116]. Parathyroid hormone has anabolic effects on bone, and it also works at the kidney to stimulate the reabsorption of calcium and enhance the synthesis of 1,25-dihydroxyvitamin D. Because spaceflight is associated with decreased parathyroid hormone levels, the use of parathyroid hormone treatment in space may make sense physiologically [117]. Parathyroid hormone might stimulate bone formation, increase vitamin D synthesis, and stimulate calcium reabsorption. The main risk of parathyroid hormone treatment is the development of osteosarcoma. While rare, this would be a devastating complication if it occurred in space.

Artificial Gravity

Knowledge of bone physiology and an understanding of the different physical, dietary, and pharmacological means to minimize bone loss are important if the crew will be exposed to weightlessness. An alternative approach to preventing bone loss in space would be to provide artificial gravity to keep the bones loaded as they are on Earth. This can be done in two ways. One is to spin the entire spacecraft and provide continuous artificial gravity (the method used in the movie *2001: A Space Odyssey*). The other is to provide a device onboard the spacecraft that rotates the crew member intermittently. Both of these methods are discussed in chapter 4. Research is underway to define the optimal level and duration of intermittent artificial gravity to be used in space. It is possible that intermittent "dosing" of artificial gravity could be effective, as short periods of standing can reduce calcium excretion during bed rest [82].

Monitoring Space-related Bone Loss

To date, most studies of calcium metabolism in space have involved the collection of samples in space and then the analysis of these samples on the ground. For long-duration flights, however, the crew medical officers and crew members will need to have information available during the flight to assess whether the countermeasure program they are using is effective. Although changes in bone mass would be a key variable, current measurement techniques, such as dual x-ray absorptiometry, are not sensitive enough to detect small changes and would not provide timely information on countermeasure effectiveness.

Two minimal requirements would be (1) the ability to measure serum calcium and electrolytes, as several of the potential drugs that could be used to prevent bone loss can alter serum calcium and electrolyte levels and (2) the ability to monitor bone resorption. Additional capabilities that would be desirable are the ability to measure bone mass (and/or density) and markers of bone formation and bone resorption in blood and urine.

Serum Calcium Measurement

Equipment already exists in the space inventory to measure serum calcium. The Portable Clinical Blood Analyzer has flown on the Space Shuttle and the International Space Station [118]. The available flight data from this device showed a decrease in ionized calcium in space, which is an unexpected finding that requires confirmation. Analyzers to measure electrolytes and other blood parameters have been part of the both the Mir and International Space Station programs, as discussed in chapter 12.

Monitoring Bone Resorption

Many biochemical markers of bone formation and bone resorption are available to monitor osteoporosis treatment [119]. Most are kits that require some laboratory support to perform. The simplest approach to monitoring bone loss is to measure urinary calcium excretion at various times throughout the flight. Urinary calcium is easy to measure and has been used as a marker in bed-rest studies of bone loss [2, 81]. Although urinary calcium excretion can be affected by diet, clinical use of repeated 24-hour calcium monitoring to follow patients who frequently form kidney stones has been successful. Consistent urinary values can be obtained over time, even without dietary controls [120]. In kidney stone patients, it has been shown that if urinary parameters can be kept within acceptable ranges, then the recurrence of a kidney stone is highly unlikely. This may be an approach that can be adopted for use in space. If urinary calcium can be kept near preflight values in space, it is very likely that significant bone resorption will not take place. Validated procedures and normative values would need to be established to make this a reliable method for use in space. Also, the addition of measurements of other markers (N-telopeptide, for example) would provide confirmation of the urinary calcium results.

Technological advances are likely to provide better analytical instruments. Miniature mass spectrometers may make it possible to measure a variety of important biochemical markers in both urine and blood. Additionally, the ability to measure bone mass or bone density using ultrasound would help determine whether the countermeasure program in space is working [121], if a method could be developed to use ultrasound on the hip rather than on the heel. Measurements of heel bone parameters using ultrasound do not correlate with hip losses in space [66].

Recommendations Based on Current Knowledge

The countermeasure program to prevent bone loss in space while weightless should be individualized. Due to a combination of genetic, physical, or environmental factors the amount of bone loss could differ markedly between crew members. Also, missions will differ. On some missions the crew members may be able to devote a considerable amount of time to countermeasures. On others, due to equipment problems or tight schedules, the adherence to the program may be spotty. These facts suggest, however, that a uniform approach for all crews may not be successful. The countermeasure program needs to be flexible. This flexibility could be provided through (1) the ability to monitor bone loss and (2) a variety of potential interventions that could be targeted to a particular crew member. With this method the crew members would be monitored during the flight, and the countermeasure program could be adjusted to keep bone loss to a minimum either by adjusting the exercise program, changing the diet, or adding drugs if necessary.

In general, however, drugs should be avoided. Particularly on long-duration voyages, the crew may not have the resources to deal with a significant side effect or adverse reaction. A case of erosive esophagitis due to alendronate, or thrombophlebitis due to a SERM, for example, would present a major problem in space. Nonetheless, the countermeasure program should have the option to use drugs. If monitoring shows that urinary calcium levels are high and not coming down, treating this problem with drugs may be preferable to the risk of kidney stone formation and significant bone loss. A variety of dietary and physical factors are important for maintaining the skeleton in space. Below are recommendations for helping ensure that kidney stone formation or bone loss does not affect a space mission:

1. Crew selection. Individuals with hereditary hypercalciuria would be difficult to maintain on a long-duration spaceflight. Also, a crew member with idiopathic osteoporosis or a low preflight bone mass could develop clinically significant osteoporosis as a consequence of long-duration weightlessness exposure. People with recurrent calcium oxalate stone disease would also be at high risk for developing a kidney stone in space. Individuals with clearcut clinical conditions like these would not be good candidates for long-duration spaceflight. Whether to select individuals based on the results of genetic tests, as opposed to demonstrated clinical findings, is a controversial area (discussed in chapter 12). For a very long-duration voyage, such as a trip to Mars, taking a conservative approach (using the results of genetic tests, even if their significance is not fully known) may make sense.

2. Dietary intervention. The best diet for bone health is a low salt, high calcium diet where the protein comes from vegetable sources. Animal protein creates an acid load that must be neutralized, and high sodium intake increases calcium excretion. The intake of elemental calcium should be maintained at 1000–1200 mg of elemental calcium a day. Calcium has been shown to help minimize bone loss in immobilization studies. Also, the availability of dietary calcium will minimize the need to use skeletal calcium to maintain homeostasis. The low light environment on the spacecraft necessitates that 600 IU of vitamin D be taken daily to ensure adequate vitamin D stores. The crew members also should not become deficient in vitamin K. Whether supplying vitamin K above recom-

mended daily allowances would provide added benefits for bone formation is not known. A diet that has a significant amount of soy-based foods could also make sense.

3. Exercise. The exercise program should focus on high impact loading of the lower extremities. Squats, leg abduction, and leg adduction exercises should be performed under load. Spinal extension under load should be performed to help maintain the back musculature. Treadmill exercise should be optimized to maintain aerobic fitness, and intense resistive and impact exercise should be used to maintain bone mass. Simple isometric exercises of postural muscles using minimal equipment (straps, Thera-Bands) have been demonstrated in weightlessness on the KC-135 aircraft. An exercise program that includes multiple intense activations throughout the day using exercises like these could potentially be more effective than one concentrated exercise period per day. Vibration could also be an attractive countermeasure that requires further study. A method of keeping the astronaut in contact with the vibration plate (perhaps using bungees or lower body negative pressure) would have to be devised.

4. Drug interventions. The use of potassium citrate is low risk and has been employed for kidney stone prevention both on Earth and on the International Space Station. Taking this compound on a long-duration spaceflight would make sense. The drugs that could be used fall into two main categories: those to reduce bone resorption, and those to increase bone formation. If an antiresorptive drug is selected, there are two possible approaches to administration. One is to take it orally, the other is intravenous injection. An intravenous injection of zoledronate, for example, could be given prior to a spaceflight and the effects would last a year. The problem with this approach is that the long-term effects of giving bisphosphonates to otherwise healthy people are not known, and the crew member is being exposed to a drug without first demonstrating whether bone mass could be maintained in weightlessness using other methods. An oral bisphosphonate could be administered only as needed and stopped when no longer necessary. This route of administration, however, carries the potential risk of erosive esophagitis. Recombinant human parathyroid hormone (teriparatide or hPTH) could also be considered to increase bone formation in space. Although generally well tolerated, there have been rare reports of osteosarcoma when using this drug.

5. Artificial gravity. Intermittent artificial gravity has the potential to be more effective than exercise and does not expose crew members to drug side effects. Whether to provide continuous artificial gravity would be an engineering decision made early on during the design of the mission. Artificial gravity, depending on how it is used, can cause problems with motion sickness (discussed in chapter 9).

6. Monitoring. Urinary calcium excretion and serum calcium levels should be monitored to provide feedback on the effectiveness of the countermeasure program. The greater the monitoring capability onboard (such as the ability to measure bone breakdown markers), the more flexible the countermeasure program can be. For those times when the crew is on the Moon (one-sixth of Earth's gravity) or Mars (one-third of Earth's gravity), monitoring will still be required because it is not clear whether the gravitational loading on the Moon and Mars will be adequate to maintain bone mass. A good monitoring program, however, will detect any losses and allow the crew to implement countermeasures.

References

1. Lebedev, V., *Diary of a Cosmonaut: 211 Days in Space*, D. Puckett and C.W. Harrison, eds. 1988, PhytoResource Research, College Station, TX.
2. Whedon, G.D., et al., Mineral and nitrogen metabolic studies on *Skylab* orbital space flights. Transactions of the Association of American Physicians, 1974. 87: 95–110.
3. Whedon, G.D., Disuse osteoporosis: physiological aspects. Calcified Tissue International, 1984. 36 (Suppl 1): S146–50.
4. LeBlanc, A., et al., Bone mineral and lean tissue loss after long duration spaceflight. Bone, 1996. 11: S323.
5. Grigoriev, A.I., et al., Clinical and physiological evaluation of bone changes among astronauts after long-term space flights. Aviakosmicheskaia i Ekologicheskaia Meditsina, 1998. 32(1): 21–25.
6. Linenger, J.M., *Off the Planet*. 2000, McGraw-Hill, New York.
7. Tilton, F.E., J.J. Degioanni, and V.S. Schneider, Long-term follow-up of *Skylab* bone demineralization. Aviation, Space, and Environmental Medicine, 1980. 51: 1209–13.
8. Wilmet, E., et al., Longitudinal study of the bone mineral content and of soft tissue composition after spinal cord section. Paraplegia, 1995. 33(11): 674–77.
9. Weaver, C.M., A. LeBlanc, and S.M. Smith, Calcium and related nutrients in bone metabolism, in *Nutrition in Spaceflight and Weightlessness Models*, H.W. Lane and D.A. Schoeller, eds. 2000, CRC Press, Boca Raton, FL, pp. 179–96.
10. Lemann, J. and M.J. Favus, The intestinal absorption of calcium, magnesium and phosphorus, in *Primer on the Metabolic Bone Diseases and Disorders of Mineral Metabolism*, M.J. Favus, ed. 1999, Lippincott Williams and Wilkins, Philadelphia, pp. 63–67.
11. Holick, M.F., Vitamin D: Photobiology, metabolism, mechanism of action, and clinical applications, in *Primer on the Metabolic Bone Diseases and Disorders of Mineral Metabolism*, M.J. Favus, ed. 1999, Lippincott Williams and Wilkins, New York, pp. 92–98.
12. Bushinsky, D.A., et al., Decreased potassium stimulates bone resorption. American Journal of Physiology, 1997. 272(6 Pt 2): F774–80.
13. Sebastian, A., et al., Improved mineral balance and skeletal metabolism in postmenopausal women treated with potassium bicarbonate. New England Journal of Medicine, 1994. 330(25): 1776–81.
14. Drummer, C., et al., Effects of elevated carbon dioxide environment on calcium metabolism in humans. Aviation, Space, and Environmental Medicine, 1998. 69(3): p. 291–98.
15. Messier, A.A., et al., Calcium, magnesium, and phosphorus metabolism, and parathyroid-calcitonin function during prolonged exposure to elevated CO_2 concentrations on submarines. Undersea Biomedical Research, 1979. 6(Suppl): S57–70.
16. Bushinsky, D.A., Acidosis and bone. Mineral and Electrolyte Metabolism, 1994. 20(1–2): 40–52.
17. Coe, F.L., J.H. Parks, and J.R. Asplin, The pathogenesis and treatment of kidney stones. New England Journal of Medicine, 1992. 327: 1141–52.
18. Boyden, L.M., et al., High bone density due to a mutation in LDL-receptor-related protein 5. New England Journal of Medicine, 2002. 346(20): 1513–21.
19. Judex, S., L.R. Donahue, and C. Rubin, Genetic predisposition to low bone mass is paralleled by an enhanced sensitivity to signals anabolic to the skeleton. FASEB Journal, 2002. 16(10): 1280–82.
20. Frost, H.M., Bone "mass" and the "mechanostat": a proposal. Anatomical Record, 1987. 219(1): 1–9.
21. Nilsson, B.E. and N.E. Westlin, Bone density in athletes. Clinical Orthopaedics and Related Research, 1971. 77: 179–82.
22. Uusi-Rasi, K., et al., Long-term recreational gymnastics, estrogen use, and selected risk factors for osteoporotic fractures. Journal of Bone and Mineral Research, 1999. 14(7): 1231–38.
23. Taaffe, D.R., et al., High-impact exercise promotes bone gain in well-trained female athletes. Journal of Bone and Mineral Research, 1997. 12(2): 255–60.

24. Huddleston, A.L., et al., Bone mass in lifetime tennis athletes. Journal of the American Medical Association, 1980. 244(10): 1107–9.
25. Bevier, W.C., et al., Relationship of body composition, muscle strength, and aerobic capacity to bone mineral density in older men and women. Journal of Bone and Mineral Research, 1989. 4(3): 421–32.
26. Orozco, P., and J.M. Nolla, Associations between body morphology and bone mineral density in premenopausal women. European Journal of Epidemiology, 1997. 13(8): 919–24.
27. Biering-Sorensen, F., H. Bohr, and O. Schaadt, Bone mineral content of the lumbar spine and lower extremities years after spinal cord lesion. Paraplegia, 1988. 26(5): 293–301.
28. Biering-Sorensen, F., H.H. Bohr, and O.P. Schaadt, Longitudinal study of bone mineral content in the lumbar spine, the forearm and the lower extremities after spinal cord injury. European Journal of Clinical Investigation, 1990. 20(3): 330–35.
29. Krebs, D.E., et al., Hip biomechanics during gait. Journal of Orthopaedic and Sports Physical Therapy, 1998. 28(1): 51–59.
30. Schultheis, L., The mechanical control system of bone in weightless spaceflight and in aging. Experimental Gerontology, 1991. 26(2–3): 203–14.
31. Strickland, E.M., et al., In vivo acetabular contact pressures during rehabilitation. Part I: Acute phase. Physical Therapy, 1992. 72(10): 691–99.
32. Anderson, F.H., et al., Androgen supplementation in eugonadal men with osteoporosis: effects of six months' treatment on markers of bone formation and resorption. Journal of Bone Mineral Research, 1997. 12: 472–78.
33. Bikle, D.D., et al., The molecular response of bone to growth hormone during skeletal unloading: regional differences. Endocrinology, 1995. 136(5): 2099–109.
34. Marcus, R., Recombinant human growth hormone as potential therapy for osteoporosis. Bailliere's Clinical Endocrinology and Metabolism, 1998. 12(2): 251–60.
35. Bateman, T.A., et al., Histomorphometric, physical, and mechanical effects of spaceflight and insulin-like growth factor-I on rat long bones. Bone, 1998. 23(6): 527–35.
36. Dempster, D.W., et al., Anabolic actions of parathyroid hormone on bone. Endocrine Reviews, 1993. 14: 690–709.
37. Canalis, E., J.M. Hock, and L.G. Raisz, Anabolic and catabolic effects of parathyroid hormone on bone and interactions with growth factors, in *The Parathyroids*, J.P. Bilezikian, M.A. Levine, and R. Marcus, eds. 1994, Raven Press, New York, pp. 65–82.
38. Deal, C., The use of intermittent human parathyroid hormone as a treatment for osteoporosis. Current Rheumatology Reports 2004. 6(1): 49–58.
39. Quattrocchi, E., and H. Kourlas, Teriparatide: a review. Clinical Therapeutics 2004. 26(6): 841–54.
40. Thomas, T., et al., Ineffectiveness of calcitonin on a local-disuse osteoporosis in the sheep: a histomorphometric study. Calcified Tissue International, 1995. 57(3): 224–28.
41. Hantman, D.A., et al., Attempts to prevent disuse osteoporosis by treatment with calcitonin, longitudinal compression and supplementary calcium and phosphate. Journal of Clinical Endocrinology and Metabolism, 1973. 36(5): 845–58.
42. Tham, D.M., C.D. Gardner, and W.L. Haskell, Clinical review 97: Potential health benefits of dietary phytoestrogens: a review of the clinical, epidemiological, and mechanistic evidence. Journal of Clinical Endocrinology and Metabolism, 1998. 83(7): 2223–35.
43. Potter, S.M., et al., Soy protein and isoflavones: their effects on blood lipids and bone density in postmenopausal women. American Journal of Clinical Nutrition, 1998. 68(6 Suppl). 1375S–79S.
44. Gennari, C., et al., Effect of ipriflavone—a synthetic derivative of natural isoflavones—on bone mass loss in the early years after menopause. Menopause, 1998. 5(1): 9–15.
45. Iwamoto, J., T. Takeda, and Y. Sato, Effects of vitamin K2 on osteoporosis. Current Pharmaceutical Design 2004. 10(21): 2557–76.
46. Heer, M., Nutritional interventions related to bone turnover in European space missions and simulation models. Nutrition, 2002. 18(10): 853–56.
47. Caillot-Augusseau, A., et al., Space flight is associated with rapid decreases of undercarboxylated osteocalcin and increases of markers of bone resorption without changes in

their circadian variation: observations in two cosmonauts. Clinical Chemistry, 2000. 46(8 Pt 1): 1136–43.

48. Rubin, J., et al., Formation of osteoclast-like cells is suppressed by low frequency, low intensity electric fields. Journal of Orthopaedic Research, 1996. 14(1): 7–15.

49. Flieger, J., et al., Mechanical stimulation in the form of vibration prevents postmenopausal bone loss in ovariectomized rats. Calcified Tissue International, 1998. 63(6): 510–14.

50. Rubin, C., et al., Anabolism. Low mechanical signals strengthen long bones. Nature, 2001. 412(6847): 603–4.

51. Bosco, C., et al., Adaptive responses of human skeletal muscle to vibration exposure. Clinical Physiology, 1999. 19(2): 183–87.

52. Rubin, C., et al., Prevention of postmenopausal bone loss by a low-magnitude, high-frequency mechanical stimuli: a clinical trial assessing compliance, efficacy, and safety. Journal of Bone Mineral Research, 2004. 19(3): 343–51.

53. Jee, W.S. and Y. Ma, Animal models of immobilization osteopenia. Morphologie, 1999. 83(261): 25–34.

54. Garland, D.E., et al., Osteoporosis after spinal cord injury. Journal of Orthopaedic Research, 1992. 10(3): 371–38.

55. Otter, M.W., et al., Does bone perfusion/reperfusion initiate bone remodeling and the stress fracture syndrome? Medical Hypotheses, 1999. 53(5): 363–68.

56. Lau, E.M. and J. Woo, Nutrition and osteoporosis. Current Opinion in Rheumatology, 1998. 10(4): 368–72.

57. Bourland, C., et al., Food systems for space and planetary flights, in Nutrition in space-flight and weightlessness models, H.W. Lane and D.A. Schoeller, eds. 2000, CRC Press, Boca Raton, FL, pp. 19–40.

58. Rubin, J., et al., Mechanical strain inhibits expression of osteoclast differentiation factor by murine stromal cells. American Journal of Physiology, Cell Physiology, 2000. 278(6): C1126–32.

59. Greenfield, E.M., Y. Bi, and A. Miyauchi, Regulation of osteoclast activity. Life Sciences, 1999. 65(11): 1087–102.

60. Roodman, G.D., Cell biology of the osteoclast. Experimental Hematology, 1999. 27(8): 1229–41.

61. Hofbauer, L.C., Osteoprotegerin ligand and osteoprotegerin: novel implications for osteoclast biology and bone metabolism. European Journal of Endocrinology, 1999. 141(3): 195–210.

62. Grigoriev, A.I., B.V. Morukov, and D.V. Vorobiev, Water and electrolyte studies during long-term missions onboard the space stations SALYUT and MIR. Clinical Investigator, 1994. 72(3): 169–89.

63. Smith, S.M., et al., Collagen cross-link excretion during space flight and bed rest. Journal of Clinical Endocrinology and Metabolism., 1998. 83: 3584–3591.

64. Caillot-Augusseau, A., et al., Bone formation and resorption biological markers in cosmonauts during and after a 180-day space flight (Euromir 95). Clinical Chemistry, 1998. 44: 578–585.

65. Vico, L., M.H. Lafage-Proust, and C. Alexandre, Effects of gravitational changes on the bone system in vitro and in vivo. Bone, 1998. 22(5 Suppl): 95S-100S.

66. Lang, T., et al., Cortical and trabecular bone mineral loss from the spine and hip in long-duration spaceflight. Journal of Bone and Mineral Research, 2004. 19(6): 1006–12.

67. Vico, L., et al., Effects of long-term microgravity exposure on cancellous and cortical weight-bearing bones of cosmonauts. Lancet, 2000. 355(9215): 1607–11.

68. Davis, B.L., et al., Ground reaction forces during locomotion in simulated microgravity. Aviation, Space, and Environmental Medicine, 1996. 67(3): 235–42.

69. McCrory, J., et al., In-shoe force measurements from locomotion in simulated zero gravity during parabolic flight. Clinical Biomechanics (Bristol, Avon), 1997. 12(3): S7.

70. McCrory, J.L., J. Derr, and P.R. Cavanagh, Locomotion in simulated zero gravity: ground reaction forces. Aviation, Space, and Environmental Medicine, 2004. 75(3): 203–10.

71. Holick, M.F., Vitamin D and bone health. Journal of Nutrition, 1996. 126: 1159S-64S.

72. Heller, H.J., et al., Pharmacokinetics of calcium absorption from two commercial calcium supplements. Journal of Clinical Pharmacology, 1999. 39(11): 1151–54.
73. Heller, H.J., The role of calcium in the prevention of kidney stones. Journal of the American College of Nutrition, 1999. 18(5 Suppl): 373S-78S.
74. Martini, L.A. and R.J. Wood, Should dietary calcium and protein be restricted in patients with nephrolithiasis? Nutrition Reviews, 2000. 58(4): 111–17.
75. Schneider, V.S. and J. McDonald, Skeletal calcium homeostasis and countermeasures to prevent disuse osteoporosis. Calcified Tissue International, 1984. 36(Suppl 1): S151–44.
76. Hall, S.J., The biomechanics of the human lower extremity, in *Basic Biomechanics*, S.J. Hall, ed. 1995, Mosby, New York, pp. 208–42.
77. Hodge, W.A., et al., Contact pressures in the human hip joint measured in vivo. Proceedings of the National Academy of Sciences USA, 1986. 83(9): 2879–83.
78. Lindh, M., Biomechanics of the lumbar spine, in *Basic Biomechanics of the Musculoskeletal System*, M. Nordin and V.H. Frankel, eds. 1989, Lea and Febiger, Philadelphia, pp. 183–207.
79. LeBlanc, A., et al., Regional muscle loss after short duration spaceflight. Aviation, Space, and Environmental Medicine, 1995. 66(12): 1151–54.
80. LeBlanc, A., Final Report for Experiment E029: Life and Microgravity Spacelab, in *Life and Microgravity Spacelab (LMS) Final Report*, J.P. Downey, ed. 1998, NASA, NASA-Marshall Space Flight Center. 361–98.
81. Issekutz, B.J., et al., Effect of prolonged bed rest on urinary calcium output. Journal of Applied Physiology, 1966. 21(3): 1013–20.
82. Vernikos, J., et al., Effect of standing or walking on physiological changes induced by head down bed rest: implications for spaceflight. Aviation, Space, and Environmental Medicine, 1996. 67(11): 1069–79.
83. Oganov, V.S. and V.S. Schneider, Skeletal system, in *Space Biology and Medicine*, A.E. Nicogossian and O.G. Gazenko, eds. 1996, American Institute of Aeronautics and Astronautics, Reston, VA, pp. 247–66.
84. Chow, Y.W., et al., Ultrasound bone densitometry and dual energy X-ray absorptiometry in patients with spinal cord injury: a cross-sectional study. Spinal Cord, 1996. 34(12): 736–41.
85. Taaffe, D.R., et al., Comparison of calcaneal ultrasound and DXA in young women. Medicine and Science in Sports and Exercise, 1999. 31(10): 1484–89.
86. Fisher, J.E., et al., Alendronate mechanism of action: geranylgeraniol, an intermediate in the mevalonate pathway, prevents inhibition of osteoclast formation, bone resorption, and kinase activation in vitro. Proceedings of the National Academy of Sciences USA, 1999. 96: 133–38.
87. Grigoriev, A.I., et al., Effect of exercise and bisphosphonate on mineral balance and bone density during 360 day antiorthostatic hypokinesia. Journal of Bone and Mineral Research, 1992. 7(Suppl 2): S449–55.
88. Rodan, G.A. and H.A. Fleisch, Bisphosphonates: mechanisms of action. Journal of Clinical Investigation, 1996. 97: 2692–2696.
89. Thompson, D.D., et al., Aminohydroxybutane bisphosphonate inhibits bone loss due to immobilization in rats. Journal of Bone and Mineral Research, 1990. 5(3): 279–86.
90. LeBlanc, A., et al., Alendronate as a spaceflight countermeasure-preliminary results. Bone, 1998. 23: S405.
91. Greenspan, S.L., et al., Alendronate stimulation of nocturnal parathyroid hormone secretion: a mechanism to explain the continued improvement in bone mineral density accompanying alendronate therapy. Proceedings of the Association of American Physicians, 1996. 108: 230–38.
92. Plotkin, L.I., et al., Prevention of osteocyte and osteoblast apoptosis by bisphosphonates and calcitonin. Journal of Clinical Investigation, 1999. 104(10): 1363–74.
93. Adami, S. and N. Zamberlan, Adverse effects of bisphosphonates. A comparative review. Drug Safety, 1996. 14: 158–70.
94. de Groen, P.C., et al., Esophagitis associated with the use of alendronate. New England Journal of Medicine, 1996. 335: 1016–21.

95. Gertz, B.J., et al., Studies of the oral bioavailability of alendronate. Clinical Pharmacology and Therapeutics, 1995. 58: 288–98.
96. Porras, A.G., S.D. Holland, and B.J. Gertz, Pharmacokinetics of alendronate. Clinical Pharmacology, 1999. 36(5): 315–28.
97. Peter, C.P., et al., Effect of alendronate on fracture healing and bone remodeling in dogs. Journal of Orthopaedic Research, 1996. 14(1): 74–79.
98. Ruggiero, S.L., et al., Osteonecrosis of the jaws associated with the use of bisphosphonates: a review of 63 cases. Journal of Oral and Maxillofacial Surgery, 2004. 62(5): 527–34.
99. Adami, S., et al., Treatment of postmenopausal osteoporosis with continuous daily oral alendronate in comparison with either placebo or intranasal salmon calcitonin. Osteoporosis International, 1993. 3 (Suppl 3): S21-S27.
100. Bone, H.G., et al., Dose-response relationships for alendronate treatment in osteoporotic elderly women. Alendronate Elderly Osteoporosis Study Centers. Journal of Clinical Endocrinology and Metabolism, 1997. 82: 265–74.
101. Chesnut, C.H., et al., Alendronate treatment of the postmenopausal osteoporotic woman: effect of multiple dosages on bone mass and bone remodeling. American Journal of Medicine, 1995. 99: 144–52.
102. Ray, W.A., Thiazide diuretics and osteoporosis: time for a clinical trial? Annals of Internal Medicine, 1991. 115: 64–65.
103. Wasnich, R., et al., Effect of thiazides on rates of bone mineral loss: a longitudinal study. British Medical Journal, 1990. 301: 1303–5.
104. LaCroix, A.Z., et al., Low dose thiazide prevents bone loss in older adults: results of a 3-year randomized double-blind controlled trial. Bone, 1998. 23: S151.
105. Barry, E.L.R., et al., Expression of the sodium-chloride cotransporter in osteoblast-like cells: effects of thiazide diuretics. American Journal of Physiology, 1997. 272: C109–16.
106. U.S. Pharmacopeia, U.S., *Drug Information for the Health Care Professional*, 19th ed. 1999, Micromedex, Greenwood Village, CO.
107. Field, M.J., and J.R. Lawrence, Complications of thiazide diuretic therapy: an update. Medical Journal of Australia, 1986. 144: 641–44.
108. Yendt, E.R., and M. Cohanim, Prevention of calcium stones with thiazides. Kidney International, 1978. 13: 397–409.
109. Delmas, P.D., et al., Effects of raloxifene on bone mineral density, serum cholesterol concentrations, and uterine endometrium in postmenopausal women. New England Journal of Medicine, 1997. 337(23): 1641–47.
110. Draper, M.W., et al., Antiestrogenic properties of raloxifene. Pharmacology, 1995. 50(4): 209–17.
111. Mundy, G., et al., Stimulation of bone formation in vitro and in rodents by statins. Science, 1999. 286(5446): 1946–49.
112. Meier, C.R., et al., HMG-CoA reductase inhibitors and the risk of fractures. Journal of the American Medical Association, 2000. 283(24): 3205–10.
113. Wang, P.S., et al., HMG-CoA reductase inhibitors and the risk of hip fractures in elderly patients. Journal of the American Medical Association, 2000. 283(24): 3211–16.
114. Edwards, C.J., D.J. Hart, and T.D. Spector, Oral statins and increased bone-mineral density in postmenopausal women. Lancet, 2000. 355(9222): 2218–19.
115. Chan, K.A., et al., Inhibitors of hydroxymethylglutaryl-coenzyme A reductase and risk of fracture among older women. Lancet, 2000. 355(9222): 2185–88.
116. Ma, Y., et al., Parathyroid hormone and mechanical usage have a synergistic effect in rat tibial diaphyseal cortical bone. Journal of Bone and Mineral Research, 1999. 14(3): 439–48.
117. Holick, M.F., Microgravity-induced bone loss—will it limit human space exploration? Lancet, 2000. 355(9215): 1569–70.
118. Smith, S.M., et al., Assessment of a portable clinical blood analyzer during space flight. Clinical Chemistry, 1997. 43: 1056–1065.

119. Garnero, P. and P.D. Delmas, Biochemical markers of bone turnover. Applications for osteoporosis. Endocrinology and Metabolism Clinics of North America, 1998. 27(2): 303–23.
120. Parks, J.H., M. Coward, and F.L. Coe, Correspondence between stone composition and urine supersaturation in nephrolithiasis. Kidney International, 1997. 51: 894–900.
121. National Osteoporosis Society, The use of quantitative ultrasound in the management of osteoposisis in primary or secondary care. NOS, Camerton, Bath, UK.
122. Breslau, N.A., Calcium homeostasis, in *Textbook of Endocrine Physiology*, J.E. Griffin and S.R. Ojeda, eds. 2000, Oxford University Press, New York, pp. 357–92.
123. Boyce, B.F., et al., Recent advances in bone biology provide insight into the pathogenesis of bone diseases. Laboratory Investigation, 1999. 79(2): 83–94.

2

Psychosocial Support: Maintaining an Effective Team

Introduction

As a group, astronauts function at a high level and are relatively free of psychological or psychiatric problems. Significant psychiatric illnesses, such as schizophrenia, have not been a problem for space missions. Astronauts are screened for major psychiatric illnesses before selection and would not be assigned to crews if they could not function effectively in a team. Nevertheless, the social isolation and confinement of long-duration spaceflight present significant psychological challenges. Psychosocial problems appear to have played a part in long-duration missions terminated for medical reasons (table 2-1) [1, 2].

Long-duration spaceflight can test any individual's psychological well-being. Factors such as confinement, under- or overwork, sleep loss, and monotony can combine to worsen interpersonal tensions or even lead to frank depression. Conflicts can arise with ground control, with a resulting loss of trust and teamwork. A chronic dispute between or among crew members can destroy team functioning. Suppressed anger or frustration can erupt unexpectedly and create potentially hazardous situations. In the past, a variety of psychological events have occurred in both space and in Antarctica, and these events have had a major impact on the missions.

Both crews and ground controllers need to be aware of the potential damage that psychosocial problems can produce on a mission. They must be able to recognize the signs of psychological problems both in themselves and in others. They need tools to manage conflicts and to repair relationships after a dispute. They need to able to recognize and treat depression, anxiety, and other clinical problems. This chapter reviews the main psychosocial problems that occur in spaceflight and presents some preventive measures and treatments.

Psychosocial Issues Relevant to Spaceflight

The *Psychiatric Diagnostic and Statistical Manual IV* (DSM IV) [3] lists numerous possible psychiatric diagnoses. Most of these have little applicability to spaceflight. Individuals with significant psychiatric problems or personality disorders would most likely not make it through the screening processes to fly in space. There are a few key areas, however, that are important for space missions: interpersonal conflict, depression, and anxiety disorders. Other clinical psychiatric diagnoses would be expected to occur very infrequently. For example, post-traumatic stress disorder could develop in reaction to a loss or near disaster on a Mars or lunar mission.

Interpersonal Conflict

Interpersonal conflict is a fact of life. Often conflict can be beneficial, but it can also lead to strong emotions like anger. On Earth, disengaging from the other person for while to gain perspective and "cool off" can dissipate anger and other emotions. In space, however, crew members must interact even if they are in conflict. This means that each person must be able to deal with conflict effectively and not let conflict interfere with good team functioning and a successful mission.

Table 2-1. Some Significant Psychosocial Events in Spaceflight and Antarctica

Mission	Occurrence	Year	Reference
IGY 1958 Antarctica	Evacuation due to psychosis	1957	[26]
IBEA Antarctica	Evacuation due to probable depression	1980	[17]
Soyuz 21-Salyut 5	Interpersonal conflict may have played a role in evacuation	1976	[1, 2]
Soyuz T14-Salyut 7	Depression may have contributed to evacuation	1985	[15, 16]
Soyuz TM2-Mir	Evacuation, interpersonal differences a possible contributor	1987	[2]

This table lists some times in recent Antarctic expeditions and space voyages when a crewmember was removed from the expedition or the flight. Although details about these events are spotty, the available evidence suggests that psychosocial factors have been a major contributor to long-duration space mission termination.

Person–person conflict

One of the best insights into what it is like to experience long-duration spaceflight comes from the book *Diary of a Cosmonaut* by Valentin Lebedev [4]. At one point in the mission he notes: "We don't understand what's going on with us. We silently walk by each other, feeling offended. We have to find some way to make things better"(p. 139). Although he doesn't elaborate on what the conflicts might be, clearly, the atmosphere on the station was tense and unpleasant. Over time, the tension affected both mood and performance.

On some crews, the situation can deteriorate. Interpersonal issues reportedly played the major role in the decision to terminate the Russian long-duration mission *Soyuz 21* in 1976 [1, 2]. In analog environments, breakdowns due to conflict can also occur. In a Russian chamber isolation study, two groups were isolated in adjoining chambers. One group (4 Russian men) spent 240 days in the first chamber. A group of Russian and German men (3 Russians and 1 German) spent 110 days in the second chamber and then were replaced by a group consisting of 3 men (from Austria, Russia, and Japan) and a Canadian woman. When the all-Russian group and the new group got together for a New Year's Eve party, two of the Russians had a fistfight and one tried to kiss the Canadian woman. After this incident the Japanese man left the study, and relations among the others were strained for the remainder of the isolation period [5]. The event also received considerable negative press coverage in Canada. Although this isolation study was not typical, it did serve to highlight two possible sources of conflict: cross-cultural differences and male-female relationships.

Overall, interpersonal conflict on a long-duration space mission has the potential to terminate missions and is arguably more of a risk to a mission than any other medical or physiological factor. Although the majority of groups entering space will be able to handle conflict well and develop an atmosphere of mutual respect, acceptance and camaraderie, it is also possible for conflicts to degenerate into feuds and power struggles that can jeopardize the mission.

Person–control center conflict

Relationships between the crew and ground control can easily become strained. Although it is possible for crew members in conflict to communicate freely and have a shared understanding of their situation, this is often not possible with ground control. It is difficult for the ground control center to fully appreciate the situation on the spacecraft, and it is easy for the crew to view ground control as out of touch with the mission. Some degree of tension in the ground-control-crew relationship is inevitable (and, as is discussed below, can be beneficial), but a complete breakdown can compromise the safety of the mission.

On the Russian *Mir 23* space station mission, the relationship between the commander and the ground was so poor that at some points the commander was shouting at ground control over the air-to-ground loop [6]. The lack of understanding between ground control and the crew of *Mir 23* contributed to the collision of a unmanned *Progress* module into the station, which almost led to an evacuation of the station [6, 7]. After the collision, the commander lost sleep and developed serious heart irregularities [7].

On the *Apollo 7* flight, an adversarial relationship developed between the commander and ground control [8, 9]. The commander sharply criticized the ground controllers' work and was curt and disdainful in his communications. Although the mission ended successfully and it was the commander's last flight, the other crew members were not as lucky. They were never assigned to other missions. On the *Skylab 4* mission, considerable tension developed between ground control and the crew over the issue of scheduling. On this flight the ground control center was providing a detailed crew plan for each day. The plan, which was impossible for the crew to execute in the time allotted, did not offer the crew flexibility in how to arrange their tasks. As a result, tension developed. Mission control became concerned that tasks were not being completed; the crew needed flexibility to get the work done in their environment. Ultimately, the crew cut off communications [10]. This precipitated an in-depth discussion between mission control and the crew that established mutual understanding and gave the crew more control over their own schedule. One lasting benefit of this has been an appreciation of the importance of allowing the crew to organize as much of their schedule as possible.

Relations between ground control and crew can be strained deliberately. Sometimes crew cohesion and functioning can be enhanced if the crew has a common adversary, which can be ground control. Just as workers in the field might complain about the head office or deployed soldiers might suspect the competence and ability of headquarters, astronauts can easily develop animosity toward ground control [11]. This is not necessarily bad if it is the only way to maintain crew cohesion at a difficult time on the flight and if ground control has the wisdom to understand what is happening. If, however, confidence and respect are completely lost, the results can be hazardous.

Depression

Chronic isolation can lead to depression [12]. Also, chronic frustration or a perceived inability to change or improve one's situation can lead to depression [13]. Finally, low light levels are a factor in the depression associated with seasonal affective disorder. All of these factors can occur in long-duration spaceflight. As a result, it is possible for a crew member to become depressed because of underwork, overwork, isolation, separation from family, conflict, environmental factors, or dissatisfaction with the mission.

It is possible that a case of depression terminated a space mission. In November 1985 the crew of the *Salyut 7* space station returned to Earth after 56 days in space, 160 days earlier than planned. One of the crew members had been reported as not sleeping or eating well and lacking motivation. The problem had been noted a few weeks before the return and had been a cause for frequent communication between the crew and the ground [14]. After the flight, some reports offered appendicitis as the reason for the return; others reported prostatitis. Statements by the crew members at the time, however, suggested the reason was in part psychological [15 688]. Reports from recent sources suggest that depression may have been the cause [16]. Depression has been the cause of evacuation from an Antarctic mission [17].

Depression is extremely common on Earth, and when accompanied by strong suicidal thoughts, it can be fatal. Although good treatments for depression exist, the main problem is diagnosis. Changes in mood are normal and often understandable, so it is often hard to discern when a depressed mood has progressed to clinical depression. Within a small crew, the gradual worsening of a fellow crewmate's mood may not be perceived until it has progressed to the point where it is interfering with performance. On a space mission, where the contribution of each person is important to the welfare of the crew, having a person affected by depression can be a hazard.

Anxiety Disorders

Data from Antarctic missions show that psychosocial problems constitute 4–5% of total morbidity on the missions, with anxiety being a common complaint [18]. On long-duration submarine missions, the most frequently occurring psychopathology is anxiety attacks [19]. Spaceflight is hazardous, so some degree of apprehension about major events such as launch, landing, or extravehicular activity is normal. An anxiety disorder exists when worries and concerns about the mission or other life events starts to interfere with normal function. Astronauts are screened at selection for anxiety disorders and experience situations that could provoke anxiety throughout their training. As a result, anxiety disorders are rare within the astronaut population. Nevertheless, on a long-duration flight that is not going well, the crew member may be fatigued, working with a crewmate that they have a conflict with, or in a situation where they have lost confidence in ground control. In this kind of environment, it may be hard to retain a balanced perspective, and new events on the mission may be anxiety provoking. From accounts published after the *Mir 23* mission, it seems that a combination of sleep deprivation, equipment malfunctions, and poor ground control support led to anxiety and physical problems in the commander of this mission [6, 7].

Just as for depression, effective anxiety treatments exist, but the main problem is diagnosis. Rightly or wrongly, a psychological problem, such as depressed mood or anxiety, is often viewed as a sign of weakness among astronauts. The selection process reinforces this. A high degree of self-awareness among the crew members is therefore required to acknowledge that a problem exists.

Asthenia

In the Russian space program, the term "asthenia" is used to describe a set of psycho-physiological changes that commonly occur among cosmonauts. Asthenia is defined as "an abnormal state marked by weakness, increased tendency to fatigue, irritability and disorders of attention and memory" [20, p. 419]. In the Russian classification, asthenia is described as having three stages. In the first stage, there is irritability, fatigue in the evening, and an increase in emotional lability. In the second stage, there is a feeling of fatigue accompanied by sleep disturbances. Performance mistakes can be seen at this stage. The third stage seems to be the same as clinical depression, with emotional depression, frequent conflicts, and performance mistakes [20, 21]. This classification has proven to be useful in the Russian space program because it summarizes a common set of symptoms that have occurred during long-duration spaceflights (typically greater than 4 months long).

In the United States asthenia (or neurasthenia) is not considered a distinct diagnosis, but it is classified as an "undifferentiated somatoform disorder" in the text revision of the DSM-IV [22]. Neurasthenia was first described by the American neuropsychiatrist George Miller, who ascribed neurasthenia to nervous exhaus-

Table 2-2. Diagnostic criteria for neurasthenia

A. Either of the following must be present:
  1. Persistent and distressing complaints of feelings of exhaustion after a minor mental effort (such as performing or attempting to perform everyday tasks that do not require unusual mental effort),
  2. persistent and distressing complaints of feelings of fatigue and bodily weakness after minor physical effort.

B. At least one of the following symptoms must be present:
  1. Feelings of muscular aches and pains,
  2. Dizziness,
  3. Tension headaches,
  4. Sleep disturbances,
  5. Inability to relax,
  6. Irritability,

The patient is unable to recover from the symptoms in criterion A 1 or 2 by means of rest, relaxation, or entertainment. The duration of the disorder is at least 3 months.

Most commonly used exclusion clause: This disorder does not occur in the presence of organic emotionally labile disorder, postencephalitic syndrome, postconcussional syndrome, mood disorders, panic disorder, or generalized anxiety disorder.

Reprinted with permission from: *The ICD-10 Classification of Mental and Behavioural Disorders, Diagnostic Criteria for Research*, World Health Organization, 1993 [46].

tion, with depletion of the "stored nutrients" in the nerve cells [22]. The depletion resulted from stress. Miller may have been prescient: today the depletion hypothesis states that prolonged stress lowers the levels in neurotransmitters in neurons. Depletion of brain amines (dopamine, serotonin) does produce anxiety and depression symptoms. The diagnostic criteria for neurasthenia are shown in table 2-2. Many of the signs and symptoms of neurasthenia overlap with depressive and anxiety disorders, making it difficult to diagnose. Nevertheless, several of the symptoms of neurasthenia, such as fatigue, difficultly concentrating, and lack of improvement after sleep, fit with symptoms that are seen in space. In the Russian program, some psychopharmacologic agents are administered prophylactically to prevent the cognitive symptoms [20].

Other

The DSM IV lists a variety of psychoses, neuroses, and personality disorders that are all possible complicating factors for a spaceflight. They are all very unlikely, however, due to preselection screening and the long training periods preceding flights. Any major psychiatric diagnosis would be likely to be identified well before a long-duration flight.

Approaches to Psychosocial Issues

As is true for most medical problems that can arise in space, the key factor for psychosocial problems is prevention. Once a psychosocial problem exists, treatment can be difficult, and, in some cases, as mentioned above, the mission has to be terminated. Preventing psychosocial problems is difficult because psychosocial states cannot be followed as reliably as, for example, calcium loss from bone. Prevention relies on selecting out those who have psychological problems and then developing a high level of awareness among crew members so that they can detect a problem early.

Selection

There is a natural tension in selecting astronauts for long-duration missions. On the one hand, individuals who are achievement oriented and competitive will be attracted to the astronaut corps. These qualities are important to foster a dynamic, productive program. On the other hand, however, excessive competitiveness or a consuming interest in achievement can be wearing on others during a long-duration mission. The selection process is directed toward finding well-integrated people who are accomplished and have significant individual achievements, but who also can work well with others to accomplish team goals.

Screening or "select out"

Screening out individuals with significant psychiatric problems was shown to be critical during one early Antarctic mission in the 1950s. One of the crew sent to establish the station had a psychotic episode while at the pole, which created a tough situation for the entire crew. The individual had to be sedated and isolated until a rescue could be arranged [23]. This event prompted the establishment of more aggressive screening procedures for significant psychiatric illness. Selection for the astronaut corps focuses on identifying significant psychiatric diagnoses or personality disorders. The process involves a battery of paper-and-pencil psychological tests and a psychological interview [24]. This is a "select out" process, as it identifies factors that should keep a particular person out of the astronaut corps. The process is not designed to identify those individuals who would be well suited for long-duration space missions.

Choosing people who will do well or "select in"

Once an individual is in the astronaut corps, several methods exist to select that person for a long-duration mission. One method could be called the "administrative/experienced-based" method. In this approach an individual (e.g., the head of the astronaut corps) or a committee (e.g., experienced astronauts) makes their best assessment of crew composition based on their knowledge of the potential crew members and other factors. This is a common-sense approach that allows factors such as seniority, national origin, and so on, to be included in the selection process. The drawback of this approach is that it can easily be perceived as overly political and open to prejudice.

Another method could be called the "self-select" method. In this approach a group of potential candidates for a mission is given the opportunity to rate who they think they would most like to fly with. The results from this assessment are then used to put together a crew of people who have indicated that they would like to work together.

One other approach could be called the "field test" approach. In this method, a group of people is placed into an analog environment (e.g., camping trip, isolation chamber) or given a test (e.g., the Homeostat test used in Russia [25]), and an expert observer notes their interactions. Crew selection is then based on the input from the observer. These different methods are not mutually exclusive, and elements of all of them can be incorporated into a combined selection procedure. For example, a group of candidates could be placed in an analog setting, then each member of the group could be given the opportunity to rate who they would like to fly with. A psychological expert and an administrator/experienced astronaut could review these ratings.

Testing select-in strategies scientifically is extremely difficult because it could take years and many subjects before a statistically meaningful comparison between different selection techniques could be produced. Considering the variability among people and missions, it is unlikely that any selection procedure could be validated based on data from space missions. Analog environments, such as Antarctica, submarines, or Special Forces deployments do offer the potential to study different selection strategies. From studies in Antarctica, a series of general selection principles have been formulated [26]. Successful Antarctic winterover personnel are flexible, sociable, and competent at their work. People who have high achievement needs and are easily bored will not do well. The criteria identified by Stuster [26] for selecting long-duration flyers are listed in table 2-3.

Table 2-3. Recommended personnel selection criteria for long-duration missions.

1. Develop a personnel selection program based on the behavioral principle that the best predictor of future performance is past performance.

2. Avoid over-reliance on psychological tests of personality and evaluations of blood chemistry (e.g., cortisol, catecholamines). Personality tests can provide unreliable indicators of performance, and differential blood chemistry might not result in differential behavior.

3. Emphasize actual performance of relevant behavior:
   a. Technical competence and task motivation
   b. Biographical data indicators of relevant skills and traits
   c. Interview responses
   d. Peer and supervisor ratings
   e. Performance and adjustment during high fidelity simulations

4. Identify technically competent candidates, but select individuals who exhibit the following appropriate social skills and behavioral traits:
   a. Social compatibility, or likeability
   b. Emotional control
   c. Patience
   d. Tolerance (low irritability)
   e. Self-confidence (without being egotistic, arrogant or boastful)
   f. Subordination of one's own interests to the goals of the team.
   g. Agreeableness and flexibility
   h. Practicality and hard-working attitude

   To the above scientifically established traits the following items identified anecdotally might be added:
   a. Tactfulness in interpersonal relations
   b. Effective conflict resolution skills
   c. Sense of humor
   d. The ability to be easily entertained.

5. Conduct high-fidelity simulations of planned expeditions, as the final step of the selection process, for formal evaluation of the relevant performance of candidate crew members before final assignments are made.

Adapted from Stuster [26], with permission.

Evaluating cultural differences

The trend in space exploration is toward international cooperation and collaboration. As a result, crews are composed of individuals from different nations with varied cultural backgrounds. This diversity can be a great strength for a mission, but in some circumstances it can also be a weakness. Studies have shown that groups composed of individuals with shared attitudes and values tend to be compatible and work well together [27]. As noted by Stuster, "it is important to note that the most successful (i.e. remarkable) expeditions have been conducted by relatively homogeneous groups or groups that have been organized specifically on the basis of compatibility" [28, p. 176]. Cultural backgrounds can affect the perception of appropriate leadership, proper role responsibilities, adequate communication, and fair decision making. Work habits and expectations about privacy, personal space, and working relationships have cultural components. Customs associated with personal hygiene and with personal family events, such as death, also differ among cultures.

In times of crisis, crewmates may understand the predicament and interpret more accurately the behavior of the crew member whose culture is more familiar to them. They may judge the behavior of a crew member from another culture as being a sign of disrespect or a personality conflict, when in fact the behavior reflects the person's culture. Such divisions can escalate what originated as a cultural misunderstanding into an individual personality conflict. Conflict resolution becomes much more difficult if the affected crew members' styles of expressing themselves are misinterpreted due to intercultural differences.

In the International Biomedical Expedition to Antarctica (IBEA), the international crew did not come together as a unified group and instead separated into individual national groups [17]. The NASA-Mir series of flights also created significant tensions between Russia and America as the crews and ground-control teams worked through differences in language, command styles, and objectives [6, 7]. Relationships were often strained.

To put these problems in perspective, however, both the IBEA and NASA-Mir programs were ultimately completed successfully. Also, these two programs shared a few characteristics. Preselection of the participants was minimal in both programs. The crews did not have much time to work together beforehand, and questions about the scientific and logistical aspects of the missions had not been fully worked out when the projects began. The lessons learned from these programs affected the planning of future missions. By close attention to selection and training, crews from a variety of cultures work together successfully on projects like the International Space Station and Antarctic research stations.

Gender in crew selection

As is true with cultural differences, mixed-gender crews offer both advantages and potential problems (see also chapter 10). Numerous examples exist of successful mixed gender expeditions. The addition of women to Antarctic crews is believed to have resulted in more productive missions with improved behavior [28]. With mixed-gender crews, however, there is greater potential for jealousies to develop and for intimate relationships to form. Even more dangerous for a long-duration mission is the breakup of an intimate relationship, which could affect not only the mood and behavior of the participants, but also the interactions within the entire crew. Instances of promiscuity, sexual harassment, dissolved marriages, and disruptive relationships have all occurred in Antarctic missions, with predicable negative effects on group functioning [28]. Also, because a significant percentage of the population is homosexual, single-gender missions do not necessarily preclude intimate relationships.

For long-duration space missions, different recommendations on how to approach mixed-gender crews have been made. One sensible approach is to select married couples, although the risk of divorce or significant conflict still exists. Ultimately, however, the success of the mission will depend on the motivation and professionalism of the crew. With careful selection and training, the crew will be fully aware of the potential problems that can arise and can work to avoid them. Also, clear, mutually agreed-upon rules for behavior are needed to ensure that the crew has a shared understanding of what is appropriate mission behavior.

Training

On short-duration spaceflights, astronauts can suppress animosities toward other crew members, and they can limit their contact with another person to professional interactions. This approach is totally unacceptable for long-duration spaceflight. To be effective on a long-duration flight, a crew needs an understanding of each other's strengths and weaknesses and a commitment to the success of the group. The crew will succeed or fail as a group, not as individuals, and so they must come together as a mutually supporting team. This requires good communication, superior conflict resolution skills, and sensitivity to cultural differences. Fortunately, training can help to improve communication, sharpen conflict resolution skills, improve teamwork, and foster understanding of different cultures.

In the aviation field, crew resource management training (CRM) is widely used to foster communication and teamwork between flight crew. Human factors studies have shown that poor communication was a key factor in many aviation accidents [29]. The addition of training has been shown to improve communication and to reinforce good practices [30]. Although it is difficult to prove if this has in turn improved safety, pilots who have received the training are perceived as more effective than those who have not [31]. When CRM-like training was implemented in a medical setting, error rates were dramatically reduced [32]. Overall, although some individuals will be better team members than others, team performance can be improved through training.

Similarly, efforts to train individuals in conflict resolution and cross-cultural sensitivity also have shown generally good results, although performing a controlled trial on a given training program is exceedingly hard to do [33]. Ultimately, the use of training is based on the common-sense notion that a well-informed and sensitive crew member will be able to handle difficult interpersonal situations tactfully and skillfully.

On the ground, training can be done through a combination of classroom and field work. In space, computer-based training materials, optimally using multimedia approaches, can be used by the crew for self-assessment and for refresher training. The most valuable form of training for the crews, however, is actual experience. Crews must have the opportunity to work together for long periods in analog environments before the flight to ensure that the crew members can relate to each other both professionally and socially.

Sleep

Lack of sleep can interact with other factors to exacerbate or precipitate psychosocial problems. The most well-publicized example of this from spaceflight is the case of *Mir 23*. On this flight, the crew was chronically sleep deprived because of their ongoing efforts to find a persistent ethylene glycol leak. This crew also had to deal with a fire, a near miss from a *Progress* spacecraft during a docking maneuver, and an actual collision with a *Progress* spacecraft. Toward the end of the mission, the ground team decided to send up a fresh crew to make repairs in the aftermath of the collision because the onboard crew was psychologically and physically drained [6, 7].

Chronic sleep loss leads to decrements in daytime performance that can jeopardize both productivity and performance [34]. On long-duration missions, there can be changes in the quality of sleep, and operational constraints can lead to chronic sleep loss [35]. Crews need to monitor their sleep and be aware when they are reaching a point where chronic sleep loss is putting them at risk for mistakes and psychosocial problems.

Medications

Typically, the use of psychoactive medications is discouraged in aviation settings. In long-duration spaceflight, however, the consequences of allowing psychosocial problems to reach a critical stage can easily outweigh the risk of drug side effects. Nevertheless, the decision of when and in what settings to use drugs must be made carefully. Also, guidelines need to be in place for dispensing drugs and recording their use.

Asthenia

The Russian medical kit includes at least four nootropics—drugs that are reported to improve cerebral blood flow and improve cognitive function (learning and memory). Some of the drugs have been reported to reduce anxiety as well. These drugs are more popular in Europe than they are in the United States, where they are not FDA approved. The four drugs used by the Russian space program are pyritinol, pantogram, phenibut, and piracetam. The most well known of the three is piracetam, which is a gamma-amino butyric acid (GABA) analogue that has been used to improve memory, enhance cognitive function, and reduce anxiety. While some studies have suggested a beneficial effect from piracetam [36], others have not [37]. In general, animal studies show that piracetam has an antiamnestic effect and can increase tolerance to hypoxia. The drug has few side effects and has been used regularly in the Russian space program. Piracetam has been given prophylactically in space to avoid the symptoms of asthenia. Phenibut is also widely used in Russia to relieve tension, anxiety, and fear. It is used in the therapy of disorders characterized by asthenia and depression.

In the United States, neurasthenia would likely be treated with selective serotonin reuptake inhibitor such as fluoxetine. Other antidepressant medications such as nefazodone and mirtazapine have also been used on the ground to treat neurasthenia. Benzodiazepines are also a possible treatment choice, as is a stimulant such as dexamphetamine (Dexedrine) [22].

Depression

On Earth, the mainstay of therapy for depression is selective serotonin reuptake inhibitors (SSRIs). The most well-known drug in this class is fluoxetine (Prozac), which is available in the U.S. medical kit on the space station. In addition to treating depression, the SSRIs may be useful for generalized anxiety. The drugs have an excellent safety record and are not addictive. Overdoses are rare and nonlethal. Side effects include insomnia and agitation (it is recommended to take the drug in the morning). Some patients may experience sedation. The SSRIs can produce gastrointestinal upset. An adequate trial of an antidepressant takes 4 weeks [38].

If fluoxetine is not effective, the U.S. medical kit also includes the tricyclic antidepressant nortriptyline, which may be a better choice for severe depression. Stimulant drugs can also be tried in depression. Dexamphetamine is available on the space station. Also, the Russian drug Syndocarb is a mild stimulant that has amphetamine-like actions.

Anxiety

The benzodiazepines are excellent drugs for anxiety, and both the Russian and American medical kits have several benzodiazepines to choose from. Diazepam, medazepam, phenazepam, temazepam, tofizopam, nitrazepam, and alprazolam have been available on the International Space Station. In addition to treatment for anxiety, they can be used to treat insomnia and seizures. Alprazolam has the advantage of being useful for anxiety combined with depression. Diazepam is long acting, which can be a problem if it is taken too frequently, but temazepam is short acting, making it a good choice for insomnia. Some drugs in the Russian onboard formulary (phenazepam, medazepam, and nitrazepam) are not FDA approved.

The benzodiazepines have a good safety profile but can cause side effects. The most common side effect is sedation, which can be undesirable in an operational setting. Other effects are weakness, amnesia, and nausea. The drugs can sometimes affect performance on motor tasks, and the crew will need to be very aware of this potential side effect [39]. Also, these drugs can lead to addiction or dependency.

The SSRIs can also be useful for anxiety. Several of the SSRIs have been approved for the treatment of generalized anxiety disorder in addition to their use in depression.

Insomnia

Sleeping medications are some of the most common drugs taken on short Shuttle flights [34] and are also needed during long-duration flights. As mentioned above, any of the benzodiazepines can be helpful for insomnia, but a short-acting benzodiazepine would be the best choice. Zolpidem (Ambien) is a very short-acting sleeping aid with minimal side effects. Diphenhydramine (Benadryl), the antihistamine, will also cause sedation and can be used as a sleeping aid. The Russian medical kit includes the herbal remedy valerian (*Valeriana officinalis*), which is used for insomnia and anxiety.

Continual use of sleeping medications can be hazardous. If a long-duration benzodiazepine such as diazepam is used, the metabolites can accumulate in the body, leading to a decrease in performance. Also, a crew member could develop dependency on the drug. In general, drugs for insomnia should be viewed as a short-term solution.

Countermeasures for Psychosocial Problems in Space

Being a member of a well-run, efficient, and high-performing team can be one of the most memorable and satisfying events of a lifetime. The goal of the countermeasure program is to ensure that the crews of long-duration flights form effective teams. If the crew is well trained, works well together, and has a high level of motivation, it will be able to endure grueling hardships. The key to forming this team is proper selection and training.

The first step is to screen potential crew members for serious psychiatric illness. A neurotic person with a personality disorder could end up being a drain on the group's resources, rather than contributing to their success. After screening, long-duration crews should be assembled carefully. Each potential crew must have a period of time, before selection is finalized, in which they can work together as team in a demanding environment. In those instances where crews were not carefully selected, such as the International Biomedical Expedition to Antarctica [17], or where they did not have time to work together, such as the SFINCSS-99 isolation study [5], significant problems occurred. Once the crew is in place, they need training in how to deal with conflict, how to recognize depression and asthenia, and how to assess their own condition accurately.

Conflict Resolution Training

Before the flight, the crew should receive classroom training in how to approach conflict where they can learn and practice the key skills. In this training, each crew member should develop an awareness of their "hot button" issues—those things that they may react to very strongly because of their own history and background. They should also know the hot button issues of their crewmates and strive not to bait or provoke their crewmates on those issues. The crew members should practice dealing with simulated conflicts and learn to avoid "win-lose" thinking, where one side has to be the clear loser. Whenever possible, they should try to find ways to resolve the conflict so that both sides gain something. They can learn that an argument over a matter of principle is not likely to be resolved and should be avoided unless it is critical for the mission. To solve conflict they need to find areas of agreement and not difference. These skills are often stressed in the business world [40].

Conflict resolution training can be computerized and provided during the flight [41]. In this way, the crew members can work through simulated conflicts or review material they had learned on the ground at times inflight when it may be timely and useful. Even though the crew member may not have someone they can talk to about how to approach a conflict they are having (the role a close friend or spouse could play on Earth), they can get ideas about different approaches to take by reviewing material presented both in video and text on a computer.

Ultimately, the goal is conflict resolution, not conflict avoidance. Suppression of anger or resentments over a long period can lead to outbursts at unexpected and sometimes inappropriate times. The crew needs to have the ability to discuss disagreements or points of tension. Sometimes this can occur at a regularly scheduled crew meeting or at some other agreed-upon gathering. At these meetings problems can be worked out without excessive emotion. Without the opportunity to do this, it is possible to have arguments and disagreements when the crew is tired or stressed. At those times the arguments can escalate quickly. Working through issues early helps prevent conflict at inopportune times.

Simulations

Settings where the crew needs to depend on each other provide an important way to establish an effective team. Outdoor survival training or living in an underwater habitat or other situations similar to spaceflight force the crew to work together. These experiences will quickly highlight any problems that may exist. In the Russian space program, a time where the crew works together closely is considered absolutely essential. This is also part of the International Space Station program.

Psychological Support Program

To help combat the sense of isolation and distance that can develop, crews are offered various ways to maintain contact with loved ones on the ground and with the world at large. Audio and video links can be arranged periodically to talk with family members. E-mail allows for crews to keep in touch with both family and friends on the ground. Special events like birthdays and anniversaries should be celebrated to provide a sense of continuity with life on the ground [2].

Packages can be delivered to the space station regularly, and these often include special videos, CDs, books, family pictures, and notes that help keep the crew member connected with family and friends. This support system has been noted in books written about long-duration missions as an extremely important factor for maintaining a stable mood [6, 7].

Problem-Solving Therapy

If depression develops despite the efforts at proper selection and training, there are some potential methods to allow the affected crew member to work through the depression using a computer-based system. Problem-solving treatment is an approach to mild and moderate depression that is used in clinics and can be performed after receiving minimal training [42]. The common-sense approach of the therapy and the fact that extensive training is not required makes problem-solving therapy suitable for implementation on a computer, which allows the crew member to work through it on his or her own.

Problem-solving therapy is a relatively new approach to treating depression and is simpler to understand and use than other therapies (e.g., cognitive-behavioral therapy). The method is based on having the subject develop and implement solutions to life problems that cause or contribute to depression. The treatment takes a seven-step approach. First there is an introduction to problem-solving therapy, followed by a time to clarify and define the problems (e.g., conflict with other crew member, monotony, overwork, etc.). The next step is to brainstorm and find solutions (schedule nonwork times, change work schedule, etc.) that are doable. The subject then chooses the preferred solution, implements it, and evaluates progress. As the problems are solved, the person is able to reestablish control over his or her life, which helps improve mood. Problem-solving therapy has been shown to be as effective as antidepressants in mild and moderate depression [42–45].

To someone who is not depressed the therapy sounds obvious. Once depression takes hold, however, feelings of hopelessness and powerlessness, as well as cognitive slowing, can severely impair the ability of the affected person to take action. The therapy provides a structure for action.

Medications

Using psychoactive drugs in space poses a significant dilemma. Benzodiazepines can produce sedation and impair the performance of motor tasks. SSRIs, though usually well tolerated, can produce insomnia, which might exacerbate fatigue. Uncontrolled use of psychotropic drugs is possible on the space station, and this has reportedly resulted in excessive medication being taken at times [20]. Intervening too late with medications, or undermedicating, can also be hazardous. Antidepressants can take weeks to have their full effect, which is a long time to live with crewmate whose performance may be impaired. To find the right time to start medication requires close cooperation between the crew member and the flight surgeon on the ground.

In addition to starting medication when a problem, like anxiety or depression, arises, medications can be taken prophylactically. Some of the nootropics and other medications in the Russian onboard formulary are administered when a crew member is asymptomatic. They are used like vitamins or nutritional supplements. Since these drugs have, by and large, good safety profiles, side effects should be rare with this approach. But the nature and magnitude of the benefit the crew members receive from these medications is not clear, making it difficult to determine if the benefit of these drugs is worth the risk of taking them.

Monitoring Psychological Well-being

Typically, in an aviation environment, where crew members can view illness as a sign of weakness, reporting problems will be limited. A crew member may want to see if the condition will improve or try to "tough out" a problem rather than discuss it with the ground. Also, even if crew members are concerned about a problem, they may not want to discuss it because of concerns about confidentiality and how it will be perceived. Because of these factors, it is important to provide two things to a crew: (1) objective, self-administered tests of performance and (2) onboard self-assessment tools.

Self-assessment

For space station flights, and even more importantly for Mars flights, the crews will need tools to assess their mental state. On the International Space Station, the crews take a cognitive test battery called WinSCAT (Spaceflight Cognitive Assessment Tool for Windows). The test evaluates memory and other cognitive functions. The crew member establishes a norm for their performance on the test, and if they score significantly outside of their preflight norms on two consecutive performances, this information is shared with the ground. This information allows the crew member and ground to discuss the reasons for the change and assess sleep, mood, and other factors.

Crew members should also have tools available to assess their mood. A variety of different tools exist to measure depression (e.g., Beck scale, Hamilton scale) and these different questionnaires may be useful to the crew. Also, the PHQ-9 (Personal Health Questionnaire 9) is useful. The PHQ-9 is a multiple-choice depression measure commonly used in conjunction with problem-solving treatment. It is part of the PRIME-MD brief evaluation of mental health problems in primary care patients, and it is based directly on the DSM-IV. The PHQ-9 asks about the occurrence and severity of each diagnostic criterion for major depression. It has high face validity and can be used as a self-administered depression measure. Questionnaires about anxiety also exist.

Self-assessment questionnaires and tests can be administered by computer. The use of computers for assessments, as described above, offers some advantages. Studies have shown that people will often provide more honest answers to a computer than they will to an interviewer. In space, if the computer is not connected to downlink, the crew member can answer the questions with complete honesty, without worry that others might view the answers.

Voice, Text Monitoring

Although assessments that crew members can take anonymously or only share under defined circumstances can be useful, as discussed above, past experience has shown that psychosocial questionnaires that do go to ground control or to an investigator are often not useful. During the NASA-Mir program, the crew members participated in a study to assess their status using the Profile of Mood States (POMS) [11]. In general, the questionnaires showed stable moods across the time periods. Subsequent publications about the actual situation on the *Mir* at various times during the program showed that there were significant and marked changes in mood in several of the participants. For a variety of reasons, the crew members were unwilling to share the actual situation on the station with others.

Because of this unwillingness, monitoring methods have been developed in the Russian space program to infer the psychosocial situation on the spacecraft indirectly. Voice communications from the spacecraft are subjected to speech analysis, and biotelemetry as well as performance data (mistakes on tasks) are examined. Medical and psychological specialists assess this information. The text of reports sent from the spacecraft is also analyzed for content, length, and word choice to help provide insights into the state of mind of the writer. The chief psychoneurologist at the control center then draws an overall conclusion from the data. These assessments are used to decide whether to move forward with demanding tasks, like extravehicular activity, or whether to institute countermeasures [20].

Recommendations Based on Current Knowledge

Although this chapter focuses on what can go wrong on a long-duration spaceflight, the larger story is that the vast majority of long-duration spaceflights are completed successfully. Crew members have great adaptive powers, and when motivation is high they can accomplish nearly any goal. Also, being a member of a well-functioning team is a satisfying experience. The goal of the countermeasure program is to make sure the teams are strong and function well. Data from the literature and past experience suggest that the following guidelines should be followed:

1. Select-out individuals with psychiatric problems. Individuals with personality disorders and other psychopathology could be very difficult to deal with on a long-duration flight.

2. Select-in based on background, self-selection, and time spent working together. Questionnaires and interviews are inadequate for selecting members of a team because someone who is highly motivated to make the trip can fake them. It is essential for a potential crew to work together as a team before the crew selection is finalized.

3. Establish ground rules. A source of stress for a crew is the need to work out significant issues (such as sexual relationships, division of tasks among national representatives, etc.) when the mission is already underway. The crew should take the time to talk through how they will deal with various situations to develop a common understanding for the major issues.

4. Train participants to recognize depression and asthenia in self and others. Once depression is firmly established, it can take significant amount of time to reverse. The crew needs to be aware that depression is a risk, and they must understand how to recognize it and seek appropriate treatment.

5. Train participants in conflict resolution. The crew will not be able to avoid conflict, but they can work to resolve successfully the conflicts that arise. Although some people are naturally better at conflict resolution than others, there are aspects of conflict resolution that can be taught.

6. Provide onboard self-assessment and training tools. Questionnaires with data about mood and feelings that are transmitted to the ground can be unreliable. The crew needs the ability to assess themselves and to be sensitive to the changes in others.

7. Monitor body mass. In addition to changes in mood and cognitive function, another sign of depression is weight loss. Any unexplained weight loss should be aggressively pursued.

8. Monitor cognitive function. Although mood may be difficult to assess, a clear decrement in cognitive function needs to be explained.

9. Assess sleep. Poor sleep is a key contributor to psychosocial problems. If the crew is chronically sleep deprived, the ground control should be able to obtain this information. Also, the crew should be aware of the possibility of getting out of synchrony with their circadian rhythms. The proper use of bright light can help resynchronize their sleep and work schedules.

10. Aim for moderation in drug usage. If medications are used too readily, problems with side effects can develop. If there is too long a wait before starting medication, the crew member may suffer unnecessarily.

References

1. Ignatius, A., Russian psychiatrist tries to make sure Russian cosmonaut stays up. *Wall Street Journal*, 1992, p. 1
2. Holland, A.W., Psychology of Spaceflight, in *Human Spaceflight: Mission Analysis and Design*, W.J. Larson and L.K. Pranke, eds. 2000, McGraw Hill, New York, pp. 155–91.

3. APA, *Diagnostic and Statistical Manual of Mental Disorders*, 4th ed. 1994, American Psychiatric Association, Washington, DC.

4. Lebedev, V., *Diary of a Cosmonaut: 211 Days in Space*, D. Puckett and C.W. Harrison, eds. 1988, PhytoResource Research, College Station, TX.

5. Gushin, V.I., J.M. Pustynnikova, and T.M. Smirnova, Interrelations between the small isolated groups with homogeneous and heterogeneous composition. Journal of Human Performance in Extreme Environments, 2001. 6(1): 26–33.

6. Linenger, J.M., *Off the Planet: Surviving Five Perilous Months Aboard the Space Station Mir*. 2000, McGraw Hill, New York.

7. Foale, C., *Waystation to the Stars: The Story of Mir, Michael and Me*. 1999, Headline Book Publishing, London.

8. Kranz, G., *Failure is Not an Option*. 2000, Simon and Schuster, New York, pp. 223–33.

9. Kraft, C., *Flight: My Life in Mission Control*. 2002, Dutton/Plume, New York.

10. Carr, G.P., Human experience in space. Cutis, 1991. 48(4): 289–90.

11. Kanas, N., et al., Crewmember and ground personnel interactions over time during Shuttle/Mir space missions. Aviation, Space, and Environmental Medicine, 2001. 72(5): 453–61.

12. Tarzi, S., et al., Methicillin-resistant *Staphylococcus aureus*: psychological impact of hospitalization and isolation in an older adult population. The Journal of Hospital Infection, 2001. 49(4): 250–54.

13. Gilbert, P., The evolution of social power and its role in depression, in *Depression: The Evolution of Powerlessness*, P. Gilbert, ed. 1992, The Guilford Press, New York, pp. 147–86.

14. Newkirk, D., *Almanac of Soviet Manned Space Flight*. 1990, Gulf Publishing Company, Houston, TX.

15. BBC Radio, Summary of World Broadcasts: Savinykh's Diary on State of Health of Vasyutin, in Pravda. January 7, 1986, London.

16. Channel 4 Television, Helen Sharman's Tomorrow's World Interview. June 1995, London.

17. Rivolier, J., G. Cazes, and I. McCormick, The International Biomedical Expedition to the Antarctic: psychological evaluations of the field party, in *From Antarctica to Outer Space: Life in Isolation and Confinement*, A.A. Harrison, Y.A. Clearwater, and C.P. McKay, eds. 1991, Springer-Verlag, New York, pp. 283–90.

18. Lugg, D., Current international human factors research in Antarctica, in *From Antarctica to Outer Space: Life in Isolation and Confinement*, A.A. Harrison, Y.A. Clearwater, and C.P. McKay, eds. 1991, Springer-Velag, New York, pp. 31–42.

19. Weybrew, B.B., Three decades of nuclear submarine research: implications for space and Antarctic research, in *From Antarctica to Outer Space: Life in Isolation and Confinement*, A.A. Harrison, Y.A. Clearwater, and C.P. McKay, eds. 1991, Springer-Verlag, New York, pp. 103–14.

20. Myasnikov, V.I., and I.S. Zamaletdinov, Psychological states and group interactions of crew members in flight, in *Space Biology and Medicine*, A.E. Nicogossian, et al., eds. 1996, American Institute of Aeronautics and Astronautics, Reston, VA, pp. 419–32.

21. Kanas, N., et al., Asthenia—does it exist in space? Psychosomatic Medicine. Special Issue: Outerspace Research, 2001. 63(6): 874–80.

22. Sadock, B.J., and V.A. Sadock, Chronic fatigue syndrome and neurasthenia, in *Kaplan and Sadock's Synopsis of Psychiatry*. 2003, Lippincott Williams and Wilkins, New York, pp. 661–67.

23. Stuster, J., Behavioral effects of isolation and confinement, in *Bold Endeavors: Lessons from Polar and Space Exploration*. 1996, Naval Institute Press, Annapolis, MD, pp. 7–13.

24. Santy, P.A., and D.R Jones, An overview of international issues in astronaut psychological selection. Aviation, Space and Environmental Medicine, 1994. 65(10 Pt 1): 900–3.

25. Eskov, K.N., et al., Group dynamics and crew interaction during isolation. Advances in Space Biology and Medicine, 1996. 5: 233–44.

26. Stuster, J., Personnel selection criteria, in *Bold Endeavors: Lessons from Polar and Space Exploration*. 1996, Naval Institute Press, Annapolis, MD, pp. 247–70.

27. Connors, M.M., A.A. Harrison, and F.R. Akins, Small groups, in *Living Aloft: Human Requirements for Extended Spaceflight*. 1985, NASA, Washington, DC, pp. 145–86.
28. Stuster, J., Group interaction, in *Bold Endeavors: Lessons from Polar and Space Exploration*. 1996, Naval Institute Press, Annapolis, MD, pp. 164–87.
29. Wiegmann, D.A., and S.A. Shappell, Human error and crew resource management failures in Naval aviation mishaps: a review of U.S. Naval Safety Center data, 1990–96. Aviation, Space, and Environmental Medicine, 1999. 70(12): 1147–51.
30. Salas, E., et al., Team training in the skies: does crew resource management (CRM) training work? Human Factors, 2001. 43(4): 641–74.
31. Helmreich, R.L., et al., Preliminary results from the evaluation of cockpit resource management training: performance ratings of flightcrews. Aviation, Space, and Environmental Medicine, 1990. 61(6). 576–79.
32. Morey, J.C., et al., Error reduction and performance improvement in the emergency department through formal teamwork training: evaluation results of the MedTeams project. Health Service Research, 2002. 37(6): 1553–81.
33. Stevahn, L., et al., Effects of conflict resolution training integrated into a high school social studies curriculum. Journal of Social Psychology, 2002. 142(3): 305–31.
34. Dijk, D., et al., Sleep, Circadian rhythms and performance during *Space Shuttle* missions, in *The Neurolab Spacelab Mission:Neuroscience Research in Space*, J.C. Buckey and J.L. Homick, eds. 2003, NASA, Houston, TX, p. 211–22.
35. Gundel, A., V.V. Polyakov, and J. Zulley, The alteration of human sleep and circadian rhythms during spaceflight. Journal of Sleep Research, 1997. 6(1): 1–8.
36. Dimond, S.J., and E.M. Brouwers, Increase in the power of human memory in normal man through the use of drugs. Psychopharmacology (Berlin), 1976. 49(3): 307–9.
37. Abuzzahab, F.S.S., et al., A double blind investigation of piracetam (Nootropil) vs placebo in geriatric memory. Pharmakopsychiatrie Neuro-Psychopharmakologie, 1977. 10(2): 49–56.
38. Schatzberg, A.F., J.O. Cole, and C. DeBattista, Antidepressants, in *Manual of Clinical Psychopharmacology*. 2003, American Psychiatric Publishing, Washington, DC, pp. 37–157.
39. Schatzberg, A.F., J.O. Cole, and C. DeBattista, Antianxiety agents, in *Manual of Clinical Psychopharmacology*. 2003, American Psychiatric Publishing, Washington, DC, pp. 321–80.
40. Greenhalgh, L., *Managing Strategic Relationships: The Key to Business Success*. 2001, Free Press, New York.
41. Bosworth, K., et al., Using multimedia to teach conflict-resolution skills to young adolescents. American Journal of Preventive Medicine, 1996. 12(5 Suppl): 65–74.
42. Barrett, J.E., et al., The treatment effectiveness project. A comparison of the effectiveness of paroxetine, problem-solving therapy, and placebo in the treatment of minor depression and dysthymia in primary care patients: background and research plan. General Hospital Psychiatry, 1999. 21(4): 260–73.
43. Mynors-Wallis, L., Problem-solving treatment: evidence for effectiveness and feasibility in primary care. International Journal of Psychiatry in Medicine, 1996. 26(3): 249–62.
44. Mynors-Wallis, L., et al., A randomised controlled trial and cost analysis of problem-solving treatment for emotional disorders given by community nurses in primary care. British Journal of Psychiatry, 1997. 170: 113–19.
45. Mynors-Wallis, L.M., et al., Randomised controlled trial of problem solving treatment, antidepressant medication, and combined treatment for major depression in primary care. British Medical Journal, 2000. 320(7226): 26–30.
46. WHO, *The International Classification of Mental and Behavioural Disorders,* 10th ed., *Diagnostic Criteria for Research.* 1993, World Health Organization, Geneva.

Introduction

In the novel *Space* by James Michener, the fictional Apollo astronauts encounter a situation that fortunately did not occur during the real program [1]. In the book, the astronauts are on the surface of the Moon when the radiation from a solar flare arrives. They receive fatal doses. The inspiration for this scenario may have been the solar storm of August 4, 1972, which over half a day delivered a radiation dose that could have caused radiation sickness or death had it reached the crew while they were out on the lunar surface or in the thinly shielded lunar module [2]. Although fictional, the story highlights that radiation can be a major challenge on a space mission and one that has to be predicted and understood.

The Earth's mass, atmosphere, and magnetic field provide the planet's surface with considerable protection from space radiation. Offering a comparable level of protection in current and planned spacecraft presents a significant engineering challenge and would greatly increase the cost and complexity of space operations. As a result, space travel exposes crew members to ionizing radiation at levels much higher than they would experience on Earth. Also, the nature of the radiation exposure in space differs from that on Earth. Solar storms, which eject powerful bursts of high-energy protons and other particles, do not usually contribute significantly to radiation exposure on the Earth's surface. As noted above, however, a solar flare could deliver a fatal radiation dose if the crew member were outside of the spacecraft when the radiation arrived. Galactic cosmic radiation, which consists of very high-energy particles, reaches the Earth's surface only in small amounts. In interplanetary space, however, cosmic radiation is significant and hard to shield against. Because exposure to this kind of radiation is so limited on Earth, experience with its long-term effects is minimal.

Radiation exposure can be an acute and long-term problem for long-duration spaceflight, and it must be properly monitored and assessed. This chapter discusses the types of radiation experienced in space, the likely biological effects of this radiation exposure, and methods to reduce the adverse consequences.

Radiation Concepts and Terms

Ionizing Radiation

Although most biological and physiological work is concerned with objects the size of molecules and larger, the initial effects of radiation take place at the atomic level. Most radiation effects involve interactions of high-energy photons (gamma rays, x-rays) or particles (protons, neutrons, nuclei of atoms) with electrons and nuclei in atoms. The most common interaction is an ionization. Energy is imparted from the radiation to the atom, ejecting an electron and creating an ion. This ionized atom or free electron can, in turn, form free radicals (highly reactive molecules like superoxide) that damage surrounding molecules. Although any molecule can be damaged, the molecules of most concern are the DNA molecules in the nucleus. Free radicals can damage DNA, and ionizations within the DNA molecule itself can also do damage. Figure 3-1 summarizes how radiation can harm DNA molecules.

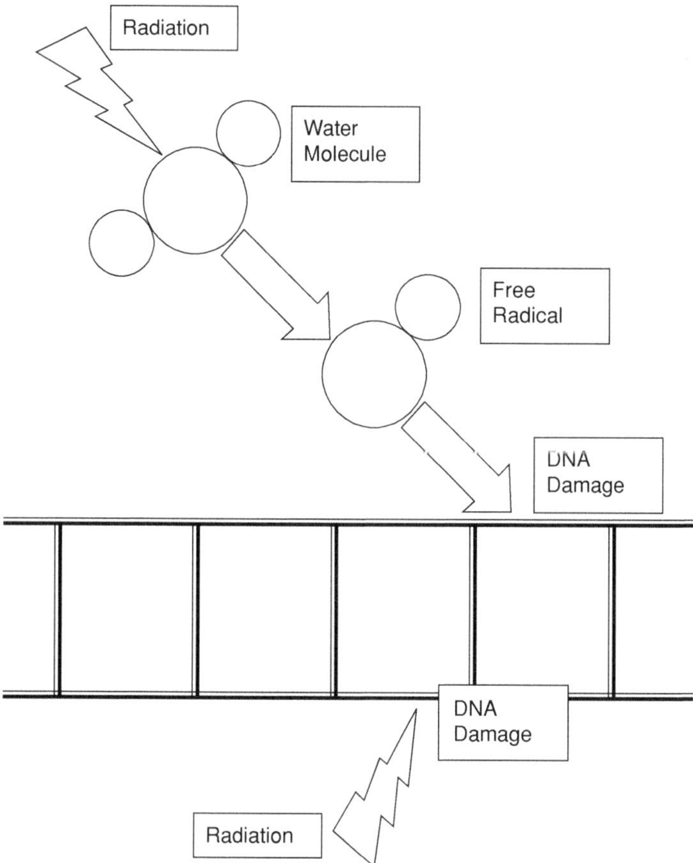

Figure 3-1. Ionizing radiation can produce damage in two ways. The radiation can create free radicals, and these in turn damage molecules in the DNA. If the atoms within the DNA are ionized, this can damage DNA directly.

There can be a difference between how photon radiation and particle radiation produce damage. Figure 3-2 illustrates the different ways radiation can deposit energy in cells. X-rays create ionizations throughout the area of irradiation and deposit small amounts of energy in multiple cells. In contrast, a particle like a neutron may deposit the same total amount of energy, but it does so all in one track and encounters just a few cells. This difference can be important when considering the effects of galactic cosmic radiation.

The higher the energy in the radiation, the more damaging it can be. Table 3-1 summarizes the different types of high-energy radiation. The energy carried by ionizing radiation is usually measured in electron volts (eV). Medical x-rays, for example, have energies measured in thousands of electron volts (KeV), while some of the particles that travel through interplanetary space (galactic cosmic rays) have energies of millions of electron volts (MeV) or more. The energy of ionizing radiation is only one measure of its effectiveness. As radiation travels through tissue, the energy

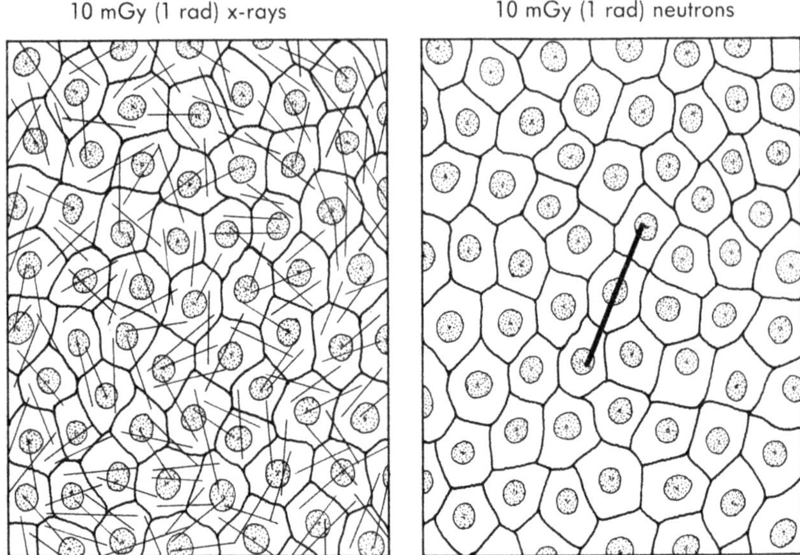

Figure 3-2. The key difference between radiation with x-rays and radiation with a high-energy particle. The left panel depicts cells irradiated by x-rays, and the right panel shows cells receiving the same dose of radiation from neutrons. On the left, when a group of cells are irradiated with x-rays, charged particle tracks traverse every cell. On the right, however, the neutrons only pass through a small proportion of cells. When the x-ray dose is increased, the average energy deposited per cell increases; when the neutron dose is increased the number of cells receiving radiation increases. Figure reprinted from Hall and Cox [9], with permission of Elsevier.

it imparts through ionizations is called linear energy transfer (LET). Compared to photons, the protons and heavy nuclei present in solar flares and galactic cosmic radiation are much larger and more massive. Particles like protons and heavy nuclei can produce many more ionizations than photons per unit distance traveled. Because of this, protons and heavy nuclei are considered high LET radiation. Also, when a heavy particle stops within the tissue, the LET increases as it slows down, imparting energy to the molecules in the area as the particle stops—like a speeding car hitting a tree. In fact, this characteristic of proton irradiation is used in some radiation treatment protocols to maximize tissue damage at the point where the protons stop. Another feature of high-energy particles is that they are more likely to interact with other nuclei. These interactions can in turn produce secondary radiation, which can also be harmful.

Solar Particle Events (Solar Flares)

One of the unique forms of radiation encountered in space is radiation from a solar particle event, or solar flare, which comes from solar storms. These powerful storms release energy that arrives at the Earth and can damage satellites and disrupt com-

Table 3-1. Characteristics of various types of ionizing radiation.

Radiation type	Energy	Fluence	Composition	Dose delivered
Chest X-ray	40–60 KeV	5×10^7 photons/cm²/sec	Photons	0.0001 Sv
Relativistic electrons	500 KeV and up	Variable 10^2–10^4 electrons/cm²/sec	Electrons	Variable
Radiation therapy (cobalt source)	1.2 MeV	10^9 photons/cm²/sec	Photons	Varies, up to 80 Sv
Solar particle event (August 1972, no shielding)	10–100 MeV	10^{10} particles/cm²/sec [4]	Mainly protons	Variable, can be 1–5 Sv
Galactic cosmic radiation (no shielding, solar minimum)	300–3000 MeV [3]	4 protons/cm²/sec; 0.4 helium ions/cm²/sec; 0.04 HZE particles/cm²/sec [4]	85% Protons 14% helium nuclei, 1% heavier nuclei	0.5 Sv/year

munications. This radiation consists mainly of protons, although some ions with a higher atomic number (such as carbon, nitrogen and oxygen) are also present [3]. When a solar flare occurs, a burst of electromagnetic energy can precede the particles and travel from the Sun to reach the Earth within 8 minutes. As soon as 10 minutes after that, energetic particles arrive at the Earth [4]. The peak particle intensities usually do not occur until 4–6 hours later [2].

Usually, a spacecraft provides adequate shielding against radiation from solar flares. Also, an exposure in interplanetary space is more concerning than an exposure in low Earth orbit because the magnetic field of the Earth provides some protection to orbiting crews. Crew members outside of a spacecraft in a spacesuit, however, would be at significant risk. Furthermore, a strong enough storm could still deliver significant doses to the crew even if they were inside the spacecraft. To be safe, crew members should be in a well-shielded environment when the energy from a solar particle event arrives. This requires some ability to predict the storms and warn the crew. Predicting and monitoring these events is a major priority to adequately shield the crew when the particles arrive. Solar storms can begin on the side of Sun facing away from Earth, and then can become a problem as the Sun rotates the storm toward the Earth or Mars-bound spacecraft. Therefore, the ability to monitor both sides of the Sun (i.e., facing Earth and facing away from Earth) would help to provide advance warning [5]. On a Mars mission, where the crew could be far from Earth in a spacecraft or on the surface of Mars, they will have to depend on reliable monitoring.

Galactic Cosmic Radiation

Galactic cosmic radiation is the background radiation in the solar system. Whereas solar flares are intermittent bursts of fairly high-energy, low-mass particles, galactic cosmic radiation is a continuous flux of very high-energy, higher mass particles. The particles in galactic cosmic radiation range from protons to the nuclei of heavier atoms such as iron. The damage this kind of radiation can do in tissue increases with the atomic number (Z) of the particle. These high-energy, high atomic number nuclei are called HZE particles. In general, an HZE particle is defined as a nucleus with an atomic number > 2 and with sufficient energy to penetrate at least 1 mm of spacecraft or spacesuit shielding [6]. If these particles stop in tissue or shielding, they can deposit large amounts of energy (i.e., they have a high LET). Galactic cosmic radiation is often referred to as high-LET radiation (as opposed to gamma rays or x-rays, which are low-LET radiations).

Galactic cosmic radiation is hard to shield against due to the high energies of the individual particles. For example, a high-energy iron nucleus encountering an aluminum shield might stop in the metal. But, in the process, the iron particle may be involved in a head-on collision with the nucleus of an aluminum atom. When this happens, the nuclei can fragment into lighter nuclei that will continue traveling. These fragments can in turn have other collisions. The net result is the production of secondary radiation that enters the spacecraft. In fact, dense materials like lead that shield effectively against gamma rays and x-rays are poor shields for galactic cosmic radiation. These dense materials create more secondary radiation than materials like hydrogen or water. Shielding strategies to protect against galactic cosmic radiation are discussed later in this chapter.

Relativistic Electrons

Belts containing radiation (the Van Allen belts) surround the Earth. The outer electron belt contains relativistic electrons, which are electrons with energies starting at about 500 KeV (see table 3-1 for energy comparison). These electrons can penetrate space suits. The risk from these electrons varies, since their flux fluctuates over time by many orders of magnitude. Nevertheless, a space station could spend up to 20% of its orbit in the relativistic electron belt every 24 hours. Over time, the cumulative dose of radiation from electrons can be significant depending on the orbit and flux.

One significant risk from electrons is performing an EVA during a highly relativistic electron event (HRE). During an HRE, the flux of electrons can reach very high levels. If this were to occur, the radiation dose delivered to the skin and eyes could put a crew member over the short-term dose limit. While exposure to these radiation belts is an important issue for operations in low-Earth orbit [2], on an interplanetary flight electrons would not be a concern.

Fluence

Fluence (or flux) is the measure of how much radiation passes through a given area per unit time. Radiation delivered for medical treatments usually has a high fluence. Solar particle events can also produce an extremely high fluence. One large solar particle event in 1972 showed 5×10^9 particles/cm^2 with energies > 30 MeV and 1.1×10^{10} particles/ cm^2 with energies > 10 MeV [7]. Galactic cosmic radiation, in contrast, is delivered at a low fluence. Table 3-1 shows the different fluence of solar flares and galactic radiation. The fluence of radiation can be important because cells

Table 3-2. Radiation exposures from various activities.

Activity	Dose
1 year in Houston	0.001 Sv [30]
1 year in Denver	0.002 Sv [30]
8-day Shuttle flight	0.0053 Sv [30]
5 month stay on Mir	0.16 Sv [30]
Dose on Mir during October 1989 solar event	0.15 Sv [2]
Trip to Moon	0.011 Sv [30]
Trip to Mars, interplanetary flight	0.50 Sv/year [3]
Surface of Mars	0.12 Sv/year [45]

have the ability to recover from ionizing radiation damage. Cellular repair mechanisms may be better able to keep up with the damage produced by a low fluence of radiation.

Absorbed Energy

The most important consideration for biologic effects, however, is not the energy of the radiation or its flux, but how much is absorbed by the tissues. The absorbed dose is the amount of energy available to do damage in tissue. This dose is measured in Gray (Gy); 1 Gy = 1 joule/kg. An older unit often seen in the literature is the rad; 100 rad = 1 Gy. A fatal single dose of radiation would be about 4–7 Gy to the entire body. Ninety days in low Earth orbit would expose the crew to approximately 0.07 Gy. The Gray, however, doesn't take into account the effectiveness of different forms of radiation. As mentioned earlier, a heavy nucleus has a much greater cross-section for interaction with tissue, and so it may have the ability to do more damage. This difference is called the relative biologic effectiveness of radiation (RBE), and it is estimated with a quality factor. Gamma rays and x-rays have a quality factor of one. Protons over a wide range of energies have an RBE very similar to x-rays [4]. The value of RBE depends on the biological effect that is being considered. For example, neutrons have an RBE for cell killing that is similar to gamma rays, but they can have a much greater RBE for cataract formation. Iron nuclei (such as might appear in galactic cosmic radiation) have an estimated RBE of 20 for cell killing. The quality factor is multiplied by the dose in Grays to give a dose in Sieverts (Sv). Sieverts are the units used to measure radiation exposures and exposure limits. The 10-year career limit for astronauts ranges from 0.4 to 3 SV, depending on sex and the age when the astronaut was first exposed [8]. Table 3-2 shows the range of exposures in Seiverts for various activities.

Dose Rate

The rate at which a radiation dose is given is also an important consideration. Although the career limit for an astronaut can be as high as 3 Sv, if this dose were given in one brief exposure, it would cause acute radiation sickness and could be fatal. This dose, however, can be spread out over a career with apparent safety. To account for this, a dose rate effectiveness factor is often incorporated into risk estimates to take into account the ability of the body to repair damage from ionizing radiation.

Radiation Biology Relevant to Spaceflight

Ionizations Involving Water and Oxygen

The ionizations caused by radiation are randomly distributed. Because water is the most common substance in tissue (composing 70–80% of tissues), many of the effects of radiation stem from the products of interactions between radiation and water molecules. The amount of energy deposited with each interaction usually is much greater than the energy of chemical bonds, so the initial effects are to break these bonds and produce very energetic ionizations. The energetic ionizations and subsequent interactions initially occur so fast that the affected molecules do not have time to diffuse. Consequently, most of the chemical events (such as oxidation/reduction reactions with DNA and proteins) can occur only after less energetic products are produced. The most prevalent reaction is the production of hydroxyl radicals (OH•) and energetic electrons (or, at lower pH, hydrogen atoms). A radical is an atom or molecule with an unpaired electron in its outer shell and is usually very reactive. The hydroxyl radical will oxidize (receive electrons) from any organic molecule at a rate equal to its collision rate with the molecule, while the energetic electrons or hydrogen atoms are almost equally powerful reductants (donate electrons). Consequently, they cause a wide range of chemical changes.

Oxygen plays a critical role in the damage caused by ionizations from radiation. If oxygen levels are low in tissue, the tissue becomes radioresistant. The presence of oxygen can increase the sensitivity to radiation almost threefold. The effects occur because the oxygen molecule combines rapidly with chemical intermediates that had the potential to undergo reverse reactions (i.e., chemical repair). This results in free radicals (such as superoxide) that cannot undergo the reverse reactions. If these free radicals interact with DNA, they can produce significant damage. It is estimated that about two-thirds of the damage caused by x-rays or gamma rays is mediated by free radicals [9]. The addition of free radical scavengers (compounds that react with the free radicals before they damage with proteins or DNA) to the tissue helps minimize radiation damage. Antioxidants are compounds that can react with reactive intermediates or diffusing free radicals, reversing or preventing damage from occurring. In addition, some enzymes (such as superoxide dismutase and catalase) can neutralize some reactive intermediates (superoxide anions and hydrogen peroxide) and therefore moderate radiation injury.

Direct Ionizing Effects

In addition to damage from reactive molecules, damage can also occur by direct interaction of ionizing radiation with key molecules, especially DNA. Because of the high LET of heavier particles, these can be especially damaging to macromolecules, destroying them by multiple ionizations within a single molecule. DNA can

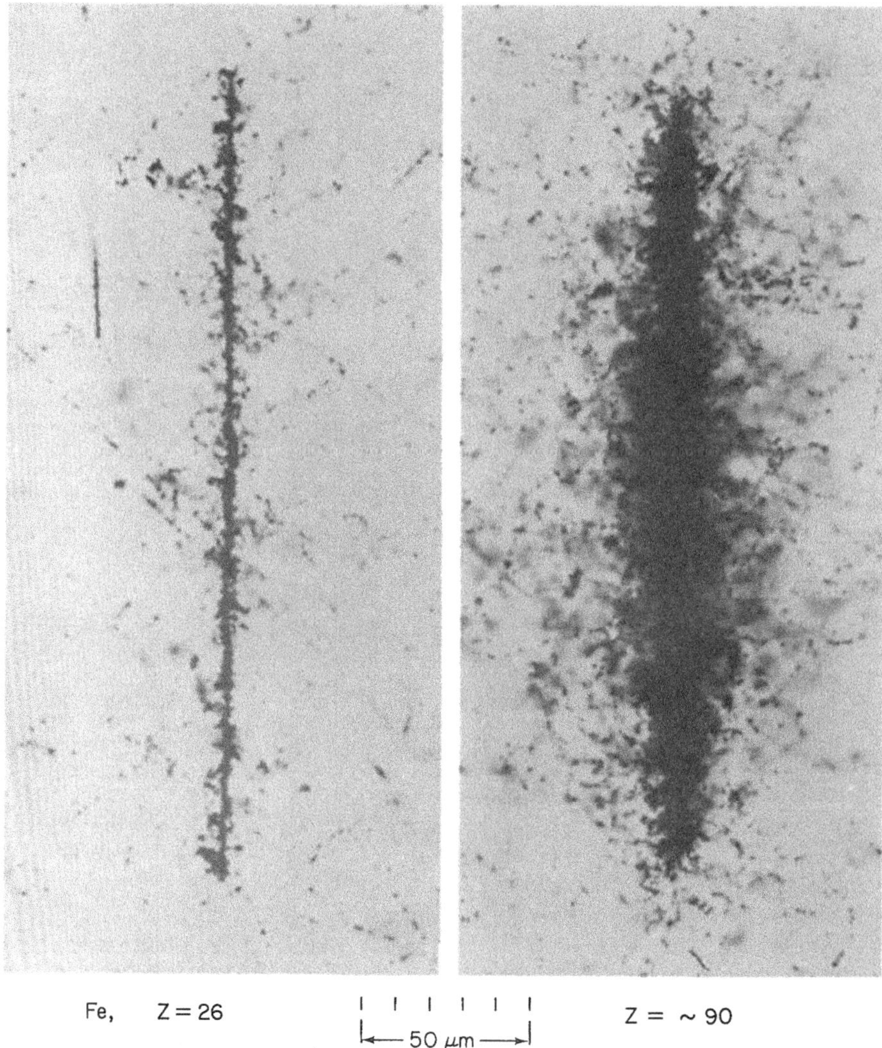

Figure 3-3. Although microlesions in tissue have been hard to demonstrate, this picture shows tracks of high-energy particles as they traverse a photographic emulsion. The panel on the left shows the track of an iron nucleus. Along the track of the ion there is an ionization track. Delta rays (electrons) are emitted to the sides. Figure reprinted from Curtis [10], with permission of Elsevier.

undergo single- and double-strand breaks as a result of an interaction with a heavy particle [9]. Heavy ions can produce an intensely ionizing track through the tissue, killing cells that lie within the path. The central track of the particle is surrounded by a "penumbra," where tissue can also be damaged by the delta rays (high-energy electrons) emitted from the ionization. Figure 3-3 shows a track of a heavy ion captured on photographic emulsion during a high-altitude balloon flight [10]. Although tracks

Table 3-3. Number of cells and cell nuclei hit by galactic cosmic rays on a simulated 3-year Mars mission.

	Retina	Hippocampus	Nucleus basalis of Meynert	Thalamus
Total number of cells	6.5×10^4	4.32×10^7	4116	1.83×10^6
Nuclear cross-sectional area (um3)	40	60	99	100
Total hits to a nucleus inside the spacecraft on a 3-year mission with a particle of $z \geq 15$	3.8×10^3	3.3×10^4	508	2.3×10^5

For this determination the shielding was estimated to be that required for a pressure vessel (1 g/cm^2 aluminum [approximately 0.4 cm thick]) plus the shielding provided by equipment (5 g/cm^2 aluminum [approximately 1.9 cm thick]). The scenario did not include a Martian surface stay. Data from Curtis et al. [11].

like this have not been directly visualized in tissue, they are thought to occur and are called microlesions. A microlesion would be expected to consist of at least four kinds of cells: cells killed by irreparable direct hits due to ionization, cells surviving but mutated as a consequence of direct hits due to ionization, cells surviving but mutated as a consequence of off-track electron damage, and unhit cells. This cell damage is of concern because killed cells might not be replaced in tissues that divide slowly (like the brain), and mutated cells might become cancerous.

The killing of cells by high-energy particles is of particular importance in the central nervous system because this cell loss could be permanent. Curtis et al. [11] used computer modeling to estimate how many cells in critical areas within the central nervous system would be hit by galactic cosmic radiation particles during a 3-year Mars mission [11]. Table 3-3 summarizes the results. Overall, ions with an atomic number ≥ 15 would hit approximately 6–12% of the entire population of neuron nuclei (depending on size and location). Also, there would be hits outside of the nucleus. Although it is not known if any functional capability would be lost due to the potential loss of brain cells due to these ion strikes, the data do show that the number of cells hit would be significant.

Immediate Dose-related Effects of Radiation

Large radiation doses, such as those from radiation therapy or exposure to a solar flare, produce immediate, dose-related effects. These effects, sometimes called deterministic or acute effects, are due to the depletion of cells, or cell functions, in a given organ or tissue. Except at the highest doses, most of the effects are due to failure of cells to reproduce. Therefore, the effects of radiation occur initially in those organ systems

Table 3-4. Direct (deterministic) effects of radiation.

Dose (Sv)	Predicted physiologic effects
0.1–0.5	No obvious effects, some minor changes in blood counts
0.5–1	Fatigue, transient reduction in lymphocyte and neutrophil counts; 5–10% experience nausea and vomiting for approximately 1 day; no deaths anticipated
1–2	50% reduction in lymphocytes and neutrophils; 25–50% experience nausea and vomiting for 1 day; no deaths anticipated.
2–3.5	75% reduction in circulating blood elements; most experience nausea and vomiting; loss of appetite, diarrhea, and minor hemorrhage also seen; death in 5 to 50% of those exposed ("bone marrow syndrome")
3.5–5.5	Nearly all experience nausea and vomiting on first day followed by fever, hemorrhage, diarrhea, and emaciation; death of 50–90% within 6 weeks; survivors convalesce for about 6 months (overlap of "bone marrow syndrome" and "GI syndrome")
5.5–7.5	All experience nausea and vomiting within 4 hours, followed by severe symptoms of radiation sickness; death of up to 100% ("GI syndrome")
7.5–10	Severe nausea and vomiting may continue into the third day; survival time reduced to less than 2.5 weeks (death may be due to effects on CNS and/or heart, primary pathophysiology is acute loss of function of the GI barriers to infection and fluid balance)
10–20	Nausea and vomiting within 1 to 2 hours; all die within 2 weeks ("CNS/Cardiac syndrome")
45	Incapacitation within hours; acute central nervous system syndrome due to radiation can be seen (disorientation, ataxia, convulsions, coma); all die within days.

Modified from Robbins and Yang [12]. This table is for whole-body exposures for the radiation dose. A local dose of the same magnitude would be much less dangerous.

whose normal functioning requires active cell reproduction, such as the bone marrow and gastrointestinal tract. The immediate clinical implications of the loss of the cellular reproductive capability depend on the tissue's function. For example, suppression of cell reproduction occurs at about the same levels of radiation exposure in both the testicles and the bone marrow, but the immediate clinical risk is associated only with the latter. The timing of the clinical effects depends on the normal reproduction of the cells. For example, when the doses are high enough to suppress reproduction in both the lining of the gastrointestinal tract and the bone marrow, the clinical consequences are seen in the gastrointestinal tract first. This is because the turnover rate of the gastrointestinal epithelium is much faster, and the bone marrow has more reserve capacity. Table 3-4 lists the main deterministic effects of radiation and the expected doses at which they occur. Nausea and vomiting are produced at the lower radiation levels. As the exposure increases, bone marrow suppression, and then central nervous system effects, are seen [12]. These deterministic radiation effects would be of concern after an unprotected exposure to the radiation experienced in a solar flare.

Long-term Radiation Effects

Central nervous system damage

Because there are few actively dividing cells in the central nervous system, the brain and spinal cord are very resistant to radiation effects. In an acute radiation exposure, central nervous system symptoms do not occur until the highest doses are given (see table 3-4). Those that do occur result from acute edema and electrolyte imbalance, not cell killing. In radiation therapy for cancer, very large doses of radiation (10–80 Gy) can be delivered to the brain [13] if the dose is spread out over weeks, so that acute effects do not occur. In contrast to tissues such as the gastrointestinal tract or bone marrow, however, the central nervous system cannot regenerate lost cells easily; therefore, radiation damage to the central nervous system that kills cells is likely to be permanent.

Damage to the central nervous system from galactic cosmic radiation is hard to predict. On one hand the dose rates are low (particularly when compared to radiation therapy), but on the other hand the heavier nuclei are more likely to kill cells compared to a similar dose of x-rays or gamma rays. On a long-duration flight, many cells could be hit by heavy nuclei, as outlined in table 3-3. Only a limited number of studies have examined the central nervous system effects of heavy ions [14].

Cataracts

Radiation exposure can lead to cataracts [15]. A threshold exposure of approximately 2 Sv has been estimated to be the approximate threshold dose of low-LET (x-rays or gamma ray) radiation for cataract formation [4]. This estimate comes from high-dose rate exposures, so it is not clear what the threshold dose would be for the low-dose rate exposure that would occur in space. Also, the dose of high-LET (heavy nuclei in galactic cosmic radiation) radiation needed to produce cataracts is not clear, although the risk of cataracts appears greater in those astronauts exposed with lens doses > 8 mSv [15]. Neutrons are very efficient in producing cataracts, presumably because of the high LET of this radiation, so it is likely that similar radiation from galactic cosmic rays also would be cataractogenic.

Reduced fertility

For a male, an acute dose to the gonads of approximately 0.5–4.0 Sv of low-LET radiation can cause temporary sterility. For women, doses as low as 1.25 Seiverts could also temporarily affect fertility. An acute dose of 0.15 Sv can reduce sperm counts. As with most radiation data, these exposures assume a single, acute exposure.

Statistical Long-term Effects of Radiation

An acute exposure to high radiation levels is very likely to cause symptoms. These effects are deterministic because they are predictable and almost universal. Long-term effects of lower level radiation exposure, such as cancer induction, may or may not occur. These effects are described best by statistics and are called the statistical long-term effects of radiation (or stochastic radiation effects). These statistical effects are thought to be due to radiation-induced changes randomly distributed in the cellular DNA throughout the body. These mutations could lead to cancer or to a genetically transmissible chromosomal defect. A long latent period exists between the exposure and the effect—cancer can occur 2–20 years after the exposure. Exposure limits are set to try to reduce the frequency of these effects. To set the limits, it is assumed that stochastic effects increase with dose and that there is no lower threshold. The main

Table 3-5. Radiation limits for space crews from the National Council on Radiation Protection set in 2000.

Limit	Bone marrow	Eye	Skin
30-day	0.25	1.0	1.5
Annual	0.5	2.0	3.0
Career	Table 3-6	4.0	6.0

The limits are based on a 3% lifetime increased risk of induced cancer. Data in Gray-equivalents, which is the organ dose multiplied by the relative biological effectiveness appropriate for the tissue and radiation quality under study. Reprinted from Townsend and Fry [8], with permission from Elsevier.

effect of concern for spaceflight is cancer prevention. The risk of producing a heritable defect due to radiation-induced mutation is probably quite low [4].

Cancer risk

Most data on cancer induction due to radiation exposure come from studies of the survivors of the Hiroshima and Nagasaki atomic bombs and from observations of patients who had radiation delivered for medical diagnosis or therapy. These and other data were reviewed by the National Council on Radiation Protection to provide guidance on acceptable radiation exposures for astronauts [8]. To make their assessment, the limit chosen was a radiation exposure that would increase the risk of cancer no more than 3% over the baseline risk. The population risk of cancer over a lifetime is 20–25% without excess radiation exposure. The current guidelines for radiation exposure and effective dose limits are presented in tables 3-5 and 3-6.

These estimates include considerable uncertainty. Most populations exposed to radiation that have been studied extensively, like the atomic bomb survivors, received acute exposures. Few data exist on individuals who have been exposed chronically to low-level radiation as would occur in space. Also, the nature of space radiation, as discussed above, differs significantly from that of most Earth-based exposures. Galactic cosmic radiation is a minor concern on Earth, but a major concern in space. The relative biologic effectiveness of galactic cosmic radiation for inducing cancer is not firmly established. As recently noted by the National Research Council, most

Table 3-6. Ten-year career effective radiation dose limits.

Age at exposure (years)	Dose Limit (Sv)	
	Female	Male
25	0.4	0.7
35	0.6	1.0
45	0.9	1.5
55	1.7	3.0

The limits are based on a 3% lifetime increased risk of induced cancer. Reprinted from Townsend and Fry [8], with permission from Elsevier.

data on cancer induction due to high-energy, high-atomic-number particles come from a study on the Harderian gland of mice [4], and the relationship of these data to human cancer is not known. These data are the underpinning of the estimates about relative biologic effectiveness, and those estimates in turn affect the exposure limits. The exposure limits are educated guesses based on the best available information and could easily be revised upward or downward based on new findings.

To determine if occupational radiation exposure increases cancer risk, epidemiological studies have been carried out among both astronauts and pilots. Astronauts receive significant levels of radiation in orbit, and pilots, with their prolonged high-altitude exposures, also receive high lifetime doses of galactic cosmic radiation. At the NASA-Johnson Space Center, a longitudinal study of astronaut health compares astronauts with a matched population. Two publications from this study have reported cancer rates in the astronaut population [16, 17]. The first, which included data collected up to 1991, showed no increased risk of cancer, but a definite increase in death due to accidents in the astronaut population. In the 1998 study, astronauts had a higher age-specific cancer mortality than the comparison group, but the difference was not significant. Both groups had lower age-specific cancer rates than the general population.

Studies have shown an increased risk of acute leukemia and melanoma in pilots with more than 5000 hours of flight time [18]. Rafnsson et al. [19] also showed an increase in melanoma in a cohort of airline pilots flying international routes, but no increase in other cancers. It could not be determined if the excess risk of melanoma correlated with sun exposure in these groups. Blettner et al. [20] recently reviewed the data on increased cancer risk in pilots and flight attendants. Some studies have shown an increased cancer risk, but others have not, and the issue remains unresolved. In view of the considerable background level of cancer, the relatively small number of individuals involved in the statistical assessments, and the unknown differences in susceptibility to cancer in the population that was studied, it is difficult to draw firm conclusions from these data.

Living at high altitude also increases radiation exposure. Individuals living in Denver, Colorado, receive approximately twice the national average radiation exposure per year. Over 50 years a Denver resident would accumulate approximately 0.05 Sv more radiation than if they lived at sea level. In some areas of the Earth, the inhabitants receive close to 0.01 Sv per year. Despite this, epidemiological studies have not shown increased cancer rates for high-altitude residents [21]. Again, however, the database for measuring these effects probably is not sufficient to support reliable conclusions.

Another complicating factor is the possibility that humans could adapt to radiation exposure. Lymphocytes from individuals who are occupationally exposed to radiation are less sensitive to radiation damage than lymphocytes from unexposed individuals [22]. This effect has been called the adaptive response, and it has been demonstrated in numerous in-vitro studies [23]. The data suggest that radiation exposure can induce an increase in DNA repair capability, which in turn could lead to improved radiation tolerance. It is also possible that individuals differ in their ability to mount this response, which could be an important consideration for crew selection. Genetic screening and genetic susceptibility to cancer are covered in chapter

12. Changes in immune function in space could also influence the risk of developing cancer, and this is also discussed in chapter 12.

Radiation Hazards on Long-duration Flights

On a long-duration flight in low-Earth orbit, even though radiation protection from the atmosphere is reduced, the crew still receives protection from the Earth's magnetic field. Also, the mass of the Earth blocks half of the galactic cosmic radiation. The trapped radiation belts (Van Allen belts), however, provide an extra radiation source not present on the ground. On an interplanetary flight, the crew would receive no protection from the Earth's magnetic field or mass. Trapped radiation, however, would not be a problem. On Mars, the crew would again have some protection. Mars would block half of the galactic cosmic radiation (i.e., half of the radiation would have to pass through Mars to reach the crew). Mars lacks a significant atmosphere or magnetic field, so no additional protection beyond the mass of the planet itself would be offered. The main radiation risks on long-duration flights are an ongoing low-dose-rate exposure combined with the possibility of a short, high-dose-rate exposure.

High-Dose-Rate Exposure

As mentioned earlier, the most worrisome radiation exposures are those that occur while the crew is outside of the spacecraft performing an extravehicular activity. Protons of 10 MeV can penetrate nearly three-quarters of the area of the space suit. Protons with energies of 25 MeV or greater can penetrate the visor, the most heavily shielded part of the suit. Electrons, such as would be found in the trapped radiation belts, can penetrate some areas of the suit at energies of 0.5 MeV and greater. For the August 1972 solar particle event, this would translate into peak dose ranges to the blood-forming organs of 0.15–0.3 Sv/hour (depending on the assumptions made). Worst-case estimates of the dose to blood-forming organs for a 6-hour extravehicular activity during a solar particle event range from 0.06 to 1.2 Sv [2]. Although the low end of this range would not be of concern, the high-end estimates could produce radiation sickness.

Within the shielded spacecraft, radiation doses increase significantly during a solar particle event. As can be seen in table 3-2, the dose received by the crew on *Mir* during the October 1989 solar particle event was comparable to the radiation that would be received over 5 months of orbital spaceflight without a solar storm. A severe solar storm can deliver as much radiation in a few hours as the crew might otherwise receive over several months.

Low-Dose-Rate Exposure

On an interplanetary flight, galactic cosmic radiation becomes a significant concern. Inside the spacecraft, about 64% of the dose of cosmic radiation is due to protons, followed by helium (15%), oxygen (4.4%), carbon (3.2%), and iron (1.9%) nuclei

[24]. Estimates of dose are in the 0.5 Sv/year range, although this would vary depending on how heavily the spacecraft is shielded. On Earth, few high-weight atomic particles reach the surface, so this level of exposure to galactic cosmic radiation is unique to spaceflight. Although cells are very good at repairing radiation damage, some cells will be killed by the exposure to heavy nuclei, and over time the numbers of cells lost could become significant. Also, the risk of cancer due to this kind of radiation exposure (low-dose-rate, continuous exposure to heavy nuclei) is very difficult to estimate.

Countermeasures for Radiation Hazards

Shielding

Countermeasures for radiation exposure are simple because adequate shielding can reduce the exposure to any desired level. Two main shielding approaches exist: passive and active. Passive shielding works by inserting a mass of shielding material (such as aluminum or water) in between the radiation source and the crew. Active shielding works by creating a magnetic field that can deflect the radiation, just as the Earth's magnetic field provides protection.

Passive shielding is often described as having a "thickness" measured in grams per square centimeter. To get the actual thickness of shielding necessary to provide a given level of protection, divide the thickness by the density of the material. A thin layer of a dense material like lead might provide the same protection as a thick layer of a less dense material like water. While this is generally true, other factors also affect the choice of material, particularly when shielding against galactic cosmic radiation. In medical applications, healthcare workers use lead or some other dense material to block x-rays from medical procedures. For space applications, however, this approach would not be ideal. The very high-energy particles in galactic cosmic radiation will interact with the nuclei in a material like lead and produce secondary radiation (such as neutrons and gamma rays). This secondary radiation, while not as energetic as that from the galactic cosmic radiation, is still quite damaging. For example, at a shielding thickness of 20 g/cm^2, the yearly dose of radiation inside a spacecraft shielded with lead is 2.25 times greater than if it were shielded with water and 9 times greater than if liquid hydrogen were used [25]. One shielding strategy for high-energy particles is to alternate heavy and light materials to provide protection against both high and low LET radiation.

Low-density materials that contain a high percentage of hydrogen molecules (such as polyethylene or water) are ideal choices as shielding materials for interplanetary flight. In some proposed Mars craft-designs, the water for the mission would surround the habitable areas. Materials high in hydrogen produce less secondary radiation. More dense materials, like aluminum, may be useful for a "storm shelter"—a small, but highly shielded area that could be used for protection during a solar particle event.

Passive shielding strategies have limits. In interplanetary space, the annual radiation dose in the spacecraft would remain >0.25 Gy, even with a thickness of aluminum > 30 cm. This reflects the difficulty in shielding against galactic cosmic radiation. Also, the thicker the spacecraft, the more difficult it is to launch. For example, a cylinder of aluminum 6 cm thick, 396 cm wide (13 feet), and 1219 cm long (40 feet) could fit in the payload bay of the Space Shuttle. It would weigh 24,212 kg, however, which is almost the entire payload capacity of the Space Shuttle system. On the Moon or Mars, where soil there could be used to provide protection, approximately 5–10 m of soil would be required to bring radiation doses down to terrestrial levels.

Active shielding offers another approach. A magnetic field is established around the spacecraft. The engineering and power constraints of this approach have made it impractical in the past. New technological developments, however, could revive the idea. High-temperature superconducting coils could produce magnetic fields with much less power and mass than was required in the past. Also, the energy needed to produce the shield decreases approximately as the third power of the radius of the coil used to generate the field. An additional benefit to a large-radius coil is that the minimum mass needed to produce a given field also decreases as the radius increases [26]. If a very large-radius coil (1–10 km) could be deployed around the spacecraft while the mission was underway, this would require very little power to provide magnetic protection against radiation.

Antioxidants

Radiation causes biological damage by producing free radicals and other reactive molecules in tissues. Antioxidants help minimize the damage from the reactive molecules formed. The term "antioxidant" suggests these compounds work against oxygen, but they function primarily against oxidative damage from reactive molecules, which does not necessarily involve oxygen directly. Oxygen increases the amount of oxidative damage, but not the type of damage that occurs. The antioxidants react with damaged molecules, repair them chemically, or react with chemical intermediates before they damage key biological molecules [27]. Several different types of antioxidants exist. Naturally occurring or nutritional compounds such as cysteine, glutathione, vitamin C, vitamin E, selenium, and superoxide dismutase are one class. Drugs, such as amifostine, diethyldithiocarbamate, and tempol are another. Table 3-7 lists compounds that could potentially be used to combat the adverse effects of radiation exposure.

Nutritional and naturally occurring antioxidants

Cysteine is an amino acid containing a sulfur atom in a thiol group. This was the first compound shown to provide protection against ionizing radiation [27]. Cysteine is part of the tripeptide glutathione (along with glutamine and glycine). Glutathione is also an antioxidant and is important in several detoxification reactions [28]. Cysteine is toxic, however, at doses effective for radiation protection and it is not active orally [29]. Also, administering glutathione orally is ineffective. Efforts have been made to circumvent these problems by synthesizing other compounds containing thiol groups. Those compounds are discussed below.

Other dietary antioxidants that may provide some degree of radiation protection are vitamin E, vitamin C, vitamin A (and its precursors beta-carotene and other carotenoids), and the trace elements copper, manganese, zinc, and selenium [30]. Typi-

Table 3-7. List of compounds that might be useful to minimize radiation damage.

Compound	Studied in humans?	Studied in animals?	Proposed human dose	Notes
Cysteine compounds (cysteine, cysteamine, N-acetyl-L-cysteine, ribose-cysteine)	Yes	Yes	140 mg/kg followed by 70 mg/kg every 4 hours × 12 doses of N-acetyl-L-cysteine (Mucomyst) is given for acetaminophen overdose; dose for radiation protection not known	Can be given as the prodrug, ribose-cysteine [46, 47]
Vitamin E	Yes [48]	Yes [49]	30 mg/day alpha-tocopherol/day [48]	Human study gave volunteers an antioxidant mixture for 4 months then radiated the lymphocytes in vitro
Vitamin C	Yes [48]	Yes	150 mg/day ascorbic acid [48]	Human study gave volunteers an antioxidant mixture for 4 months then radiated the lymphocytes in vitro
Vitamin A	Yes	Yes	3 mg retinol acetate [48]	
Beta-carotene	Yes [48, 50]	Yes [51]	15–40 mg/day	Children who had been exposed to radiation at Chernobyl received supplement and lipid peroxidation products in the blood were reduced [50]

(continued)

cally, however, these naturally occurring dietary antioxidants protect only against lower doses or lower dose rates of radiation [27]. These compounds do have the benefit of low toxicity, and they can be taken orally and dosed daily. Antioxidants are discussed in more detail in chapter 8.

Superoxide dismutase is an enzyme that catalyzes the breakdown of superoxide (a free radical formed in tissue), converting it to hydrogen peroxide. Superoxide dismutase must be administered parentally and cannot be used as a nutritional supplement. Drugs that have superoxide dismutase-like activity are discussed below.

Drugs

Amifostine (also know as WR-2721) emerged from extensive testing of antiradiation compounds at the Walter Reed Medical Center. This drug is used clinically to reduce the side effects of radiation and chemotherapy, and like cysteine it has a thiol group. The usual dose is 740–910 mg/m^2. The main side effect of the drug is hypotension [31].

Table 3-7. (*continued*)

Compound	Studied in humans?	Studied in animals?	Proposed human dose	Notes
Selenium	Yes	Yes	70 µg (male), 55 µg (female)	Recommended dietary allowance from the National Research Council [30]
Glutathione	Yes [28]	Yes	1500–3000 mg/ square meter IV	Low bioavailability when given orally in the rat [52]; may be most effective to protect the kidneys, liver and peripheral nerves
Superoxide dismutase	Yes	Yes	10,000 µg/kg IV given in studies of closed head injury [32]	Not studied for radiation protection. Used in studies for amyotrophic lateral sclerosis, head injury, bronchopulmonary dysplasia
Amifostine	Yes	Yes	100–900 mg/m^2 IV before radiation	IV administration, short duration of action, causes nausea/vomiting/ hypotension
Tempol	No	Yes	Not known	IV administration, short duration of action causes hypotension, seizures [34]

There are no solid data on the long-term use of any of these compounds indicating that they prevent damage from continuous low-level radiation.

There are no data on its long-term use as a radioprotectant. This drug could be useful in the space program if an acute, high dose of radiation was expected or had recently occurred. The dose of the drug to use in that instance is not firmly established.

Various forms of superoxide dismutase have been prepared to determine if the exogenously administered compound could be used to prevent oxidative damage in different settings. Superoxide dismutase has been combined with polyethylene glycol and administered within liposomes [32, 33]. It must be administered parentally, and it is rapidly eliminated from the circulation (although this is improved with the liposomal preparation). No data exist on the long-term administration of superoxide dismutase to prevent cumulative radiation damage. The dose to use in the case of an acute exposure to radiation, such as would occur in space, has not been established.

Other compounds have superoxide like activity. Nitroxides are stable free radical compounds that possess superoxide dismutase activity and have been shown to protect against the toxicity of reactive oxygen species both in vitro and in vivo [34, 35]. Tempol is a cell-permeable hydrophilic nitroxide that protects against oxidative stress. Topical application of Tempol may help prevent hair loss due to radiation therapy. Numerous animal studies have examined the effects of Tempol in minimizing the effects of oxidative stress. Tempol is not currently available for human use, and no data exist on long-term use. It is possible that Tempol might be useful for prevention of radiation damage during an acute exposure.

Another well-studied antioxidant is *N*-acetyl-L-cysteine (NAC); a sulfhydryl donor capable of reducing oxidized intracellular glutathione, thereby restoring its effectiveness. NAC is most commonly used as an antidote for acetaminophen poisoning, but it also prevents lung damage and respiratory distress syndrome in septic shock. NAC can be administered orally, but there are no studies assessing long-term administration. NAC has been shown in animal studies to help minimize the effects of ionizing radiation [36, 37]. Similar to amifostine, NAC could be considered for use in space if a high dose of radiation was expected or had recently been received.

Monitoring Radiation Exposure

Radiation Monitoring

To allow research into the long-term effects of radiation and to determine the effectiveness of radiation protection, measures of radiation dose are needed. A dosimeter is a device that can provide information about radiation dose. Physical dosimeters, such as film badges, provide information on dose and dose rate for a variety of different kinds of radiation. A variety of different devices exist to measure radiation inside a spacecraft and on a person. The problem with dosimeters, however, is that they do not provide any information on the biological effects of the radiation. This has led to the search for biological markers of radiation dose and for a way to measure the biological effects of radiation—called biodosimetry.

Biodosimetry

Radiation exposure produces chromosomal aberrations in lymphocytes in peripheral blood. Chromosomal aberrations have been shown to increase after radiation exposure [38], and individuals who live in areas with high background radiation have more chromosomal aberrations than those who do not [39]. The presence of chromosomal aberrations may also be correlated with the subsequent development of cancer [40, 41]. Biodosimetry uses biological markers of radiation exposure, such as chromosomal aberrations, to track the effects of the radiation. Chromosomal aberrations are measured in blood samples before and after spaceflight. An in-vitro calibration is done before flight with blood samples to determine the relationship between radiation dose and the frequency of chromosomal aberration. With this technique the frequency of chromosomal aberrations can be scored and compared to preflight levels. The radiation dose is inferred from the preflight dose–response curves relating radiation dose to chromosomal aberrations. Initial studies with this technique during long-duration flights have been promising [42–44].

Several markers also exist for oxidative stress, but their validity and usefulness is controversial. Superoxide dismutase and glutathione levels decrease with oxidative stress, and other makers, such as malonaldehyde, indicate that lipid peroxidation has taken place. In theory, knowledge of these levels could be useful because if antioxidant defenses are reduced, the crew member might be more susceptible to radiation damage. At present, however, there is not sufficient information to determine whether levels of antioxidants and other markers of oxidative stress could be useful in long-duration spaceflight.

Recommendations Based on Current Knowledge

The main approach to preventing radiation damage in long-duration spaceflight is adequate shielding. Developing new propulsion systems could also reduce exposure by limiting time in interplanetary space. At present, however, providing a level of shielding in interplanetary flight comparable to the protection offered by the Earth is not practical, and dramatic improvements in propulsion will take time. For Mars exploration, it is likely that crews will be exposed to levels of radiation higher than terrestrial levels. The following measures might help minimize radiation damage in space.

1. Shielding. A spacecraft should have a "storm shelter," a thickly shielded area where crews can go during solar storms. While passive shielding is the most practical short-term approach to radiation exposure in space, even very thick shields will not be able to bring radiation levels in the spacecraft to terrestrial levels. Active shielding concepts should also be explored.

2. Antioxidants. An antioxidant mixture containing vitamins E, C, A, beta-carotene, and selenium would be reasonable to take. As discussed in chapter 8, although it is important to avoid deficiency of these compounds, there are no data to support mega-dose supplementation. Also, there are no data showing that taking antioxidants regularly will reduce the long-term effects of long-duration, low-level radiation exposure. Nevertheless, these compounds have minimal side effects and are easy to take. Because extravehicular activity involves a high skin dose of radiation, antioxidant supplementation before extravehicular activity and the use of an antioxidant skin cream should be considered (i.e., a cream containing compounds such as vitamin E, vitamin A and/or alpha-lipoic acid). The best composition for the cream should be studied.

3. Drugs for unusually high exposures. For those situations where a high level of radiation might be expected or could be sustained (such as from a solar flare), the crew should have access to drugs like amifostine to minimize the damage. A protocol to deal with a significant radiation exposure should be established. This would require the ability to measure peripheral blood counts and administer antibiotics.

4. Monitoring. Physical dosimeters can provide the crew members with an ongoing record of their radiation exposure. Biodosimetry, however, provides more meaningful information on biological effects. Developing technology to provide measurements of oxidative stress would also be worthwhile and is discussed in chapter 8.

5. Selection. As discussed in chapter 12, individuals who show deficiencies in DNA repair or who do not have an adaptive response to radiation may be at higher risk for developing cancer as a result space radiation exposure. This is a complex and poorly studied area, but one that will be important for Mars exploration.

References

1. Michener, J.A., *Space*. 1998, Fawcett Books, New York.
2. National Research Council, *Radiation and the International Space Station: Recommendations to Reduce Risk*. 2000, National Academy Press, Washington, DC.
3. Letaw, J.R., Radiation biology, in *Fundamentals of Space Life Sciences*, S.E. Churchill, ed. 1997, Krieger, Malabar, FL, pp. 11–18.
4. National Research Council, *Radiation Hazards to Crews of Interplanetary Missions*. 1996, National Academy Press,Washington, DC.
5. Feynman, J., and A. Ruzmaikin, Problems in the forecasting of solar particle events for manned missions. Radiation Measurements, 1999. 30(3): 275–80.
6. Fry, R.J., Radiation effects in space. Advances in Space Research, 1986. 6(11): 261–68.
7. Robbins, D.E., et al., Ionizing radiation, in *Space Biology and Medicine*, Joint U.S./Russian Publication, A.E. Nicogossian, et al., eds. 1996, American Institute of Aeronautics and Astronautics, Reston, VA, pp. 365–93.
8. Townsend, L.W., and R.J. Fry, Radiation protection guidance for activities in low-Earth orbit. Advances in Space Research, 2002. 30(4): 957–63.
9. Hall, E.J., and J.D. Cox, Physical and biologic basis of radiation therapy, in *Moss' Radiation Oncology*, J.D. Cox, ed. 1994, Mosby, St. Louis, MO, pp. 3–66.
10. Curtis, S.B., Radiation physics and evaluation of current hazards, in *Space Radiation Biology and Related Topics*, C.A. Tobias and P. Todd, eds. 1974, Academic Press, New York, pp. 21–114.
11. Curtis, S.B., et al., Cosmic ray hit frequencies in critical sites in the central nervous system. Advances in Space Research, 1998. 22(2): 197–207.
12. Robbins, D.E., and T.C. Yang, Radiation and radiobiology, in *Space Physiology and Medicine*, A.E. Nicogossian, C.L. Huntoon, and S.L. Pool, eds. 1994, Williams and Wilkins, Philadelphia, pp. 167–93.
13. Kun, L.J., The brain and spinal cord, in *Moss' Radiation Oncology*, J.D. Cox, ed. 1994, Mosby, St. Louis, MO, pp. 737–81.
14. Vazquez, M.E., Neurobiological problems in long-term deep space flights. Advances in Space Research, 1998. 22(2): 171–83.
15. Cucinotta, F.A., et al., Space radiation and cataracts in astronauts. Radiation Research, 2001. 156(5 Pt 1): 460–66.
16. Peterson, L.E., et al., Longitudinal study of astronaut health: mortality in the years 1959–1991. Radiation Research, 1993. 133(2): 257–64.
17. Hamm, P.B., et al., Risk of cancer mortality among the Longitudinal Study of Astronaut Health (LSAH) participants. Aviation, Space, and Environmental Medicine, 1998. 69(2): 142–44.
18. Gundestrup, M., and H.H. Storm, Radiation-induced acute myeloid leukaemia and other cancers in commercial jet cockpit crew: a population-based cohort study. Lancet, 1999. 354(9195): 2029–31.
19. Rafnsson, V., J. Hrafnkelsson, and H. Tulinius, Incidence of cancer among commercial airline pilots. Occupational and Environmental Medicine, 2000. 57(3): 175–79.
20. Blettner, M., B. Grosche, and H. Zeeb, Occupational cancer risk in pilots and flight attendants: current epidemiological knowledge. Radiation and Environmental Biophysics, 1998. 37(2): 75–80.
21. Mason, T.J., and R.W. Miller, Cosmic radiation at high altitudes and U.S. cancer mortality, 1950–1969. Radiation Research, 1974. 60(2): 302–6.
22. Barquinero, J.F., et al., Occupational exposure to radiation induces an adaptive response in human lymphocytes. International Journal of Radiation Biology, 1995. 67(2): 187–91.
23. Wolff, S., The adaptive response in radiobiology: evolving insights and implications. Environmental Health Perspectives, 1998. 106(Suppl 1): 277–83.
24. Zaider, M., Microdosimetric-based risk factors for radiation received in space activities during a trip to Mars. Health Physics, 1996. 70(6): 845–51.
25. Letaw, J.R., R. Silberberg, and C.H. Tsao, Radiation hazards on space missions outside the magnetosphere. Advances in Space Research, 1989. 9(10): 285–91.

26. Sussingham, J.C., S.A. Watkins, and F.H. Cocks, Forty years of development of active systems for radiation protection of spacecraft. Journal of the Astronautical Sciences, 1999. 47(3): 165–75.
27. Weiss, J.F., and M.R. Landauer, Radioprotection by antioxidants. Annals of the New York Academy of Sciences, 2000. 899: 44–60.
28. Hospers, G.A., E.A. Eisenhauer, and E.G. de Vries, The sulfhydryl containing compounds WR-2721 and glutathione as radio- and chemoprotective agents. A review, indications for use and prospects. British Journal of Cancer, 1999. 80(5–6): 629–38.
29. Roberts, J.C., et al., Thiazolidine prodrugs of cysteamine and cysteine as radioprotective agents. Radiation Research, 1995. 143(2): 203–13.
30. Pence, B.C., and T.C. Yang, Antioxidants: radiation and stress, in *Nutrition in Spaceflight and Weightlessness Models*, H.W. Lane and D.A. Schoeller, eds. 2000, CRC Press, Boca Raton, FL, pp. 233–51.
31. Hensley, M.L., et al., American Society of Clinical Oncology clinical practice guidelines for the use of chemotherapy and radiotherapy protectants. Journal of Clinical Oncology, 1999. 17(10): 3333–355.
32. Young, B., et al., Effects of pegorgotein on neurologic outcome of patients with severe head injury. A multicenter, randomized controlled trial. Journal of the American Medical Association, 1996. 276(7): 538–43.
33. Jadot, G., et al., Clinical pharmacokinetics and delivery of bovine superoxide dismutase. Clinical Pharmacokinetics, 1995. 28(1): 17–25.
34. Hahn, S.M., et al., Evaluation of the hydroxylamine Tempol-H as an in vivo radioprotector. Free Radical Biology and Medicine, 2000. 28(6): 953–58.
35. Hahn, S.M., et al., Hemodynamic effect of the nitroxide superoxide dismutase mimics. Free Radical Biology and Medicine, 1999. 27(5–6): 529–35.
36. Neal, R., et al., Antioxidant role of N-acetyl cysteine isomers following high dose irradiation. Free Radical Biology and Medicine, 2003. 34(6): 689–95.
37. Reliene, R., E. Fischer, and R.H. Schiestl, Effect of N-acetyl cysteine on oxidative DNA damage and the frequency of DNA deletions in atm-deficient mice. Cancer Research, 2004. 64(15): 5148–53.
38. Limoli, C.L., et al., Genomic instability induced by high and low LET ionizing radiation. Advances in Space Research, 2000. 25(10): 2107–17.
39. Ghiassi-Nejad, M., et al., Long-term immune and cytogenetic effects of high level natural radiation on Ramsar inhabitants in Iran. Journal of Environmental Radioactivity, 2004. 74(1–3): 107–16.
40. Baria, K., et al., Chromosomal radiosensitivity as a marker of predisposition to common cancers? British Journal of Cancer, 2001. 84(7): 892–96.
41. Bonassi, S., M. Neri, and R. Puntoni, Validation of biomarkers as early predictors of disease. Mutation Research, 2001. 480–481: 349–58.
42. Yang, T.C., et al., Biodosimetry results from space flight *Mir-18*. Radiation Research, 1997. 148: S17–S23.
43. Durante, M., et al., Risk estimation based on chromosomal aberrations induced by radiation. Radiation Research, 2001. 156(5 Pt 2): 662–67.
44. Greco, O., et al., Biological dosimetry in Russian and Italian astronauts. Advances in Space Research, 2003. 31(6): 1495–503.
45. Nicogossian, A.E., and D.E. Robbins, Characteristics of the space environment, in *Space Physiology and Medicine*, A.E. Nicogossian, C.L. Huntoon, and S.L. Pool, eds. 1994, Williams and Wilkins, Philadelphia. pp. 50–62.
46. Carroll, M.P., et al., Efficacy of radioprotective agents in preventing small and large bowel radiation injury. Diseases of the Colon and Rectum, 1995. 38(7): 716–22.
47. Rowe, J.K., et al., Protective effect of RibCys following high-dose irradiation of the rectosigmoid. Diseases of the Colon and Rectum, 1993. 36(7): 681–88.
48. Gaziev, A.I., et al., Effect of vitamin-antioxidant micronutrients on the frequency of spontaneous and in vitro gamma-ray-induced micronuclei in lymphocytes of donors: the age factor. Carcinogenesis, 1996. 17(3): 493–99.

49. Felemovicius, I., et al., Intestinal radioprotection by vitamin E (alpha-tocopherol). Annals of Surgery, 1995. 222(4): 504–8 [discussion 508–10].
50. Ben-Amotz, A., et al., Effect of natural beta-carotene supplementation in children exposed to radiation from the Chernobyl accident. Radiation and Environmental Biophysics, 1998. 37(3): 187–93.
51. Slyshenkov, V.S., et al., Protection by pantothenol and beta-carotene against liver damage produced by low-dose gamma radiation. Acta Biochimica Polonica, 1999. 46(2): 239–48.
52. Grattagliano, I., et al., Effect of oral glutathione monoethyl ester and glutathione on circulating and hepatic sulfhydrils in the rat. Pharmacology and Toxicology, 1994. 75(6): 343–47.

4

Muscle Loss: A Practical Approach to Maintaining Strength

Introduction

On June 19, 1970, *Soyuz 9* landed after a record-breaking 18-day flight. The cosmonauts had orbited the Earth in a confined capsule, which provided little room to exercise. When the hatch was opened, the weak crew had great difficulty getting out of the capsule. They were reported to still be weak and recovering 10 days later [1]. Although a variety of physiological changes undoubtedly contributed to their condition (orthostatic intolerance, loss of balance), their weakness was most likely due to muscle atrophy in space.

Since then, long-duration space crews have followed aggressive exercise protocols designed to minimize the loss of muscle strength and function. Nevertheless, despite a 6-day-a-week, 2-hour-a-day exercise program on the International Space Station, crews still return to Earth with significant muscle loss. In addition, the lack of muscular activity in space can contribute to bone loss because the pull of muscles on bone is one factor that prevents demineralization.

To maintain high performance on missions and to minimize rehabilitation back on Earth, crews must work to maintain muscle strength. For the future, if crews return to the Moon or go to Mars, they will need to be physically fit when they arrive to meet the challenges there. Exercise is an important component of any countermeasure plan, but it takes time and increases the use of oxygen, water, and food. Crews need efficient and effective measures to maintain muscle strength. This chapter outlines the physiology of muscle loss and what measures the crews can take to prevent or minimize it.

Muscle Physiology Relevant to Spaceflight

Not all muscles are affected equally by weightlessness. The muscles that maintain posture and stability while upright on Earth are the ones that show the most dramatic changes in space.

Antigravity Muscles

On Earth, standing and walking on two legs present physiological challenges. The force of gravity must be countered to stay upright without falling. Fortunately, the skeletal, muscular, and balance systems have evolved to meet this need. The line of gravity of an upright human goes through the base of the spine (figure 4-1, table 4-1), so that the body is fairly well balanced when standing on two legs. Nevertheless, the center of gravity is in front of the ankles.

The slightly forward center of gravity means that the muscles in the back of the leg, particularly the soleus, need to be active periodically to prevent a person from falling forward. The result is that, in general, the posterior calf muscles (the soleus and gastrocnemius) are more active than the anterior leg muscles (tibialis anterior) during quiet standing [2]. The muscles' activity levels vary because a slight forward-to-back swaying is also present. The majority of the thigh muscles, except for the hamstrings (biceps femoris), are not continuously active with standing, but are activated with swaying.

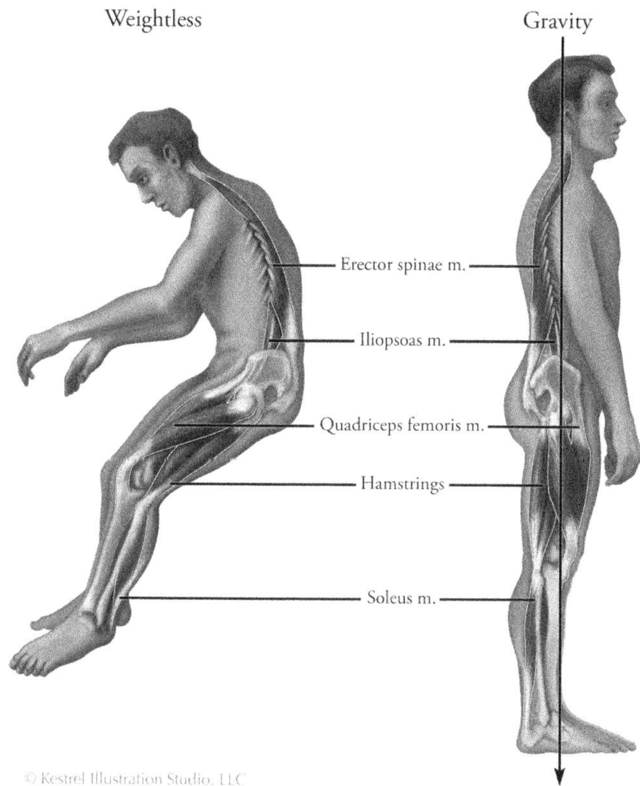

Figure 4-1. The main muscle groups used in posture. In weightlessness they are unloaded. © Kestrel Illustrations, LLC.

Muscular activity is needed to keep the spine straight (a function of the erector spinae muscle) and to balance forces around the hip and in the back. The iliopsoas muscle is frequently active in the upright posture and seems to work to stabilize the hip joint [3]. The long intrinsic muscles of the back and the abdominal muscles work to stabilize the spine as a whole [2]. The neck muscles also are active in the erect

Table 4-1. MRI-determined changes in muscle volume on landing day for different muscle groups.

Muscle group	% Loss after long-duration flight
Back (erector spinae and intrinsics)	−10.9
Iiopsoas	−20.0
Quadriceps	−12.1
Hamstrings	−15.7
Soleus	−19.6

Data from Leblare et al. [78].

posture. The muscles of the upper extremity, however, do not show much activity with changes in posture (except for the trapezius, which elevates the shoulder girdle). Although many muscles are activated intermittently while standing, surprisingly few muscles are continuously active.

The muscles involved in maintaining stability in gravity are called the antigravity muscles. By and large, these are the muscles that work to straighten the spine and extend the hip, knee, and ankle [4]. Maintaining posture is just one role of the antigravity muscles and may not be the most demanding. Moving from the lying or sitting position to standing requires powerful movements, which antigravity muscles (such as the quadriceps) provide [2]. With walking, the muscles around the hip joint must contract forcefully to keep the hip from slouching every time a foot leaves the ground. Muscles throughout the legs and spine are activated with walking and running. This allows for smooth movements and keeps the body stable. Figure 4-1 shows the main postural muscles (the antigravity muscles).

In space, the body tends to float in a neutral posture with the knees and hips bent and the spine slightly rounded. The control of movement is shifted toward the upper extremities, hands, and fingers, while the lower extremities are used mainly for stability. As a result, the powerful muscle contractions that are part of daily life on Earth are not needed in space.

Muscle Types

The antigravity muscles are called upon to provide powerful and sometimes long-lasting contractions. These demands differ from the ones placed upon muscles involved in sprinting or other intense, but short, exercises. It is not surprising, then, that muscles contain different types of fibers that are optimized for the conditions the muscles are likely to face. Many antigravity muscles, like the soleus, have a high concentration of type I muscle fibers [5]. These type I fibers are named after the type I myosin heavy chains (MHC) that predominate within these cells [6]. These fibers, also known as slow-twitch muscle fibers, do not develop force quickly, but instead are designed to provide a steady, fatigue-resistant contraction. These fibers have many mitochondria, good aerobic capacity, and high levels of myoglobin [7]. The muscles with an elevated concentration of these fibers also have a high capillary density. Type 1 fibers are obviously ideal for postural muscles, and, as can be seen in table 4-2, many of the postural muscles contain a high percentage of slow-twitch fibers. These fibers are significant in space because they seem to be the most sensitive to inactivity, immobilization, and weightlessness [4].

The other types of muscle fibers are type II or fast-twitch fibers. These fibers have high shortening velocities (for providing rapid power), but they fatigue easily. Compared to type I fibers, they have fewer mitochondria and less myoglobin [7]. These fibers are further subdivided into type IIA and type IIX, based on the type of MHC expressed in the fiber. The type IIA fibers have sometimes been called fast, fatigue-resistant fibers because they use both oxidative and anaerobic metabolism. The type IIX fibers are sometimes called fast-fatigable fibers because they rely extensively on glycolysis. Skeletal muscles usually have a mix of fiber types, with slow-twitch fibers predominating in endurance muscles (like the soleus) and fast-twitch prevalent

Table 4-2. Percentage of slow–twitch fibers in a given muscle.

Muscle	% Slow twitch
Gastrocnemius	55
Vastus lateralis	45
Soleus	90
Triceps	35
Biceps	55
Deltoid	60

Type I (slow–twitch) fibers are the most sensitive to disuse atrophy, suggesting that muscles with a high concentration of these fibers would show significant losses with disuse. The soleus has a very high percentage of slow–twitch fibers and does show substantial reductions in size and strength in microgravity. Data from Saltin and Gollnick [5].

in muscles like the triceps in the arm, which is often called upon for rapid, but not sustained, power. Training or inactivity can change the balance between the fiber types within a given muscle. During immobilization or inactivity, some type I fibers can transform into type II fibers [8].

Muscle Atrophy

Two main factors can cause muscles to lose mass and strength. Lack of activity will decrease protein synthesis within the muscle, and inadequate caloric intake (or stress) will enhance the breakdown of muscle so that the amino acids can be used for energy. Several factors can influence muscle atrophy, such as disuse, undernutrition, stress (psychological or physical), oxidative stress, and hormones.

Inactivity and disuse

Muscle must be used to maintain its structure and function. The protein in muscle is not static, but constantly in flux. Typically, the rates of protein synthesis and breakdown in muscle are balanced to meet daily demands. Loading, such as provided by weight lifting, can shift the balance between protein synthesis and protein breakdown in muscle, leading to an increase in fiber cross-sectional area and hypertrophy of muscle [9]. This effect is readily seen in bodybuilders. If, however, the loading of the muscles is reduced, protein synthesis in the muscle fiber declines. Because the half-life for degradation of protein within skeletal muscle is less than 7 days [6], atrophy can progress rapidly as synthesis falls.

Immobilization, chronic shortening, denervation, and inactivity will all lead to muscle atrophy. Through mechanisms that are not fully understood, the muscle senses the reduction in activity and reduces protein synthesis. Protein degradation is elevated transiently as the muscle remains unloaded. Eventually, a new steady state is established where both protein synthesis and degradation are reduced relative to the normal level, and the muscle stabilizes at a lower mass [6]. Atrophy typically reduces the cross-sectional area of individual fibers, but not the number of fibers.

Muscle fibers that express the type I MHC isoform (the type I muscles) seem to be the most sensitive to unloading and show the greatest reductions in size during atrophy [10]. Unloading can also affect the balance between type I and type II fibers. In some slow-twitch fibers, the protein composition of the fiber is transformed during atrophy. A portion of the slow myosin is degraded and replaced by faster MHC isoforms (chiefly IIX) [6]. This creates a new type of fiber, called a slow/fast hybrid. The net result is that a muscle that has atrophied will have a smaller size and less strength but will contract faster.

Several factors can influence the amount of atrophy that takes place in an inactive muscle. A muscle that is placed under chronic stretch when immobilized will atrophy less than if it had been immobilized at its normal length. Immobilization at a shortened length will increase atrophy [11]. If the inactive muscle can be moved passively in a way that that reproduces the length changes seen in normal use, this can also help slow atrophy [12].

Activity can prevent or minimize atrophy. Animal studies have been done using a spinal cord isolation model in which the muscle is innervated but inactive. These studies have shown that without any activation the soleus atrophies to 33% of its original size, but then remains stable at that level. A fairly small amount of activity (9 minutes of static [isometric] activity) can boost the amount of muscle retained to 64%. In primates, continuous electromyogram (EMG) recordings show that although the soleus is the most active of the lower limb muscles, it is inactive 90.5% of the time [13]. Overall, animal studies suggest that fairly short periods of activation under load can maintain muscle mass and function [12].

Undernutrition and stress

Skeletal muscle serves as a protein reservoir that is mobilized in stressful states or with undernutrition. In stress, protein turnover is increased, meaning that both protein synthesis and protein breakdown are elevated, but the increase in breakdown exceeds the increase in synthesis. The result is muscle mass loss [14]. In undernutrition, protein synthesis is decreased and breakdown is increased. There is a regulatory system that may enhance the capacity for protein breakdown in both of these cases [15]. Undernutrition and stress are important because they can add to the muscle disuse problem in space. Disuse of muscle combined with a stress response, undernutrition, or both can be a powerful combination to lose lean body mass. This kind of muscle loss is seen frequently in hospitalized patients.

Oxidative stress

Although oxygen is essential for life, it is also a very reactive molecule. The use of oxygen in the body results in the production of reactive oxygen species (e.g., superoxide, hydrogen peroxide), which can damage DNA, proteins, and membrane lipids. "Oxidative stress" refers to the balance between oxidants (such as reactive oxygen species) and antioxidants. Antioxidants are compounds that minimize the damage from reactive oxygen species. Antioxidants work by reacting with the damaged molecules, chemically repairing them, or by reacting with intermediates before they damage key biological molecules. Naturally occurring or nutritional antioxidants include compounds such as cysteine, glutathione, vitamin A, vitamin C, vitamin E, selenium, and superoxide dismutase.

Several studies in rats suggest that oxidative stress (a shift toward oxidation in the balance between reactive oxygen species and antioxidants) can exacerbate muscular atrophy [16–19]. Oxidative stress may enhance muscle breakdown by increasing the activity of the ubiquitin-proteasome pathway (a major pathway for breakdown of muscle protein) [20]. Supplementation with antioxidants may minimize muscle atrophy [21], but this has not been seen in all studies [22].

Hormonal influences

Hormonal factors play an important role in muscle atrophy and hypertrophy. At the level of the muscle cell, insulinlike growth factor-1 (IGF-1) has been shown to have a powerful effect within muscle fibers. IGF-1 initiates a cascade of events within the muscle cell that increases protein translation and transcription and markedly increases protein synthesis [6]. IGF-1 is produced both in the liver and directly in the muscle itself. The administration of IGF-1 can reduce or eliminate atrophy due to denervation [23].

Growth hormone, secreted by the pituitary gland, stimulates the liver to produce systemic IGF-1. Conversely, a drop in systemic IGF-1 levels can stimulate the release of growth hormone [24]. The growth hormone induced production of IGF-1 is also enhanced by thyroid hormone [10]. Growth hormone given exogenously to animals reduced atrophy when combined with exercise [11].

Testosterone also can have a profound effect on muscle. When testosterone was given to older men, an improvement in protein balance was demonstrated [25]. The improvement resulted from a decrease in muscle protein breakdown, since protein synthesis was unchanged. The ubiquitin-proteasome system, which plays an important role in protein breakdown, showed reduced activity after testosterone administration, suggesting that testosterone's main effect is on protein degradation. Other studies have shown that testosterone increases protein synthesis [26]. Testosterone increases concentrations of IGF-1 and reduces the concentration of an IGF inhibitory protein [24]. At present it seems that testosterone may have effects on both protein synthesis and breakdown in muscle. When testosterone is given to men with low testosterone levels, muscle mass and strength are increased [24]. Testosteronelike drugs (anabolic steroids) have been used to enhance physical performance and to reduce the muscle mass loss seen in burn and trauma patients.

Cortisol is elevated by a variety of physical and psychological stresses and is thought of as the classic stress hormone. Although the exact role of cortisol in stress is not known, it does mobilize amino acids and fats from cells, which can be used to synthesize other compounds or produce glucose. This is accomplished in part by decreasing protein synthesis and increasing protein breakdown in muscle [27]. Sustained elevations in cortisol levels lead to a catabolic state.

Effect of Bed Rest on Muscle

The *Soyuz 9* mission mentioned earlier showed what can happen to muscle mass and strength in weightlessness if there is not sufficient intervention. On that flight,

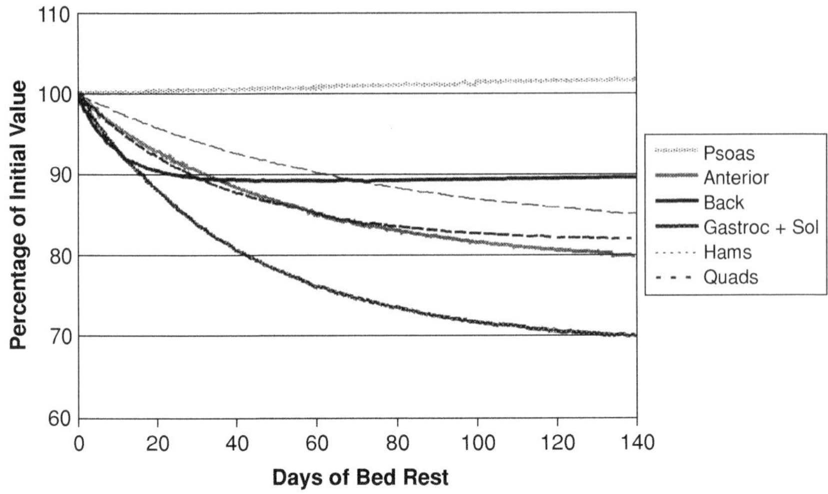

Figure 4-2. Sites and magnitude of muscle loss during bed rest. MRI-determined muscle volumes from bed rest studies ranging from 5 to 17 weeks were combined and fit with a mathematic function. The graphs give a general sense of the magnitude, location, and time course of muscle loss during bed rest. The gastrocnemius and soleus together shrink to approximately 70% of their original size. Of interest is the minimal loss in the psoas. Graph modified from LeBlanc et al. [31].

the countermeasure program was inadequate, and the 18-day mission was long enough to produce significant atrophy. Since that time, however, all spaceflights have included some kind of exercise countermeasure to prevent muscle loss. The data on muscle loss collected during or after a space mission show the combined effects of weightlessness, exercise, stress, and nutrition. For example, one study showed significant losses in calf, quadriceps, and back muscles after an 8-day spaceflight [28], but another showed no change in calf strength and fiber composition after 17 days in space [29]. Published reports often do not take into account the unique operational or psychological stresses that might have characterized a particular flight or groups of flights (such as the Shuttle/Mir program). These stresses have an impact on the physiological changes.

Because of the difficulties with spaceflight data, analogs of weightlessness are used to provide controlled studies on muscle loss. The most commonly used analog is bed rest. With horizontal or head-down tilt bed rest, gravity is not eliminated, but the postural muscles are unloaded as they would be in space. Also, muscle activity is minimal, in contrast to the situation on the spacecraft where the postural muscles are unloaded but considerable muscular activity may still take place. For animal studies, the most commonly used analog of weightlessness exposure is hindlimb suspension. In this model the hindlimbs of the rat are off the ground, although the rat can move about using the front limbs. These two models, bed rest in humans and hindlimb suspension in animals, provide most of the analog data for spaceflight exposure.

Table 4-3. Strength reductions in various muscle groups from different bed rest studies.

Muscle group	Strength loss (35-day study)	Strength loss (42-day study)	Strength loss (120-day study)
Handgrip		0%	
Elbow flexors	−7%	−9%	
Elbow extensors	2%	−7%	
Knee flexors	−8%	−8%	
Knee extensors	−19%	−8%	
Ankle flexors	−8%	−13%	−39%
Ankle extensors	−25%	−21%	−34%

Both flexors and extensors lose strength, although the largest reductions are in the ankle extensors. With longer bed-rest durations, upper body strength shows further decrements, although typically it is less affected than the lower body (data not shown). Data from Convertino [33].

In a prolonged bed rest study, subjects lie either supine or with the body tilted head-down 6°. Contraction of the postural muscles is minimized, but muscular activity is still present because the subjects can roll on their sides and adjust themselves in the bed. Bed rest causes a net decrease in protein synthesis [30] and an increase in nitrogen excretion, indicating a loss of body protein. Much of this loss is from muscle. Figure 4-2 shows the effect of bed rest on muscle volume using combined data from several bed rest studies. To create this graph, LeBlanc and colleagues [31] combined data on MRI-determined muscle volume from bed rest studies of different lengths [31]. They collected data on 11 men and 5 women who had participated in bed rest studies of 5–17 weeks duration. The data were then fit to a curve using a curve-fitting program. The result is an approximation of the time course and magnitude of muscle volume loss during bed rest. The graph focuses on the lower limb and back muscles because these muscles are most affected by bed rest. The upper body muscles also atrophy and lose strength, but this is much less marked than in the lower body [32, 33].

The data show that, as expected, the ankle flexors (soleus and gastrocnemius) are profoundly affected. The hamstring muscles and quadriceps muscles, which are involved with posture and locomotion, also show significant losses. Interestingly, the anterior muscles of the lower leg (tibialis anterior) were significantly affected, even though they usually are not as active as the soleus in maintaining posture. The psoas muscles, which are reported to be active during standing in Earth's gravity, were relatively unaffected.

Data from other bed rest studies show the practical effect from this inactivity. Table 4-3 contains a compilation of data from three bed rest studies. These data show that ankle, knee, and elbow flexors and extensors all lose strength during bed rest [33, 34]. Overall, the loss in strength is greatest in the ankle flexors, which is consistent with the slow-twitch soleus muscle's sensitivity to disuse.

Fortunately, the bed rest data do not seem to be confounded by changes in the hormonal milieu. Fourteen days of bed rest had no effect on serum cortisol, insulin, IGF-1, or testosterone levels [30]. Also, as bed rest studies are typically done with metabolic control, the likelihood that undernutrition contributed to the findings is low. Therefore, bed rest studies provide a set of expectations for the changes that might be seen in space. Bed rest is not a faithful reproduction of weightlessness, but this is compensated by the rigorous control of confounding factors that is possible in a bed rest study.

After bed rest, too much activity on the atrophied muscles can lead to muscle soreness. Studies on athletes have shown that vigorous exercise, particularly eccentric exercise (i.e., where the muscle is working as a brake and active while it is lengthening) performed with untrained muscles will lead to muscular soreness, stiffness, tenderness, and reduced muscle strength. These symptoms take time to develop after the exercise, so the syndrome is typically called delayed-onset muscle soreness. At the ultrastructural level, damage within the muscle fibers can be seen [9]. This kind of muscle soreness also may occur after spaceflight.

Effect of Spaceflight on Muscle

Most published spaceflight data are grouped results on muscle volume and strength collected after landing. These data are useful to highlight major trends, but they should be interpreted with some caution. Diet, exercise, and stress cannot be rigorously controlled on most flights, so the postflight results reflect an amalgam of all these effects.

Muscle volume

Comprehensive data on muscle loss in space are summarized in table 4-1. After long-duration spaceflights, muscle volume loss is seen in the ankle flexors, ankle extensors, knee flexors, knee extensors, and back muscles. A comparison of these data with what might be expected from bed rest studies is instructive. The loss in the back muscles and psoas in space exceeds what might be expected from bed rest data (comparing losses in table 4-1 with those in figure 4-2). The losses in the anterior leg, quadriceps, and hamstrings are similar to what might be expected in bed rest. The changes in the gastrocnemius and soleus volume may be slightly less than what is seen with bed rest. Overall, however, the muscle-volume changes in space warrant concern, considering that the space data were collected in the setting of an active countermeasure program, while the bed rest data were not.

Hormonal changes

One reason for the pre- and postflight muscle volume differences could be hormonal, but studies conducted thus far are inconclusive. The most comprehensive biochemical study performed in space was done during the Skylab program in the 1970s. The mean data from the three Skylab missions showed no significant changes in adrenocorticotropic hormone (ACTH), growth hormone, or plasma cortisol in flight [35]. Data from other flights support the finding that ACTH and growth hormone do not seem to be altered during spaceflight [36]. IGF-1 also has been measured in space and is not changed [36]. Data on urinary cortisol, however, have shown a trend to be elevated, suggesting that elevated cortisol could contribute to muscle loss in space [35, 36]. The elevation, however, is not a consistent finding [37].

Data on testosterone levels in space are sparse. One study measured testosterone levels 5 days into a flight and found they were decreased [38]. Studies done in rats have also shown decreased testosterone levels in flight [39]. Whether the changes in testosterone are a result of weightlessness or some other factor is not clear. Rat studies have shown decreases in testosterone during hindlimb unweighting [40]. Psychological stress can also reduce testosterone levels in normal men [41].

Nutrition

One common finding in many spaceflight studies is that the astronauts are often in negative energy balance; they burn more calories than they take in and lose weight [14]. This effect is first seen early in a space mission. Symptoms of motion sickness, combined with the novelty of the environment and other factors, lead to reduced dietary intake. Studies on protein metabolism early in spaceflight show that protein turnover is increased, consistent with a metabolic stress response. This combination of low dietary intake and a metabolic stress response can lead to significant protein loss early in a space mission.

Even after adapting to spaceflight, however, dietary intake can remain inadequate [14]. Measurements of protein synthesis after a few months in space show reductions, which may be related to reduced intake. The reasons for the inadequate intake are not clear, but reduced intake, particularly if it is combined with an aggressive exercise program, can set the stage for significant muscle loss. The reduced intake in space and the possible reasons for it are discussed in chapter 8.

Oxidative stress

The urinary excretion of products from DNA and lipid peroxidation are not increased in space but are markedly increased during the recovery period [42]. It is possible that a reduction in energy intake in space could also reduce the effectiveness of the antioxidant system. After spaceflight, oxidative stress is increased, making this an important consideration for recovery [42].

Summary

On average, the muscle volume losses seen in space appear much greater than what would be expected from bed rest studies. Several factors could contribute to this difference, including possible increases in cortisol, decreases in testosterone, and inadequate dietary intake. Oxidative stress may also be a factor, but this has not been studied extensively. Also, as is discussed in other chapters, operational demands and psychological stress can play a significant role, depending on the mission. A crew member who develops mild depression also could be expected to have inadequate dietary intake, which would have an adverse effect on muscle. These data are balanced by the fact that some crew members return with only minimal reductions in muscle mass and strength [29, 33], suggesting that countermeasure programs can be effective.

Approaches to Muscle Loss in Space

Although weightlessness is common to all spaceflights, the level of stress, nutrition, and exercise varies. The response of muscle to spaceflight may differ substantially between individuals and between flights. Therefore, any countermeasure program may have to be adaptable and individualized to maintain the right level of function at the lowest cost of oxygen, food, and water.

Aerobic Exercise

The inactivity of spaceflight can lead to aerobic deconditioning. Good aerobic capability is important for performing work and can at times be critical for demanding work in a spacesuit (see chapter 5). Aerobic training also reduces the risk of cardiovascular disease, increases insulin sensitivity, and can improve mental health [9]. Regular exercise training increases the effectiveness of the antioxidant system. On a space mission, however, aerobic exercise also increases the use of food, water, and oxygen. Greenleaf et al. [43] estimated that for a long-duration space voyage, each 30-minute/day reduction in exercise training time would save 110,869 kcal and 91 l of water per year [43]. So, although maintaining aerobic conditioning is worthwhile, extensive endurance training is wasteful of resources. Also, aerobic conditioning is not optimal training for maintaining muscle mass.

Studies have suggested that 30 minutes of exercise at 50% or greater of maximal oxygen uptake will improve aerobic fitness in untrained individuals and may be considered the minimum level of aerobic exercise needed on Earth to produce a training effect [9]. Slightly more may be needed in weightlessness because of the lack of other aerobic activity (walking, etc.) in space.

Stretching

One simple intervention that can be used to minimize atrophy is to maintain the muscle as much as possible in the stretched condition. Chronic passive stretch has been shown to help maintain muscle mass during unloading in animal studies [44, 45]. In the Russian space program, a special suit is used to take advantage of this effect. The Penguin suit (also marketed commercially as the Adeli suit) is a snug fitting, full-length, long sleeved jumpsuit made with elastic inserts at the collar, waist, wrists, ankles, and along the vertical sides of the suit. The inside of the suit contains a system of elastic cords, straps, and buckles that can be used to adjust the fit and tension of the suit. One set of elastic elements load the body along the long axis with an adjustable force of 15–40 kg. Other elastic elements make it possible to adjust the position of the limbs. The angle of the major joints, like the knee and ankle can be set, allowing, for example, the foot to be dorsiflexed, which would stretch the soleus. The relative contribution of the suit to a countermeasure program is hard to determine, but the approach makes sense physiologically.

Strength Training

Resistance exercise leads to increased protein synthesis, fiber hypertrophy, and increased strength. It is a natural choice to counteract muscle loss in space. The details of the resistance exercise program, such as the types of exercises, how frequently they are performed, and at what intensity, are important to obtain the maximum benefit.

Human bed rest studies and rat hindlimb suspension studies have shown a few basic principles. One is that the intensity of the exercise stimulus is probably more important that the duration. In hindlimb suspension studies it was found that a small amount of high-loading-bearing activity per day (as little as 6 minutes) could preserve a significant amount of soleus mass. Also, short intermittent bouts of exercise done throughout the day could be as or more effective than a single long bout [11]. An intense bout of resistance training every other day during a bed rest study was sufficient to maintain protein synthesis and prevent the loss of muscle strength [46].

Types of resistance exercise can be divided into concentric, eccentric, and isometric [9]. Concentric exercise is when the muscle is active and shortens. During a biceps curl, for example, the biceps muscle is undergoing a concentric contraction when the weight is being lifted. Eccentric exercise is when the muscle is active while it is being lengthened (i.e., when it is serving as a brake). If, after completing a curl, the biceps brachii muscle is active while it lowers the weight slowly, it is contracting eccentrically. Activation of the muscle when it neither lengthens nor shortens is an isometric contraction. For example, trying to hold a weight in the hand at a constant position would require an isometric contraction from the biceps.

In a weight-training program on Earth, the typical recommendations are to perform the exercises at a level where five to six repetitions can be done at the chosen load (also called five to six repetitions maximum or RM). A good mix of exercise types is 15%, eccentric, 10% isometric, and 75% concentric for usual weight training. For spaceflight, the kinds of muscles that need the most work are the ones involved in extending the ankle, knee, and back. An exercise that involves starting in a squatting position and moving under load to a standing position would activate most of the major postural muscles. Muscles in the hip also require activation. The current exercise program followed on the International Space Station, which includes the appropriate types of exercises, is listed in table 4-4.

Electrical Stimulation

Volitional activity activates muscle and increases protein synthesis in muscle. Activation by an externally supplied electrical current will also stimulate protein synthesis and prevent the decline in oxidative enzymes that occurs with disuse [47, 48]. Electrical stimulation also is used frequently to reduce rehabilitation after immobilization [49]. In immobilized patients, the application of electrical stimulation promoted the growth of muscle [47]. In this study the muscle was stimulated 10 hours a day for 9 days. In bed rest studies, shorter periods of stimulation have been used. In one bed rest study, the knee extensors, knee flexors, ankle extensors, and ankle flexors of one leg each received four, 5-minute bouts of stimulation. Stimulation was given twice a

Table 4-4. Exercise countermeasure program in use on the International Space Station.

Exercise	Day 1	Day 2	Day 3	Day 4	Day 5	Day 6
Deadlift	X		X		X	
Bent over rows	X		X		X	
Straight leg deadlift	X		X		X	
Squat	X		X		X	
Heel raises	X		X		X	
Shoulder press		X				X
Rear raises		X				
Front raises		X				
Hip abduction		X				X
Hip adduction		X				X
Bicep curls				X		
Tricep kickbacks				X		
Upright rows				X		
Hip flexion				X		
Hip extension				X		
Lateral raises						X
Front raises						X

The resistance exercises are performed on a device that has pulleys and a shoulder harness to provide the resistance. Early in the mission the goal is 12–15 repetitions maximum (RM) for the resistance exercises, progressing to 4–6 RM later in the mission. The exercise program is as follows: aerobic conditioning—2 days/week, 1 hour/session; resistive exercise—6 days/week, 1 hour/session; interval training—4 days/week, 1 hour/session; extravehicular training—as needed.

day during a 3 days on, 1 day off schedule. The other leg served as a control. Muscle strength and size was significantly better in the stimulated leg [50].

In the Russian space program, electrical stimulation of muscle has been in use for many years. The Tonus-2 and Tonus-3 electrical stimulators flew on the *Salyut* and *Mir* space stations, respectively. The contribution they made to the countermeasure program, however, is not known precisely.

Artificial Gravity

Since the daily exposure to gravity experienced on Earth is enough to prevent muscle atrophy, it is reasonable to expect that introducing Earth-level gravity in a space vehicle would eliminate muscle atrophy due to disuse of postural muscles. What is not known, however, is how effective lower levels of artificial gravity would be (or how effective the one-sixth or one-third of Earth's gravity on the Moon and Mars, respectively, would be). The minimum level of gravitational loading needed to maintain postural muscles has not yet been determined.

Although continuous artificial gravity would likely be a very effective way to prevent muscle loss, it also is the most difficult to achieve. Various devices have been proposed to provide intermittent artificial gravity. Short centrifuges could be used as sleepers to provide a gravity gradient during sleep [51]. Another possibility is a space cycle, where crew members pedal a cycle around an axis, thereby generating a centrifugal force [52]. A short-arm (4–6 m diameter) meter centrifuge also could be used. Studies in animals suggest that different organ systems require different levels and durations of artificial gravity. Using a rat hindlimb suspension model, Zhang and colleagues [53] showed that approximately 4 hours a day of 1 G was needed to maintain soleus mass. Increasing the centrifugation to 2.6 G did not provide additional benefit [53]. In the 1960s, studies on humans using intermittent centrifugation were performed to evaluate intermittent artificial gravity as a countermeasure. These studies suggested that four short (11.2 minute) daily exposures on a short-arm centrifuge were effective at preventing muscle loss [54]. Further study is needed to define the optimal level and duration of intermittent artificial gravity that should be used in space.

Medications and Other Interventions

Because several factors can modify muscle atrophy, a wide range of potential interventions exist to slow or prevent muscle loss.

Antioxidants

Studies in rats have suggested that antioxidants may play a role in muscle disuse atrophy [16, 18, 21]. In a study of rats with one leg immobilized, those given vitamin E injections showed significantly less atrophy than a control group. This was not found, however, in rats where the atrophy was produced by hindlimb unweighting [22]. No human studies exist at present, so no firm recommendations can be made about giving antioxidants to prevent muscle atrophy in space. Antioxidants will likely be given in space to help with radiation damage, so any positive effect on muscular atrophy would be a welcome extra benefit.

Growth hormone

At present, there is no evidence that growth hormone is reduced in space or that space-related muscle atrophy results from growth hormone deficiency. Supplementation with growth hormone, however, could perhaps minimize the muscle changes in space. Studies in older adults suggest that administering exogenous growth hormone can be beneficial but also can produce side effects. When growth hormone is given to older adults, lean body mass increases, although functional ability and strength are not improved [24]. The incidence of side effects (glucose intolerance, carpal tunnel syndrome) is high with growth hormone supplementation [55]. This side effect profile makes growth hormone undesirable for use in space.

Growth factors

IGF-1 plays a key role in maintaining muscle. Recent evidence suggests that IGF-1 may be an important messenger in the pathway where mechanical signals are converted into biological signals to increase protein synthesis [6]. IGF-1 injected into denervated mouse muscle sustained muscle diameter, muscle weight, and strength compared to normal muscle [23]. In a study on human burn victims, IGF-1 (complexed with its principal binding protein, IGF-1 binding protein-3 [IGFBP-3]) was given intravenously at doses of 1, 2, and 4 mg/kg per day [56]. Protein synthesis in the leg was improved with IGF-1. IGF-1, however, also has side effects. The insulinlike effect causes hypoglycemia. Neuropathies have also developed during IGF-1 treatment. The administration of IGF-1 along with its binding protein IGFBP-3 helps minimize side effects [57], but in space, where the ability to deal with a side effect is limited, any adverse effects are unwelcome.

Atrophy inhibitors

The biochemical pathways involved in protein degradation during muscle atrophy are also potential targets for countermeasures. In muscle cells, proteins that are marked for destruction are linked to ubiquitin. Once a protein is ubiquitin-linked, it is quickly broken down by particles within the cells called proteasomes. Recent work has uncovered enzymes (the most notable of which is atrogin-1) that link ubiquitin to specific proteins. Atrogin-1 may be a critical component in the enhanced protein breakdown that characterizes muscle wasting in space [15, 58]. At present, no clinically tested product exists that can manipulate the function of this protein breakdown pathway, but it is a promising area for the future.

Clenbuterol

Clenbuterol is a beta-2 adrenergic agonist that has been given to farm animals to increase muscle mass. Chronic administration of clenbuterol induces skeletal muscle hypertrophy by shifting the balance in protein metabolism toward synthesis [59]. The exact mechanism by which clenbuterol acts is not known [60]. In animal studies, clenbuterol has been effective in reducing muscle atrophy in the rat hindlimb suspension model [59, 61]. Clenbuterol also has been used in patients to minimize atrophy during immobilization after orthopedic procedures [62]. Clenbuterol is a beta-2 agonist, and thus its main side effects are due to this effect (tachycardia and dysrhythmias).

In the United States, clenbuterol is not approved for human use. It is not clear if other beta-2 agonists that are approved (terbutaline, albuterol) would also be useful for counteracting muscle loss. The dosing and trade off between side effects and effectiveness would need to be established for beta-2 agonists to be used against spaceflight-related muscle atrophy.

Anabolic steroids

Testosterone levels may be decreased in space, suggesting that supplementation with testosterone or testosterone-like drugs (anabolic steroids) might be effective [38]. In animal studies using hindlimb suspension and immobilization, the administration of anabolic steroids has shown mixed results. The overall conclusion is that anabolic steroids can be effective when combined with exercise [63]. Pretreatment with the anabolic steroid nandrolone before unweighting has also been shown to be effective in minimizing muscle loss due to limb unweighting in rats [63].

In humans, anabolic steroids stimulate muscle protein synthesis [26]. Anabolic steroids have been given to patients with burns or other significant stress to help minimize the catabolic state that develops [64]. Testosterone has been given chronically to elderly men to help increase lean body mass and improve function [25].

Another compound of interest that is not an anabolic steroid but is a precursor molecule for steroid synthesis is androstenedione (DHEA). Androstenedione is available over the counter as a supplement. Administration of androstenedione does not elevate testosterone levels, but it does increase estradiol concentrations. It does not appear to increase protein synthesis. Thus, the administration of androstenedione does not replicate the effects of anabolic steroids [26].

While anabolic steroids can clearly be helpful in increasing muscle mass and strength, particularly when combined with an exercise program, they also have side effects. They suppress native testosterone secretion. They also can lead to increased aggressiveness (not desirable in an isolated, confined environment), heightened libido, and prostatic enlargement. The risk of hepatic cancer also may be increased. Chronic anabolic steroid administration could not be recommended for spaceflight, but it might be useful if there were a short-term need to increase strength and performance rapidly.

Amino acids

The administration of essential amino acids, either by infusion or orally, can stimulate net protein synthesis in skeletal muscle [65]. Because protein synthesis is decreased in long-duration spaceflight [66], the administration of essential amino acids might be a strategy to boost protein synthesis and maintain muscle mass. Preliminary data from a 28-day bed rest study suggested that the administration of an amino acid/carbohydrate supplement maintained muscle fiber diameter and minimized the loss of lean body mass in the legs [67]. The administration of branched-chain amino acids during a 14-day bed rest study showed a decreased loss of nitrogen in the supplemented group, although protein synthesis did not change [68]. Overall, the use of amino acid supplementation in space is attractive because it has few side effects and could be beneficial.

Monitoring Muscle Loss and Strength

To combat muscle loss in space, good monitoring is required. Monitoring allows individual crew members to adapt their countermeasure program to address particular areas of weakness. Some areas that can be monitored are weight, muscle size, muscle function, and the overall level of activity.

Weight

Perhaps the simplest, but most important, piece of information that crew members need to know is their weight. Crew members who lose weight continuously while in flight obviously are having problems with the countermeasure program, and interventions are necessary. Weight loss can prompt crew members to study their countermeasure program and identify whether their individual problem is inadequate nutrition or exercise. Equipment to measure weight in space use either the period of oscillation of the body on a known spring, or the acceleration of the body due to a known force [69]. Measurements of skinfold thickness could also complement the weight measurements.

Anthropometric Measurements

The simplest anthropometric measurement that can provide important information is calf circumference. This measurement needs to be made reliably using consistent technique at the appropriate position. Calf circumference decreases significantly upon entering space due to the headward fluid shift that occurs [70]. A continuing reduction in calf circumference can indicate that muscle mass is being lost. This measurement should be tracked during the mission to identify developing trends.

Calf circumference is also very sensitive to differences in measurement technique (i.e., tightness and location of the measurement tape). Measuring the volume of the entire leg, as was done on the Skylab program, would provide more reliable information [71]. Leg volume measurement has been a consistent feature of the Russian space program [72]. Infrared perometry also can be used to measure limb volume [73].

Another useful technique is imaging ultrasound. High-resolution ultrasonography can be used to measure muscles, such as the quadriceps, calf, and back muscles. Studies have shown that ultrasound can provide accurate measurements of cross-sectional area and muscle layer thickness [74–76]. Because ultrasound technology is readily available on the International Space Station, these kinds of measurements could be performed and used to guide exercise and nutritional prescriptions.

Functional Tests

Measurements of muscle size and architecture provide useful information but do not assess function. A variety of testing methods can be used to assess whether a crew member has lost strength or aerobic capacity. Handgrip dynamometers have been used in space to assess grip strength before extravehicular activity. Submaximal tests on the treadmill are done periodically in space with EKG monitoring to assess functional capacity. Also, crew members can get a sense of their strength on the basis of the amount of resistance exercise they are able to perform.

Activity Monitoring

Extensive monitoring and functional assessments can represent a significant workload for the crew. Crew members will need to track their performance on various tests, which can add data management tasks to the countermeasure program. One way to minimize data entry and formal assessments is to implement an unobtrusive form of continuous monitoring.

Force transducers have been placed in shoes to note the forces that are experienced during a typical day [77]. Wrist-worn activity monitors (Actigraphs) can give a general sense of the level of activity during different times in the day and give a rough sense of activity. Wireless sensors placed in clothing or around joints could provide continuous monitoring of activity that could be analyzed by computer to produce an estimate of the level of activity that a crew member is experiencing. In the future, this kind of monitoring might be able to provide automatically a measure of the type, duration, and intensity of exercise a crew member performs each day.

Recommendations Based on Current Knowledge

The muscular atrophy that occurs in space appears to be disuse atrophy complicated by an early, in-flight stress response that is possibly exacerbated by continuing undernutrition (depending on the mission). Slightly elevated levels of cortisol may aggravate the losses. Testosterone levels also may be decreased.

The case can be made that some degree of muscle loss in space should be acceptable. Working in weightlessness does not require significant strength, and time and effort are needed to maintain muscle mass. After the mission, muscle mass can be regained. Therefore, perhaps some degree of muscle wasting can be allowed to occur, which will then be recovered through rehabilitation after the mission.

While this approach does have some merit, there are other factors to consider. One is that lost bone is not as easy to recover as lost muscle, and so exercise will be needed in space to maintain bone. Exercise can be done in a way to minimize muscle loss. Another factor is that several lines of evidence from ground-based studies suggest that the exercise needed to maintain postural muscles should not be excessive. If factors such as a reduced caloric intake and stress responses can be controlled, it is conceivable that muscle mass could be maintained with a time-efficient mix of countermeasures. In this case, rehabilitation on Earth would be minimized, and the pathway toward longer and more demanding missions (say, to Mars) would be opened. The best objective of the countermeasure program may be to completely master the muscle loss produced by spaceflight because the knowledge and experience gained doing this will be essential for planning and executing longer missions.

The bulk of animal hindlimb unweighting studies and human bed rest studies suggest some basic principles for preventing disuse atrophy [11]:

- A few repetitions at a high load seem to be more effective than many repetitions at a lower intensity.
- Multiple small bouts throughout the day may be more effective than one long session.
- Passive stretch and isometric contractions (which require minimal equipment) can help.

On the basis of this information, the following recommendations can be made:

1. Weight should be carefully monitored. If a crew member is losing weight, caloric intake and the intensity of the exercise program should be evaluated. The approach to a crew member who is not getting adequate calories but who is exercising extensively would obviously differ from the approach to a crew member who is not exercising, but getting adequate calories.

2. Periodic anthropometric measurements (leg volume, calf circumference) and functional tests are essential to measure progress. Ultrasound may provide a good method for getting accurate data on muscle size while in flight. Because atrophy can progress rapidly, assessments should be conducted more frequently

(every 1 to 2 weeks) early in the mission than they are later in the mission. Later in the mission, crews should be experienced in tracking their own progress and may not need assessment as frequently. If, however, the countermeasure program is ineffective early on, it should be corrected promptly.

3. Activity and exercise should be carefully logged and shared with the ground. In the future, an unobtrusive automated system might be able to perform this function. Data on activity will be essential to determine the appropriate course of action if a crew member is losing weight or muscle mass.

4. Crews should be encouraged to use passive stretch (with a Penguin suit or other device) and isometric exercise to stimulate the muscles outside of the regular exercise countermeasure periods. A few short contractions during the day can help stimulate protein synthesis and maintain muscle mass.

5. The current International Space Station exercise program should be followed, with the flexibility to allow the program to be tailored to the needs of each crew member, based on the results of their anthropometric measurements and functional test results. The overall guideline should be to favor intensity over repetitions.

6. Nutritional countermeasures, such as amino acid supplementation, should be tried because they are easy to implement and have minimal side effects. Also, antioxidant supplementation may be worthwhile, although further studies are needed in this area. Antioxidants will likely be taken anyway as a radiation countermeasure.

7. Hormonal manipulations with growth hormone, testosterone, or IGF-1 should be avoided because of concerns about side effects. Similarly, drugs such as clenbuterol or anabolic steroids cannot be recommended for routine use. As a contingency, anabolic steroids or a beta agonist might be considered in a situation where operational stresses or an emergency put the crew substantially behind on their countermeasure program and they have only a short time to prepare physically for landing.

8. A consistent program of hormonal and other biochemical measurements should be instituted to resolve whether there are any major hormonal changes (such has increased cortisol or decreased testosterone) that complicate muscle loss in space. At present, the data are from too small a sample to be conclusive.

References

1. Newkirk, D., *Almanac of Soviet Manned Space Flight*. 1990, Gulf Publishing, Houston, TX.
2. Basmajian, J.V. and C.J. Deluca, Posture, in *Muscles Alive*. 1985, Williams and Wilkins, Baltimore, MD, pp. 252–64.
3. Nachemson, A., Electromyographic studies of the vertebral portion of the psoas muscle. Acta Orthopaedica Scandinavica, 1966. 37: 177–90.
4. Appell, H.J., Muscular atrophy following immobilisation. A review. Sports Medicine, 1990. 10(1): 42–58.
5. Saltin, B. and P.D. Gollnick, Skeletal muscle adaptability: significance for metabolism and performance, in *Handbook of Physiology*, section 10, *Skeletal Muscle*, L.D. Peachey, ed. 1983, American Physiological Society, Bethesda, MD, pp. 555–632.
6. Baldwin, K.M., and F. Haddad, Skeletal muscle plasticity: cellular and molecular responses to altered physical activity paradigms. American Journal of Physical Medicine and Rehabilitation, 2002. 81(11 suppl): S40–51.

7. Astrand, P., et al., The muscle and its contraction, in *Textbook of Work Physiology*, P. Astrand, et al., eds. 2003, Human Kinetics, Champaign, IL, pp. 31–70.
8. Ohira, Y., et al., Gravitational unloading effects on muscle fiber size, phenotype and myonuclear number. Advances in Space Research, 2002. 30(4): 777–81.
9. Astrand, P., et al., Physical training, in *Textbook of Work Physiology*, P. Astrand, et al., eds. 2003, Human Kinetics, Champaign, IL, pp. 313–68.
10. Adams, G.R., F. Haddad, and K.M. Baldwin, Gravity plays an important role in muscle development and the differentiation of contractile protein phenotype, in *The Neurolab Spacelab Mission: Neuroscience Research in Space*, J.C. Buckey and J.L. Homick, eds. 2003, NASA, Houston, TX, pp. 111–22.
11. Edgerton, V.R., and R.R. Roy, Neuromuscular adaptations to actual and simulated spaceflight, in *Handbook of Physiology*, section 4, *Environmental Physiology*, M.J. Fregly and C.M. Blatteis, eds. 1996, Oxford University Press, New York, pp. 721–64.
12. Edgerton, V.R., et al., Adaptations in skeletal muscle disuse or decreased-use atrophy. American Journal of Physical Medicine and Rehabilitation, 2002. 81(11 Suppl): S127–47.
13. Hodgson, J.A., et al., Circadian force and EMG activity in hindlimb muscles of rhesus monkeys. Journal of Neurophysiology, 2001. 86(3): 1430–44.
14. Stein, T.P., The relationship between dietary intake, exercise, energy balance and the space craft environment. Pflügers Archiv: European Journal of Physiology, 2000. 441(2–3 Suppl): R21–31.
15. Lecker, S.H., and A.L. Goldberg, Slowing muscle atrophy: putting the brakes on protein breakdown. Journal of Physiology, 2002. 545(Pt 3): 729.
16. Lawler, J.M., W. Song, and S.R. Demaree, Hindlimb unloading increases oxidative stress and disrupts antioxidant capacity in skeletal muscle. Free Radical Biology and Medicine, 2003. 35(1): 9–16.
17. Kondo, H., M. Miura, and Y. Itokawa, Oxidative stress in skeletal muscle atrophied by immobilization. Acta Physiologica Scandinavica, 1991. 142(4): 527–28.
18. Kondo, H., et al., Mechanism of oxidative stress in skeletal muscle atrophied by immobilization. American Journal of Physiology, 1993. 265(6 Pt 1): E839–44.
19. Girten, B., et al., Skeletal muscle antioxidant enzyme levels in rats after simulated weightlessness, exercise and dobutamine. Physiologist, 1989. 32(1 Suppl): S59–60.
20. Gomes-Marcondes, M.C., and M.J. Tisdale, Induction of protein catabolism and the ubiquitin-proteasome pathway by mild oxidative stress. Cancer Letters, 2002. 180(1): 69–74.
21. Appell, H.J., J.A. Duarte, and J.M. Soares, Supplementation of vitamin E may attenuate skeletal muscle immobilization atrophy. International Journal of Sports Medicine, 1997. 18(3): 157–60.
22. Koesterer, T.J., S.L. Dodd, and S. Powers, Increased antioxidant capacity does not attenuate muscle atrophy caused by unweighting. Journal of Applied Physiology, 2002. 93(6): 1959–65.
23. Day, C.S., et al., Insulin growth factor-1 decreases muscle atrophy following denervation. Microsurgery, 2002. 22(4): 144–51.
24. Kamel, H.K., D. Maas, and E.H.J. Duthie, Role of hormones in the pathogenesis and management of sarcopenia. Drugs and Aging, 2002. 19(11): 865–77.
25. Ferrando, A.A., et al., Differential anabolic effects of testosterone and amino acid feeding in older men. Journal of Clinical Endocrinology and Metabolism, 2003. 88(1): 358–62.
26. Wolfe, R., et al., Testosterone and muscle protein metabolism. Mayo Clinic Proceedings, 2000. 75(Suppl): S55–59.
27. Guyton, A.C., and H. J.E., The adrenocortical hormones, in *Textbook of Medical Physiology*, A.C. Guyton and H. J.E., eds. 1996, W.B. Saunders, Philadelphia, pp. 957–70.
28. LeBlanc, A., et al., Regional muscle loss after short duration spaceflight. Aviation, Space, and Environmental Medicine, 1995. 66(12): 1151–54.
29. Trappe, S.W., et al., Comparison of a space shuttle flight (STS-78) and bed rest on human muscle function. Journal of Applied Physiology, 2001. 91(1): 57–64.

30. Ferrando, A.A., et al., Prolonged bed rest decreases skeletal muscle and whole body protein synthesis. American Journal of Physiology, 1996. 270(4 Pt 1): E627–33.
31. LeBlanc, A., et al., Muscle atrophy during long duration bed rest. International Journal of Sports Medicine, 1997. 18(Suppl 4): S283–85.
32. Suzuki, Y., et al., Effects of 10 and 20 days bed rest on leg muscle mass and strength in young subjects. Acta Physiologica Scandinavica, 1994. 616(Suppl): 5–18.
33. Convertino, V.A., Exercise and adaptation to microgravity environments, in *Handbook of Physiology*, section 4, *Environmental Physiology*, M.J. Fregly and C.M. Blatteis, eds. 1996, Oxford University Press, New York, pp. 815–44.
34. Bloomfield, S.A., Changes in musculoskeletal structure and function with prolonged bed rest. Medicine and Science in Sports and Exercise, 1997. 29(2): 197–206.
35. Leach, C.S., and P.C. Rambaut, Biochemical Responses of the Skylab Crewmen: an Overview, in *Biomedical Results from Skylab*, R.S. Johnston and L.F. Dietlein, eds. 1977, NASA, Washington, DC, pp. 204–16.
36. Stein, T.P., M.D. Schluter, and L.L. Moldawer, Endocrine relationships during human spaceflight. American Journal of Physiology, 1999. 276(1 Pt 1): E155–62.
37. Lane, H.W. and D.L. Feeback, Water and energy dietary requirements and endocrinology of human space flight. Nutrition, 2002. 18(10): 820–28.
38. Strollo, F., et al., The effect of microgravity on testicular androgen secretion. Aviation, Space, and Environmental Medicine, 1998. 69(2): 133–6.
39. Amann, R.P., et al., Effects of microgravity or simulated launch on testicular function in rats. Journal of Applied Physiology, 1992. 73(2 Suppl): 174S-85S.
40. Wimalawansa, S.M., and S.J. Wimalawansa, Simulated weightlessness-induced attenuation of testosterone production may be responsible for bone loss. Endocrine Journal, 1999. 10(3): 253–60.
41. Chatterton, R.T.J., et al., Hormonal responses to psychological stress in men preparing for skydiving. Journal of Clinical Endocrinology and Metabolism, 1997. 82(8): 2503–9.
42. Stein, T.P., Space flight and oxidative stress. Nutrition, 2002. 18(10): 867–71.
43. Greenleaf, J.E., Energy and thermal regulation during bed rest and spaceflight. Journal of Applied Physiology, 1989. 67(2): 507–16.
44. Goldspink, D.F., et al., The role of passive stretch and repetitive electrical stimulation in preventing skeletal muscle atrophy while reprogramming gene expression to improve fatigue resistance. Journal of Cardiac Surgery, 1991. 6(1 Suppl): 218–24.
45. Sancesario, G., et al., Active muscle length reduction progressively damages soleus in hindlimb-suspended rabbits. Muscle and Nerve, 1992. 15(9): 1002–15.
46. Ferrando, A.A., et al., Resistance exercise maintains skeletal muscle protein synthesis during bed rest. Journal of Applied Physiology, 1997. 82(3): 807–10.
47. Buckley, D.C., et al., Transcutaneous muscle stimulation promotes muscle growth in immobilized patients. Journal of Parenteral and Enteral Nutrition, 1987. 11(6): 547–51.
48. Yoshida, N., et al., Electrical stimulation prevents deterioration of the oxidative capacity of disuse-atrophied muscles in rats. Aviation, Space, and Environmental Medicine, 2003. 74(3): 207–11.
49. Halar, E.M. and K.R. Bell, Immobility. Physiological and functional changes and effects of inactivity on body functions, in *Rehabilitation Medicine Principles and Practice*, J.A. DeLisa and B.M. Gans, eds. 1998, Lippincott-Raven, New York, pp. 1015–34.
50. Duvoisin, M.R., et al., Characteristics and preliminary observations of the influence of electromyostimulation on the size and function of human skeletal muscle during 30 days of simulated microgravity. Aviation, Space, and Environmental Medicine, 1989. 60(7): 671–78.
51. Cardus, D., and W.G. McTaggart, Artificial gravity as a countermeasure of physiological deconditioning in space. Advances in Space Research, 1994. 14(8): 409–14.
52. Kreitenberg, A., et al., The "Space Cycle" self powered human centrifuge: a proposed countermeasure for prolonged human spaceflight. Aviation, Space, and Environmental Medicine, 1998. 69(1): 66–72.
53. Zhang, L.F., et al., Effectiveness of intermittent -Gx gravitation in preventing deconditioning due to simulated microgravity. Journal of Applied Physiology, 2003. 95(1): 207–18.

54. Young, L.R., Artificial gravity considerations for a Mars exploration mission. Annals of the New York Academy of Sciences, 1999. 871: 367–78.
55. Zachwieja, J.J., and K.E. Yarasheski, Does growth hormone therapy in conjunction with resistance exercise increase muscle force production and muscle mass in men and women aged 60 years or older? Physical Therapy, 1999. 79(1): 76–82.
56. Debroy, M.A., et al., Anabolic effects of insulin-like growth factor in combination with insulin-like growth factor binding protein-3 in severely burned adults. Journal of Trauma-Injury Infection and Critical Care, 1999. 47(5): 904–10 [discussion 910–11].
57. Burguera, B., Risks and benefits of insulin-like growth factor. Annals of Internal Medicine, 1994. 121(7): 549.
58. Jagoe, R.T., et al., Patterns of gene expression in atrophying skeletal muscles: response to food deprivation. FASEB Journal, 2002. 16(13): 1697–712.
59. Wineski, L.E., et al., Muscle-specific effects of hindlimb suspension and clenbuterol in mature male rats. Cells Tissues Organs, 2002. 171(2–3): 188–98.
60. Castle, A., et al., Attenuation of insulin resistance by chronic beta2-adrenergic agonist treatment possible muscle specific contributions. Life Sciences, 2001. 69(5): 599–611.
61. Dodd, S.L., and T.J. Koesterer, Clenbuterol attenuates muscle atrophy and dysfunction in hindlimb-suspended rats. Aviation, Space, and Environmental Medicine, 2002. 73(7): 635–39.
62. Maltin, C.A., et al., Clenbuterol, a beta-adrenoceptor agonist, increases relative muscle strength in orthopaedic patients. Clinical Science, 1993. 84(6): 651–54.
63. Joumaa, W.H., et al., Nandrolone decanoate pre-treatment attenuates unweighting-induced functional changes in rat soleus muscle. Acta Physiologica Scandinavica, 2002. 176(4): 301–9.
64. Wolf, S.E., et al., Improved net protein balance, lean mass, and gene expression changes with oxandrolone treatment in the severely burned. Annals of Surgery, 2003. 237(6): 801–10 [discussion 810–11].
65. Volpi, E., et al., Oral amino acids stimulate muscle protein anabolism in the elderly despite higher first-pass splanchnic extraction. American Journal of Physiology, 1999. 277(3 Pt 1): E513–20.
66. Stein, T.P., et al., Protein kinetics during and after long-duration spaceflight on MIR. American Journal of Physiology, 1999. 276(6 Pt 1): E1014–21.
67. Ferrando, A.A., D. Paddon-Jones, and R.R. Wolfe, Alterations in protein metabolism during space flight and inactivity. Nutrition, 2002. 18(10): 837–41.
68. Stein, T.P., et al., Branched-chain amino acid supplementation during bed rest: effect on recovery. Journal of Applied Physiology, 2003. 94(4): 1345–52.
69. Thornton, W.E., and J. Ord, Physiological mass measurements in Skylab, in *Biomedical Results from Skylab*, R.S. Johnston and L.F. Dietlein, eds. 1977, NASA, Washington, DC, pp. 175–82.
70. Johnson, R.L., et al., Lower body negative pressure: third manned Skylab mission, in *Biomedical Results from Skylab*, R.S. Johnston and L.F. Dietlein, eds. 1977, NASA, Washington, DC, pp. 284–312.
71. Thornton, W.E., G.W. Hoffler, and J.A. Rummel, Anthropometric changes and fluid shifts, in *Biomedical Results from Skylab*, R.S. Johnston and L.F. Dietlein, eds. 1977, NASA, Washington, DC, pp. 330–38.
72. Talavrinov, V.A., et al., Anthropometric studies of crew members of *Salyut-6 Salyut-7* [in Russian]. Kosmicheskaia Biologiia i Aviakosmicheskaia Meditsina, 1988. 22(3): 22–27.
73. Tierney, S., et al., Infrared optoelectronic volumetry, the ideal way to measure limb volume. European Journal of Vascular and Endovascular Surgery, 1996. 12(4): 412–17.
74. Bleakney, R. and N. Maffulli, Ultrasound changes to intramuscular architecture of the quadriceps following intramedullary nailing. Journal of Sports Medicine and Physical Fitness, 2002. 42(1): 120–25.
75. Narici, M. and PP. Cerretelli, Changes in human muscle architecture in disuse-atrophy evaluated by ultrasound imaging. Journal of Gravitational Physiology, 1998. 5(1): P73–74.

76. Reimers, C.D., et al., Calf enlargement in neuromuscular diseases: a quantitative ultrasound study in 350 patients and review of the literature. Journal of the Neurological Sciences, 1996. 143(1–2): 46–56.
77. McCrory, J., et al., In-shoe force measurements from locomotion in simulated zero gravity during parabolic flight. Clinical Biomechanics, 1997. 12(3): S7.
78. LeBlanc, A., et al., Muscle volume, MRI relaxation times (T2), and body composition after spaceflight. Journal of Applied Physiology, 2000. 89(6): 2158–64.

5

Extravehicular Activity: Performing EVA Safely

Introduction

In his historic first extravehicular activity (EVA), Alexi Leonov faced almost every medical risk an EVA can present. Leonov left his *Voskhod 2* capsule in March 1965 to begin the first spacewalk ever performed. All went well until he tried to get back into the capsule. His suit was pressurized to 5.9 psi (40.5 kPa), and he had difficulty maneuvering. He reduced the suit pressure to 3.7 psi (25.5 kPa), helping his mobility, but also greatly increasing his risk for decompression sickness. His efforts to get back inside his capsule generated more heat than his suit could remove, causing his body temperature to increase, his heart rate to rise, and his visor to fog. Even though his spacewalk lasted only 12 minutes, by the end Leonov was dehydrated and physically exhausted.

Since that historic spacewalk, the equipment and procedures for EVA have improved dramatically. The medical risks, however, remain the same. During EVAs, crew members' tasks can require both good muscle strength and aerobic conditioning. High workloads can lead to crew members experiencing thermal stress in their pressurized suit. Also, the pressure in the suits, although needed to provide a suitable atmosphere, requires trade-offs between flexibility and the risk of decompression sickness.

Performing EVAs successfully demands close attention to every detail. Most of the risk in EVA is from equipment failure (suit leak, micrometeorite strike, life support system failure, etc.), but there are medical issues that need careful management. Flight surgeons need to be able to recognize when workloads become excessive or when decompression sickness has occurred. This chapter reviews the medical issues raised by EVA and what actions can be taken to prevent problems.

EVA Physiology

The key physiological areas that are relevant to EVA work are exercise physiology, thermal stress, and changes due to altered barometric pressure. Cardiac arrhythmias during EVA are discussed in the chapter on cardiovascular effects (chapter 7).

Work Capacity and Strength

On *Gemini 9* in 1966, Gene Cernan's EVA had to be terminated early because he experienced fatigue, excessive heat inside his suit, and fogging of his visor. At that point in EVA design, not enough was known about the need to provide adequate handholds and attachment points to make the EVA run smoothly. As a result, Cernan was expending considerable energy to perform what appeared to be simple tasks. Working in weightlessness presents some unique challenges.

Weightlessness does make it easy to move around. A tap with a finger can provide enough acceleration to move from point to point. Once objects (like an EVA crew member) are moving, however, they have momentum. As a result, considerable effort may be needed to stop a movement once it is started. If, for example, a crew member is rotating and there is only one handhold to grasp, the crew member will have to try to stop the movement just by using the muscles of the wrist and forearm. Muscles throughout the shoulder area and upper body also will be brought into play. Too aggressive a correction can then begin movement in an unwanted direction, requiring more effort to stop. The net effect is that a simple action, like stopping a rotation, can employ considerable muscle mass and become a significant isometric workout. Without a series of well-placed handholds and footholds, an EVA crew member can expend large amounts of energy on unproductive movements.

Weightlessness also complicates the use of tools. On Earth, the reaction force created when turning a wrench can be safely ignored (e.g., when loosening a bolt, it is rare for the person to move and not the wrench). In weightlessness, however, pulling on a wrench can send the crew member flying if the body is not stabilized. Work on EVAs must be carefully planned to ensure that the body is properly stabilized. Even with adequate hand- and footholds, having good muscle strength and tone provides an extra measure of safety.

EVA tasks typically involve the upper body. The work involves using hand tools, grasping, pulling, pushing, and turning. Most of the work is done using the arms and shoulders, while the lower body provides a stable base for the work. Exposure to weightlessness reduces muscle strength and mass. Although this reduction is more severe for the lower body, the upper body is affected as well. Measurements taken after the Skylab mission in the early 1970s showed 5–10% reductions in the peak torque generated by the arm flexors and extensors [1]. Bed rest studies also show that upper body strength is reduced. Average losses in maximal strength across studies were –6% for hand grip, –8% for forearm, and –7% for arm [1]. To maintain peak performance in weightlessness, EVA crew members need to adhere to an exercise regimen.

Thermal Stress

Work in the EVA suit generates heat, and in the enclosed insulated suit there is nowhere for the heat to go. Exercise can increase heat production about 10–20 times from what it is at rest [2]. Without an effective way to remove heat from the suit, the crew member would rapidly become uncomfortably hot, leading to heat exhaustion or possibly even to heat stroke. An increase in deep body temperature of just 4°C can significantly impair physical and mental performance [2].

As mentioned previously, early EVAs encountered major problems with overheating and fogging of the visor [3]. In contrast to these earlier suits, modern EVA suits are cooled by a liquid cooling garment containing small tubes in contact with the skin. Cooling water circulates through the tubes to remove heat. The current U.S. Extravehicular Mobility Unit (EMU) suit can handle a 252 kcal/hour metabolic load continuously, with peak loads of 403 kcal/hour for 1 hour and 503 kcal/hour for 15 minutes. The Russian Orlan suit has an average metabolic load of 299 kcal/hour with a short-term capability of 600 kcal/hour [4]. As the resting metabolic rate is approximately 60–90 kcal/hour, the suits can handle loads 3–10 times those produced at rest.

Because muscular exercise can increase heat production 10–20 times over resting levels, high-intensity exercise, such as might be produced by operating at a high percentage of maximal oxygen uptake, could not be tolerated by the suit. Actual EVA experience has shown, however, that very high levels of metabolic activity have typi-

Table 5-1. Oxygen cost and heat production of various activities.

Activity	Oxygen uptake (l/minute)	Heat production (kcal/hour)
Slow walking	0.76	228
Golf	1.08	324
Cycling at 10 mph	1.78	534
Playing basketball	2.28	684
Wrestling	2.60	780

This table shows the heat production from various everyday activities. For comparison, the current U.S. extravehicular mobility suit can handle a 252 kcal/hour metabolic load continuously, with peak loads of 403 kcal/hour for 1 hour and 503 kcal/hour for 15 minutes. Data from Webb [41].

cally not been reached. One of the highest metabolic rates recorded in space (500 kcal/hour) was during a demanding EVA in the Skylab program where a strap had to be cut. On average, Shuttle EVAs have required crew members to expend 195 kcal/hour [5]. The heat produced in various activities is listed in table 5-1.

If the EVA crew member exceeds the cooling capacity of the suit, body temperature will rise. Brain function is particularly vulnerable to heat, and studies have shown diminished mental performance on various tasks as heat stress increases [2]. The most severe consequence of heat stress is heat stroke, where the body's temperature regulating system breaks down and allows body core temperature to ramp up. This would be extremely unlikely during EVA because the crews would have other warning signs of heat stress before heat stroke, and they are closely monitored by mission control. Nevertheless, the crew members have a greater ability to produce heat than the suit has capacity to take it away, and they need to be aware of this risk.

Decompression Sickness

The sea-level atmosphere is mostly nitrogen. Even though this nitrogen does not take part in metabolic processes (as oxygen does), it is dissolved throughout the blood and tissues, just as carbon dioxide is dissolved in an unopened soft drink. When a soft drink bottle is opened, thereby releasing the pressure in the bottle, some of the carbon dioxide comes out of solution and forms bubbles. Similarly, reducing the atmospheric pressure rapidly around the body (decompression) can create a situation where nitrogen bubbles begin to form in tissue and in blood. One way to reduce atmospheric pressure is to ascend in altitude. The pressure around the body is reduced during the ascent as the atmosphere becomes less dense. Because historically the earliest way to reduce pressure around the body was by ascending in altitude, the level of decompression is often cited as an altitude rather than as a pressure level (psi, kPa, etc.). Table 5-2 lists the relevant pressures/altitudes of interest for EVA work.

As pressure around the body is reduced, evidence that bubbles are being formed can be obtained in different ways. An ultrasound image can show bubbles moving through the heart. These bubbles, called venous gas emboli or VGE, are a sign of decompression stress [6]. These bubbles can be detected by Doppler ultrasound or imaging ultrasound and are usually graded to provide an index of severity. Table 5-3 shows a commonly used method to grade VGE—the Spencer grading scale. The lev-

Table 5-2. Pressures of interest for EVA work.

Event	Pressure (psi)	Pressure (kPa)	Equivalent altitude (m)	Reference
Sea level	14.7	101.3	0	
Threshold for DCS, no prebreathing, EVA activity	9.7	67	11,000	[8]
VGE first observed, no prebreathing, resting	8.3	57	15,000	[7]
5% DCS incidence, no prebreathing, resting	6.6	45	20,500	[7]
55% DCS incidence, no prebreathing, resting	6.1	42	22,500	[7]
Pressure in Russian Orlan suit	5.7	39	24,000	[4]
Pressure in U.S. EVA suit	4.3	30	30,300	[4]
Emergency pressure in Russian Orlan suit	3.9	27	32,400	[4]

Decompression refers to the reduction in the pressure around the body, which provides a gradient for nitrogen to come out of solution. The decompression level can be given as a pressure or as an equivalent altitude. This table shows the rough pressure levels where bubbles (venous gas emboli or VGE) have been seen in the bloodstream and where decompression sickness (DCS) has occurred in ground-based studies. Both the Russian and American EVA suits are at pressures where DCS would be likely to occur if no action were taken.

els range from grade 1, where an occasional bubble signal is seen, to grade 4, where bubbles are detected in both systole and diastole of every heartbeat.

Although VGE can be used as an index of decompression stress, they are not equivalent to decompression sickness (DCS). Decompression sickness refers to symptoms caused by bubbles, not just their presence, because it is possible to have bubbles moving through the heart during decompression but not to have any pain

Table 5-3. The Spencer grading system for bubbles (venous gas emboli) detected in the bloodstream using Doppler ultrasound.

Grade	Description
0	No bubbles detected
1	Occasional bubble signal, most cardiac periods bubble free
2	Many, but less than half, of the cardiac cycles contain bubbles
3	All the cardiac cycles contain single-bubble signals, but not overriding the cardiac motion signals
4	Maximum detectable bubble signal throughout systole and diastole of every cardiac period, overriding the amplitude of the normal cardiac signals

The presence of venous gas emboli (VGE) does not predict decompression sickness (DCS) and DCS can occur without VGE. Nevertheless, VGE can provide a marker of decompression stress, and higher levels of VGE are associated with more severe DCS.

or other symptoms. If bubbles expand in and around joints, they can produce pain (the bends). This joint or limb pain is known as type I decompression sickness. If the bubbles expand within or travel to the central nervous system, they can cause neurological symptoms such as stroke, vestibular symptoms, or paralysis. This kind of DCS is called type II DCS and is the most feared medical complication of EVA. Bubbles also can cause a variety of other symptoms (skin tingling, cough, etc.), but the most common symptoms are pain in and around joints, and the most serious symptoms are neurological.

Compared to earlier spacecraft like the Apollo capsule or the Skylab space station, current spacecraft operate with a sea-level atmospheric composition and pressure. As a result, in the spacecraft the crew's tissues are saturated with dissolved nitrogen. When a crew member dons a Russian Orlan suit and goes outside the vehicle, the pressure change is equivalent to ascending to 24,000 feet (7317m). For the U.S. EVA suit, the pressure change is equivalent to ascending directly to 30,300 feet (9238 m). If no other action were taken, these changes in atmospheric pressure would provide a strong driving force for nitrogen to come out of solution in tissues and form bubbles.

In Earth-based studies, altitude exposures of these magnitudes clearly cause bubbles to form. Webb et al. [7] exposed groups of normal volunteers to altitude to determine the threshold for detectable bubbles (VGE) and symptoms related to bubble formation. They first noted bubbles in the bloodstream at 15,000 feet. At 21,200 feet they had a 5% incidence of DCS, and at 22,500 feet they had a 55% incidence of DCS (table 5-2). From this study and others [5, 8], it is clear that crew members cannot go directly to the altitude in either the Russian or U.S. space suit without running a risk of DCS. Some factors (like activity) further increase the risk, while others (like removing nitrogen from the body by breathing oxygen) can reduce the risk dramatically.

Activity and DCS risk

The difference between the partial pressure of nitrogen in tissues and the partial pressure of nitrogen in the surrounding atmosphere is an important determinant of whether bubbles are formed, but it is only one of many factors. If bubble formation were an easily predictable process, prevention of DCS would be straightforward. Unfortunately, the occurrence of DCS is variable and unpredictable [9]. In diving and high-altitude operations, episodes of DCS have occurred despite ample preventive measures, and DCS has not occurred in some settings when the preventive measures were totally inadequate.

One factor that affects DCS incidence is activity. In the 1940s, Harvey [10] demonstrated that it is impossible to produce bubbles in a blood sample at the levels of decompression where DCS usually occurs. In other words, when the blood was exposed to a DCS-producing altitude in a test tube outside the body, bubbles did not form. The hypothesis advanced was that small gas nuclei (micronuclei) exist in blood and/or tissue normally, and these serve as nucleation sites for bubble formation during decompression [11]. If tissue micronuclei exist, this raises the possibility that some of the variability in response to decompression may be due to differences in the number and/or size of tissue micronuclei before decompression.

There is indirect but compelling evidence that the numbers of micronuclei can be affected by various factors. Whitaker et al. [12] and Harris et al. [13] showed that rats and frogs that had exercised at altitude displayed a greater number of vascular bubbles than those who had rested. They postulated that exercise increased the number of gas micronuclei. This exercise effect decreased with time. Exercise before decompression produced increased vascular bubbles in frogs, but the effect decreased as the time between exercise and depressurization lengthened. These results are supported by more contemporary studies by Evans and Walder [14], Vann et al. [15], and Daniels et al. [16], who used hydrostatic compression (of water breathing animals) or gas compression (of gas breathing rats) to suppress DCS, presumably by crushing the tissue nuclei. Experiments using crabs as subjects demonstrated a resistance to the formation of visible decompression gas bubbles (seen through the carapace) when the legs of the crabs were kept motionless [17]. A similar decompression when the legs were no longer immobilized produced numerous visible gas bubbles.

The first work examining exercise and the predisposition to decompression sickness in humans was performed during the World War II. Ferris et al. [18] showed that the number of men who developed DCS in their lower limbs increased while performing step (stair climbing) exercises at simulated altitude. The response was proportional to the degree of exercise. Additionally, the subjects seemed to reach a plateau where further exercise did not produce an increase DCS incidence or decrease the time to DCS appearance. Inactivity (sitting) decreases the incidence of altitude DCS markedly [18, 19]. This effect has been reconfirmed at Johnson Space Center [20]. These studies show that some factor, possibly tissue micronuclei, develops during physical activity and decays with inactivity.

The role that activity might play in DCS risk during EVA is complex. On the one hand, because EVA requires intense physical activity, it is possible that this could increase tissue micronuclei and that this in turn would lower the threshold for DCS. On the other hand, it is also possible that the lack of gravitational loading that ordinarily occurs during standing and walking on Earth would lead to a reduction in micronuclei. This has been postulated as one possible reason for the reported lower level of DCS in space than what would be expected from ground-based studies [21].

Risk of DCS during EVA

To plan preventive strategies in space, some estimate of DCS risk is essential. Waligora et al. [22] performed a study to simulate in a ground-based setting the pressure profile experienced by an EVA astronaut. They exposed 38 subjects to an altitude of 30,300 ft (4.3 psi) after they had performed a 6-hour oxygen prebreathe. The astronauts remained at altitude for 6 hours (the maximal duration of an EVA) and performed upper body work during that time. Eighteen of the 38 (47%) had detectable VGE, and 4 of the 38 developed symptoms of DCS (11%) [22]. Because the usual prebreathe time in space is only 4 hours, these data suggest that DCS symptoms would be fairly common during EVA. In practice, however, DCS has been rare in space.

No episodes of DCS have been recorded directly after an American space mission. An episode probably did occur in the Gemini program that was not reported at the time [23]. Aside from this, there have been no crew member reports of DCS, and no need for hyperbaric treatment. There are several possible reasons for the difference between flight experience and ground-based studies. One is that since the EVA suit can produce hot spots and areas of discomfort, it may be difficult for a crew member to discriminate between pain from the bends and the sensations they normally get from the suit. Another is that weightlessness reduces DCS risk either through fluid shifts, a reduction in micronuclei, or some other factor. A third possibility is overreporting in the ground-based studies. The participants may be anxious and too eager to report symptoms. As a result, they may falsely ascribe a variety of aches and pains to DCS. Finally, crew members may have symptoms but not report them.

Reporting DCS symptoms can cause problems for a crew member. To be selected for an EVA, astronauts work and train for years. If these astronauts then experience DCS symptoms during a spacewalk, they may justifiably worry that it might affect their future in the astronaut corps. Pain-only bends symptoms often resolve spontaneously with oxygen, and the crew members are breathing 100% oxygen in the suit. Symptoms may resolve spontaneously and may not be recognized as DCS. Also, as mentioned above, astronauts may not be sure if symptoms are bends or just problems due to the pressure of the suit itself. As a result, the fact that DCS symptoms are not reported after EVAs is encouraging, but this has to be interpreted with an understanding of the disincentive to report minor, resolved, or questionable symptoms.

Patent foramen ovale and other right-to-left shunts

One other complicating factor for DCS risk in space is the presence or absence of a right-to-left shunt. Although VGE can be seen often during decompression, arterial gas emboli (AGE) are rare and can be very hazardous. If a right-to-left shunt exists, it is possible that bubbles from the venous system could pass into the arterial system, travel to the brain and spinal cord, and cause neurological type II DCS. Movement of VGE to the arterial circulation has been seen in altitude decompression sickness [24]. As the possibility of type II or neurological DCS is the greatest medical concern for EVA, how aggressively to screen astronauts for right-to-left shunts remains an important question. Right-to-left shunts can occur because of an atrial septal defect, a patent foramen ovale, or an intrapulmonary shunt. Significant atrial septal defects are usually detected during astronaut selection and are disqualifying unless they are corrected. Intrapulmonary shuts are difficult to detect. Most of the discussion of right-to-left shunts centers on patent foramen ovale because this is a very common condition. Autopsy studies show an incidence of patent foramen ovale of somewhere between 15% and 27% of the population [25–27]. If the presence of a patent foramen ovale were considered to be disqualifying for EVA, this could eliminate a significant number of candidates.

Several studies have shown that patent foramen ovale is more prevalent in divers with severe DCS [27]. Because the risk of serious DCS in diving operations is small (2.28/100,000 dives in one study) the extra risk added by a patent foramen ovale is also small. While some studies have examined patent foramen ovale only, others have looked at the issue of right-to-left shunts more generally since the presence of a patent foramen ovale does not necessarily mean there is a significant shunt. In a study by Cantais et al. [28], 101 consecutive divers presenting to a hyperbaric chamber for treatment of significant DCS were studied using transcranial Doppler. Transcranial Doppler signals in the middle cerebral artery were monitored while a contrast agent

Table 5-4. Odds of severe DCS in divers with a major right-to-left shunt.

| | Shunt | | | |
Group	Yes	No	p value	Odds ratio
Control ($n = 101$)	12	89		
DCS group ($n = 101$)	49	52	<.001	8.7
Cochleovestibular DCS group ($n = 34$)	24	10	<.0001	29.7
Cerebral DCS group ($n = 21$)	13	8	<.0001	24.1
Spinal DCS group ($n = 31$)	10	21	.013	3.9
Limb bends group ($n = 15$)	2	13	NS	1.1

In this study, the investigators used transcranial Doppler to detect a shunt. A major shunt was defined > 20 Doppler "hits" within 15 seconds of injection of 20 ml of agitated oxypolygelatin. The presence of a major shunt was strongly associated with cochleovestibular and cerebral DCS. Adapted from Cantais et al. [28].

was injected intravenously. A significant number of Doppler "hits" in the middle cerebral artery was taken as evidence that a right-to-left shunt existed. A control group of divers was also studied. The results were striking. The DCS group was eight times more likely to have a significant right-to-left shunt than the control group. When the divers were divided into those who had neurological symptoms and those who did not, the odds of having a significant right-to-left shunt was much greater in those with cerebral, cochleovestibular, or spinal DCS. Those who had only limb bends did not have any greater odds of having a significant right to left shunt [28]. These data are summarized in table 5-4 and suggest a strong association between a right-to-left shunt and neurological DCS.

Although these data do suggest that a right-to-left shunt can be a significant risk factor, there are several mitigating factors. One is that AGE are only one source of neurological decompression sickness. Bubbles can form within the tissues (autochthonous bubbles), and this will be unaffected by a right-to-left shunt. Therefore, many cases of severe DCS occur in those without a detectable shunt. Another consideration is that VGE can occur commonly during decompression stress, whereas neurological decompression sickness is rare. In other words, the presence of bubbles in the bloodstream, even with a right-to-left shunt, is rarely associated with neurological decompression sickness.

In space operations, the incentive to avoid a case of neurological DCS is extremely high. Typically, neurological DCS requires treatment in a hyperbaric chamber (see below), which is not a capability that currently exists on the International Space Station. A crew member who becomes incapacitated because of neurological DCS during a spacewalk would need to be brought back to the spacecraft, stabilized, and returned to Earth for treatment. All of those steps require contingency operations, which increase the risk for a mishap or error. In the worst case, the crew member could die or be permanently disabled, or the entire crew could be put at risk trying to save the affected crew member. Because of these considerations, the possibility of neurological DCS is taken very seriously. Spaceflight operations are inherently dangerous, and the risk of neurological DCS is dwarfed by the hazard of launch and landing. Nevertheless, a significant type II DCS event during an EVA could be tragic.

Prevention of DCS

The main method of reducing DCS risk is to eliminate some of the nitrogen from the tissues before EVA. This can be done either by reducing the pressure in the cabin of the spacecraft or by breathing 100% oxygen (these methods also can be combined). In either approach, the crew member eliminates nitrogen with each breath and therefore decreases the tissue pressure of nitrogen and decreases the risk of DCS. The main question, however, is how much time the crew member should devote to denitrogenation. This decision has important practical consequences. Time used to breathe oxygen before an EVA can reduce the amount of time available for EVA.

On the International Space Station, reducing cabin pressure throughout the station is not practical. The airlock can be partially depressurized, and one option is for the crew member to sleep ("camp out") in the airlock before EVA. Currently, however, oxygen prebreathing is the preferred option to remove nitrogen from the body before an EVA. NASA requirements call for 4 hours of oxygen breathing in the EVA suit before a spacewalk [5]. This can create operational problems because a 4-hour prebreathe plus a 6-hour EVA plus preparation time can add up to a long time in the suit and a long work day. To help improve operational effectiveness, other prebreathing protocols have been evaluated.

One reason for the slow removal of nitrogen during a resting oxygen prebreathe is that blood flow to many tissues is low at rest. This limits the exchange of nitrogen. Exercise, however, increases blood flow, so exercise while breathing oxygen has been studied as a way to speed nitrogen removal. This improved blood flow, however, requires activity, which may increase tissue micronuclei and increase the risk of DCS. Ground-based studies have shown that after exercise the half-life of putative tissue micronuclei is approximately 1 hour [20], suggesting that if exercise were used early in the prebreathe, additional risk from tissue micronuclei could be minimized.

The current International Space Station protocol involves 10 minutes of arm and leg exercise at 75% of Vo_2 max; followed by 24 minutes of light exercise, during an overall 2-hour oxygen prebreathe. No episodes of DCS occurred when this protocol was used in ground-based experiments, but there was a 31% overall incidence of VGE, and a 7% incidence of grade IV VGE [29, 30]. This protocol has been adopted for use on the International Space Station. Other options include a 4-hour prebreathe without exercise or camping out in the airlock at 10.2 psi.

Treatment of DCS

On the ground, the standard treatment for pain-only DCS is compression in a hyperbaric chamber to 2.8 atmospheres absolute (ATA) while breathing oxygen. Divers who present with pain in or around a joint should have a complete neurological exam to ensure there are no neurological symptoms that might suggests type II DCS. They should be placed in the hyperbaric chamber, compressed to 2.8 ATA, and monitored for resolution of the pain. If the pain resolves within 10 minutes, they stay in the chamber to complete a course of compression combined with oxygen. A variety of different compression treatment plans, called treatment tables, have been developed over the years. A treatment table 5, for example, lasts for 135 minutes, starts at

2.8 ATA, and then backs off to 2.0 ATA. Oxygen breathing is interspersed with air breathing to minimize the risk of oxygen toxicity. This treatment table would be the choice for a diver whose symptoms resolved within 10 minutes. For more significant symptoms, the tables can go to much higher pressures (6 ATA for a treatment table 6A) and last for longer periods of time (36 hours for a treatment table 7) [31].

The tables have been refined over the years to minimize the hazardous sequelae of DCS. Although some pain-only bends will resolve spontaneously, some cases will progress to more serious symptoms. Treatment of all cases with hyperbaric compression helps to standardize the outcome. In space operations, however, a hyperbaric chamber is currently not available, so other methods need to be used.

The mainstay of therapy for DCS in space is 100% oxygen. Breathing 100% oxygen helps to remove inert gas, and the oxygen also promotes bubble dissolution in the tissues due to the "oxygen window" [32]. Although a chamber is not available, some pressure can be applied by overpressurizing the suit using the onboard bends treatment apparatus or by increasing the cabin pressure in the station. Hydration is also important because it maintains blood and plasma volume, which helps with inert gas washout. At present, there is not strong evidence to suggest the use of corticosteroids or aspirin in the treatment of DCS [33]. In many cases of pain-only bends, it is possible that oxygen and hydration combined with the pressure that can be applied in the suit and cabin would be enough to help relieve the symptoms.

If more serious DCS develops, lidocaine can be administered. Lidocaine has been shown to preserve brain function in patients who undergo surgery on the left heart valves. As patients who have this kind of surgery often have gas emboli that reach the brain, the situation is analogous to DCS. In one study, lidocaine was given using a 1 mg/kg bolus over 5 minutes, followed by 240 mg over the first hour and 120 mg over the second hour, and then 60 mg/hour thereafter [34]. Patients who received lidocaine showed significantly fewer deficits in postoperative neuropsychological test performance.

Another potential treatment option in space is perfluorocarbon emulsions. Perfluorocarbons are a class of compounds that have very high solubility for gases. This not only makes them ideal as blood substitutes, but the high solubility of both oxygen and nitrogen in perfluorocarbon raises the possibility that these compounds could increase the rate of delivery of oxygen and the elimination of nitrogen, thereby improving the outcome in decompressions sickness. Intravenous perfluorocarbon emulsions have been shown to improve survival in rats [35, 36] and hamsters [37] after decompression. Studies in swine also have shown improved survival after decompression when perfluorocarbons are administered. Perfluorocarbons are considered an adjunctive treatment worthy of further research by the Undersea and Hyperbaric Medical Society [38]. A perfluorocarbon emulsion is provided in very small particles (~ 0.1 micron). These small particles increase the surface area for absorption of gas. Perfluorocarbons are taken up by the reticuloendothelial system and are removed from the body by vaporization through the skin and lungs. One disadvantage of this treatment is that after the administration of a perfluorocarbon the individual can develop flulike symptoms. At present, there are no perfluorocarbons available for human use to treat DCS, but this could be an attractive option for spaceflight use in the future.

Radiation Exposure

The overall risks of radiation in space are discussed in chapter 3. In low-earth orbit astronauts are exposed to protons, electrons, and galactic cosmic radiation (GCR). During an EVA, astronauts may receive increased doses of radiation to the skin and eye from trapped particles, especially during transits through the South Atlantic Anomaly. Severe solar weather conditions can produce high levels of radiation due to solar particle events or disturbed Earth magnetic field conditions. High-energy protons, such as would be common in a solar flare, can penetrate nearly three-quarters of the area of the space suit. Protons with energies of 25 MeV or greater can penetrate the visor, the most heavily shielded part of the suit. The trapped radiation belts, which spacecraft travel through on some orbits, can penetrate some areas of the suit at energies of 0.5 MeV and above. The key factor in avoiding an excessive radiation dose is proper planning and monitoring. The crews should avoid EVA during solar flares and minimize time in trapped radiation belts.

Countermeasures for EVA-related Problems

For optimum performance and safety, several measures can be taken to keep strength and endurance high and the risk of decompression sickness low.

Physical Training

Because the majority of EVA tasks involve the upper body, and because weightlessness reduces upper body strength, a regular exercise countermeasure program is essential. The current International Space Station countermeasure program includes a variety of exercises. Grip strength is maintained by working with a range of devices, such as a finger exerciser, foam ball, web of elastic straps, and grip master. Forearm strength is maintained by a set of exercises using a rod with dual grips that allows for supination, pronation, flexion, and extension. These exercises should be done in a rotating schedule as an EVA approaches.

The arms and shoulders are exercised using both a cycle ergometer and a resistance exercise device. Both resistance and endurance exercise are performed 6 days a week with a rotating set of exercises. Upper body strength and endurance training is increased in the 2 weeks preceding an EVA. Details on exercise training are given in chapter 4.

Decompression Sickness Prevention

Compared to previous space stations, the Shuttle program, and the Apollo lunar program, the International Space Station program faces greater challenges to reduce DCS risk. For the Apollo program, the spacecraft was at low pressure with a high oxygen atmosphere. As a result, crew members were fully denitrogenated when they arrived at the moon, making concern about DCS unnecessary. Similarly, the *Skylab* space station operated at a low pressure, so that moving between the EVA suit and the station did not involve a major pressure change. In future lunar or Mars exploration, use of a lower pressure habitat would simplify EVA operations, although this would involve trade-offs in other areas.

The Russian program has had space stations in orbit since the 1970s, with an active EVA program. Although the Russian stations have operated at sea-level pressure, the Russian suits operate at a higher pressure than the U.S. suits (39 vs. 30 kPa). As a consequence, prebreathe times have been short (1 hour), and there have not been any reported cases of DCS in the Russian program.

The lower-pressure U.S. suit offers greater movement flexibility than the Russian design. When the U.S. suit was used on Shuttle flights, its safety could be enhanced by reducing the Shuttle cabin pressure to 10.2 psi (71 kPa) for 24 hours before EVA, thereby allowing for a prolonged denitrogenation period [39]. For International Space Station operations, the same low-pressure suit is being used, but the ability to use a reduced cabin pressure to help with denitrogenation is not readily available (except perhaps in the airlock). The 2-hour prebreathe protocol that incorporates exercise has been effective in ground-based trials. Ground-based testing, however, has shown a significant level of VGE using this protocol, and it is not known if the amount of VGE in operational use is lower, higher, or the same as in ground-based testing.

Screening

Astronaut selection includes certain criteria to exclude those who may be at high risk of DCS. Current guidelines state that an individual with a history of type II DCS would not be qualified for selection, and neither would an individual with an uncorrected atrial septal defect or patent ductus arteriosus. The question of whether patent foramen ovale should also be disqualifying for EVA has been controversial. It is possible that the focus in space operations should not be on patent foramen ovale, as many of these are small, but on right-to-left shunts more generally. In other words, the concern in EVA is not whether a patent foramen ovale is present, but whether a significant right-to-left shunt exists.

Screening methods that use transcranial Doppler to look for evidence of echo contrast in the middle cerebral artery test for right-to-left shunts rather than for a particular anatomic abnormality. In the absence of solid data on how prevalent significant VGE are during actual EVA operations, screening for right-to-left shunts may make sense if it can be done without significant risk. There are no data showing that screening for a right-to-left shunt would improve safety in flight, but there is also no proof that it is safe for a crew member with a significant right-to-left shunt to perform EVA using the current International Space Station protocol.

Fitness

Animal studies have shown that exercise conditioning reduces the risk of DCS [40]. Obesity may increase the risk of DCS, perhaps because of the increased solubility of nitrogen in fat. The association of obesity with DCS is not clear, but it makes sense for crew members to do physical training in space to remain fit for EVA. The crew members would be participating in an exercise program anyway to help prevent bone and muscle loss.

Bubble monitoring

One problem with evaluating the countermeasures to prevent DCS is that the main end point to judge success or failure—the number of DCS cases—is unreliable in space. Crew members may be unable to tell whether a particular discomfort they are experiencing represents DCS, and, as mentioned previously, there are disincentives to reporting equivocal cases. As a result, there could be a bias toward underreporting, so that the first indication that the countermeasure program is not working might well be a severe case of DCS. In this situation, the use of Doppler or ultrasound monitoring to provide an index of decompression stress would be reasonable. In-suit Doppler monitoring was recommended in a National Academy of Sciences report [39].

An in-suit Doppler system presents technical challenges. The suit is filled with 100% oxygen, so any electrical system must meet stringent engineering criteria to be placed within the suit. The communication systems in the suit must meet these stringent standards, so the safety standards are possible to achieve. Communication ability while in the suit is an absolute requirement, however, whereas the value of Doppler monitoring could be debated. Careful testing would be needed to ensure that the hazards of monitoring do not exceed the risks that are being prevented.

Decompression Sickness Treatment Options

If a case of pain-only DCS should occur, the protocol followed in diving practice can be used. The affected crew member should receive repressurization, hydration, a careful exam, and 100% oxygen. Although a hyperbaric chamber is not present, it is still possible to provide some extra pressure. The EVA suit has a limited hyperbaric capability afforded by a special backpack adapter component (the bends treatment adapter). This can be used in the airlock. It allows pressurization of the suit to 57.2 kPa, which, in combination with an elevated cabin pressure, can provide a 160 kPa (1.6 ATA) treatment pressure. While not the standard 2.8 ATA that would be given in a DCS treatment, this amount of pressure may be adequate in many cases.

If type II DCS should occur, the crew will need to consider evacuation of the affected crew member. In addition to 100% oxygen and extra pressure, lidocaine should be given (1 mg/kg loading, followed by 240 mg over the first hour and 120 mg over the second hour, and then 60 mg/hour thereafter). If the crew member should become unconscious or unresponsive, then pressurization in the suit is not possible because access to the crew member will be needed. In the future, it is possible that perfluorocarbon emulsions might be useful in this setting, although more study is needed.

Radiation Protection

As was mentioned in chapter 3, an antioxidant mixture containing vitamin E, C, A, beta-carotene, and selenium would be a reasonable dietary supplement. Because EVA involves a high skin-dose of radiation, antioxidant supplementation before EVA and the use of a skin cream containing antioxidants should be considered.

Recommendations Based on Current Knowledge

The main risks of EVA are not medical. The crew member must depend on the suit and supporting equipment, and failures in these areas can be fatal. Nevertheless, there are medical risks. A crew member can overwhelm the cooling capacity of the suit with vigorous work. Weightlessness exposure will reduce muscle strength and mass, and crew members may be presented with challenging physical tasks. Although serious DCS is rare in diving and altitude operations and is usually associated with a violation of standard practice, it can still happen and is an unpredictable risk. A serious case of DCS in space could be devastating for the affected crew member and for the overall program. Ground-based testing of the current denitrogenation protocol shows that it works, but that 31% of subjects have VGE and 7% have grade 4 VGE. Without monitoring, it is difficult to know if the VGE in actual operational use corresponds with the ground studies or is better or worse. Based on current knowledge, the following recommendations can be made:

1. An upper body training program is essential to maintain strength, particularly for grasping, pulling, and pushing. Exercises focusing on grip strength and forearm motions are particularly important. Periodic assessments are needed before EVA to alert the crew member to any strength changes and to allow time for the exercise program to be modified to correct any deficits.

2. Doppler monitoring should be implemented. An in-suit Doppler would present safety concerns requiring a judgment about whether the information on VGE is worth the extra risk. Another option is to perform Doppler monitoring immediately after the EVA, in the 79% nitrogen, 21% oxygen International Space Station atmosphere. Although not as complete as in-suit monitoring, measurements after EVA may be useful and could provide data on the degree of decompression stress seen in actual operations. A significant level of VGE detected in the spacecraft would indicate that the decompression stress may be more than predicted and that prebreathe times may need to be prolonged.

3. Use the highest pressure suit that will be suitable for the task. While the mobility of the U.S. EMU suit may be desirable for some tasks, a higher pressure suit has been used successfully for space station maintenance in the Russian program for many years. This suit does sacrifice some mobility, but it has an established track record and adds an extra measure of safety. The ability to regulate the pressure in the suit is also important because suit pressure could be reduced in settings where DCS risk is low.

4. Have a protocol for emergency EVA that allows for rapid response while still keeping DCS risk as low as practicable.

5. In the absence of data on the severity of decompression stress in actual operations, a screening program for right-to-left shunts should be considered. Ground-based data suggest that VGE will be present, sometimes at very high levels. Significant VGE combined with a significant undetected right-to-left shunt could lead to tragic consequences. Although screening may seem to be unfair to individuals already trained in EVA who have not had problems, at present the program is not equipped to handle a severe type II DCS case. This requirement could be relaxed as operational experience is gained and further data on VGE in flight are acquired.

6. An antioxidant mixture should be taken before EVA, and a topical antioxidant preparation should be considered.

References

1. Greenleaf, J.E., et al., Exercise-training protocols for astronauts in microgravity. Journal of Applied Physiology, 1989. 67(6): 2191–204.
2. Astrand, P.O., et al., Temperature regulation, in *Textbook of Work Physiology: Physiological Bases of Exercise*. 2003, Human Kinetics, Champaign, IL, pp. 395–432.
3. Shayler, D.J., Survival in space, in *Disasters and Accidents in Manned Spaceflight*, 2000, Springer Praxis, Chichester, UK, pp. 245 58.
4. McBarron, J.W., et al., Individual systems for crewmember life support and extravehicular activity, in *Space Biology and Medicine*, A.E. Nicogossian, et al., eds. 1996, American Institute of Aeronautics and Astronautics, Reston, VA, pp. 275–330.
5. Horrigan, D.J., et al., Extravehicular activities, in *Space Biology and Medicine*, A.E. Nicogossian, et al., eds. 1996, American Institute of Aeronautics and Astronautics, Reston, VA, pp. 533–46.
6. Nishi, R.Y., A.O. Brubakk, and O. Eftedal, Bubble detection, in *Bennett and Elliot's Physiology and Medicine of Diving*, A.O. Brubakk and T.S. Neuman, eds. 2003, Saunders, New York, pp. 501–29.
7. Webb, J.T., A.A. Pilmanis, and R.B. O'Connor, An abrupt zero-preoxygenation altitude threshold for decompression sickness symptoms. Aviation, Space, and Environmental Medicine, 1998. 69(4): 335–40.
8. Kumar, K.V., J.M. Waligora, and D.S. Calkins, Threshold altitude resulting in decompression sickness. Aviation, Space, and Environmental Medicine, 1990. 61(8): 685–89.
9. Francis, T.J., and D.F. Gorman, Pathogenesis of the decompression disorders, in *The Physiology and Medicine of Diving*, P.B. Bennett and D.H. Elliot, eds. 1993, W.B. Saunders, New York, pp. 454–480.
10. Harvey, E.N., Physical factors in bubble formation, in *Decompression Sickness*, J.F. Fulton, ed. 1951, W.B. Saunders, New York, pp. 90–114.
11. Hempleman, H.V., History of the decompression procedures, in *The Physiology and Medicine of Diving*, P.B. Bennett and D.H. Elliot, eds. 1993, W.B. Saunders, New York, pp. 342–375.
12. Whitaker, D.M., et al., Muscular activity and bubble formation in animals decompressed to simulated altitudes. Journal of General Physiology, 195. 28: 213–23.
13. Harris, M., et al., The relation of exercise to bubble formation in animals decompressed to sea level from high barometric pressure. Journal of General Physiology, 195. 28: 241–51.
14. Evans, A., and D.N. Walder, Significance of gas micronuclei in the aetiology of decompression sickness. Nature, 1969. 222(190): 251–52.
15. Vann, R.D., J. Grimstad, and C.H. Nielsen, Evidence for gas nuclei in decompressed rats. Undersea Biomedical Research, 1980. 7(2): 107–12.
16. Daniels, S., et al. Micronuclei and bubble formation: a quantitative study using the common shrimp, *Crangon crangon*. in *Underwater Physiology VIII: Proceedings of the Eighth Symposium on Underwater Physiology*, Bachrach, A.J. and Matzen, M.M., eds. 1984, Undersea Medical Society, Bethesda, MD, pp. 147–57.
17. McDonough, P.M., and E.A. Hemmingsen, Bubble formation in crabs induced by limb motions after decompression. Journal of Applied Physiology, 1984. 57(1): 117–22.
18. Ferris, E.B., et al., The importance of straining movements in electing the site of the bends. 1943, U.S. National Research Council Committee on Aviation Medicine, Washington, DC.
19. Gray, J.S., and R.L. Masland, Studies on altitude decompression sickness: II The effects of altitude and exercise. Journal of Aviation Medicine, 1946. 17: 483–93.
20. Dervay, J.P., et al., The effect of exercise and rest duration on the generation of venous gas bubbles at altitude. Aviation, Space, and Environmental Medicine, 2002. 73(1): 22–27.

21. Conkin, J., and M.R. Powell, Lower body adynamia as a factor to reduce the risk of hypobaric decompression sickness. Aviation, Space, and Environmental Medicine, 2001. 72(3): 202–14.
22. Waligora, J.M., D.J. Horrigan, and J. Conkin, The effect of extended O_2 prebreathing on altitude decompression sickness and venous gas bubbles. Aviation, Space, and Environmental Medicine, 1987. 58(9 Pt 2): A110–12.
23. Collins, M., *Carrying the Fire*. 1974, Farrar, Straus and Giroux, New York.
24. Pilmanis, A.A., F.W. Meissner, and R.M. Olson, Left ventricular gas emboli in six cases of altitude-induced decompression sickness. Aviation, Space, and Environmental Medicine, 1996. 67(11): 1092–96.
25. Hagen, P.T., D.G. Scholz, and W.D. Edwards, Incidence and size of patent foramen ovale during the first 10 decades of life: an autopsy study of 965 normal hearts. Mayo Clinic Proceedings, 1984. 59(1): 17–20.
26. Schneider, B., et al., Diagnosis of patent foramen ovale by transesophageal echocardiography and correlation with autopsy findings. American Journal of Cardiology, 1996. 77(14): 1202–9.
27. Foster, P.P., et al., Patent foramen ovale and paradoxical systemic embolism: a bibliographic review. Aviation, Space, and Environmental Medicine, 2003. 74(6 Pt 2): B1–64.
28. Cantais, E., et al., Right-to-left shunt and risk of decompression illness with cochleovestibular and cerebral symptoms in divers: case control study in 101 consecutive dive accidents. Critical Care Medicine, 2003. 31(1): 84–88.
29. Vann, R.D., et al. Design, Trails and Contingency Plans for Extravehicular Activity From the International Space Station, in *Proceedings of the Bioastronautics Investigators' Workshop*. 2001, Universities Space Research Association, Division of Space Life Sciences, Galveston, TX.
30. Gernhardt, M.L., J. Conkin, and P.P. Foster, Design and testing of a 2-hour oxygen prebreathe protocol for space walks from the International Space Station. Undersea Hyperbaric Medicine, 2000. 27(Suppl): 12.
31. Joiner, J.T., NOAA Diving Manual, 4th ed. 2001, Best Publishing, Flagstaff, AZ.
32. Moon, R.E., Treatment of decompression sickness and arterial gas embolism, in *Diving Medicine*, A.A. Bove, ed. 1997, W.B. Saunders, Philadelphia, PA, pp. 184–204.
33. Moon, R.E., and D.F. Gorman, Treatment of the decompression disorders, in *Bennett and Elliot's Physiology and Medicine of Diving*, A.O. Brubakk and T.S. Neuman, eds. 2003, Saunders, New York, pp. 600–50.
34. Mitchell, S.J., O. Pellett, and D.F. Gorman, Cerebral protection by lidocaine during cardiac operations. Annals of Thoracic Surgery, 1999. 67(4): 1117–24.
35. Lutz, J., and G. Herrmann, Perfluorochemicals as a treatment of decompression sickness in rats. Pflugers Archives, 1984. 401(2): 174–77.
36. Spiess, B.D., et al., Treatment of decompression sickness with a perfluorocarbon emulsion (FC- 43). Undersea Biomedical Research, 1988. 15(1): 31–37.
37. Lynch, P.R., et al., Effects of intravenous perfluorocarbon and oxygen breathing on acute decompression sickness in the hamster. Undersea Biomedical Research, 1989. 16(4): 275–81.
38. Adjunctive Therapy ad hoc Subcommittee, *UHMS Guidelines for Adjunctive Therapy for Decompression Illness*. 2002, Undersea and Hyperbaric Medical Society, Bethesda, MD.
39. Commission on Engineering and Technical Systems, Extravehicular activity, robotics and supporting technologies, in *Engineering Challenges to the Long Term Operation of the International Space Station*. 2000, National Academy Press, Washington, DC, pp. 18–23.
40. Francis, T.J.R., and S.J. Mitchell, Pathophysiology of decompression sickness, in *Bennett and Elliott's Physiology and Medicine of Diving*, A.O. Brubakk and T.S. Neuman, eds. 2003, Saunders, New York, pp. 530–56.
41. Webb, P., Work, heat and oxygen cost, in *Bioastronautics Data Book*, J.F. Parker and V.R. West, eds. 1973, NASA, Washington, DC, pp. 847–80.

6

Balance: Neurovestibular Effects of Spaceflight and Their Operational Consequences

Introduction

A day or so after returning from Earth orbit, an astronaut drove his car onto a long, tightly curved ramp at the intersection of two interstate highways. About halfway through the curve he needed to pull over because he was having a strong sensation of tilt. The slight sideways G forces he had experienced in the turn had been interpreted by his space-adapted brain and nervous system as a significant body tilt. The most likely explanation was that his balance system had adapted to weightlessness and had not yet readapted to Earth's gravity.

Astronauts are unsteady shortly after returning from space. Walking heel-to-toe, standing still with eyes closed, and balancing on a rail can all present major challenges postflight [1]. When tested with their eyes closed, crew members usually incorrectly estimate sideways (roll) tilt of their bodies [2, 3]. These changes in balance and perception can be hazardous. Sudden roll movements of the Shuttle due to wind gusts may be misperceived as being of much greater magnitude than they actually are. If a postflight emergency should occur, the crew may be clumsy and unstable as they work through emergency procedures. An emergency that required the crew to function effectively in the dark could be particularly hazardous because vision helps compensate for changes in the vestibular system. After flight, however, the balance system improves rapidly. Crew members show improvement within hours of landing and gradually return to normal over the next several days [4].

To fly safely and function effectively in emergencies, crews need to be aware of the vestibular problems they are likely to have. Also, if crews go to the Moon or Mars, they will want to begin work there as soon as possible after landing and will need to know how to adapt rapidly to those gravitational fields. Existing data suggest that crew members will be able to adapt to new environments. Nevertheless, they need to be aware of the altered vestibular function they will experience until they have fully adapted. This chapter outlines the physiology of balance, the changes to balance that occur in space, and how crew members can manage these changes.

The Control of Balance

The balance or vestibular system plays an important role in walking, running, navigation, driving, and operating aircraft. The system integrates input from the eyes, the inner ears, and position sense to stabilize the body and allow for smooth movements [5]. Figure 6-1 is a block diagram of the system. Sensors (the eyes, the acceleration sensors in the inner ear, and pressure sensors in joints and muscles) provide information about movement and body position. The semicircular canals sense nonsustained body and head angular accelerations. The otoliths are the main gravity sensors; they indicate the static tilt of the body and detect linear acceleration. Vision offers information on body tilt and motion. Position sense (proprioception) provides feedback on body movement and data on the location of the arms and legs in space. The semicircular canals and proprioception offer rapid, high-frequency stabilization, while the otoliths and eyes work at lower frequencies [6]. Typically, these senses present complementary information to the integrative centers in the brain, which then synthesize

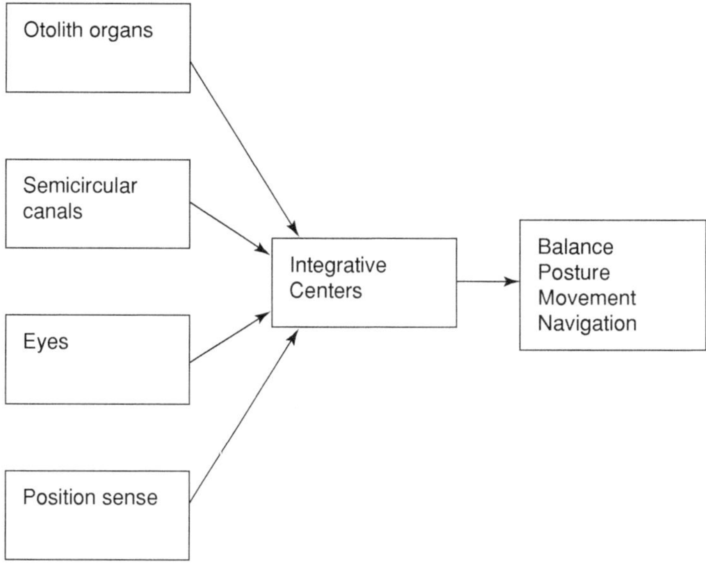

Figure 6-1. A block diagram of the vestibular system. The brain uses information from a variety of sources to synthesize a sense of body orientation and movement. Also, the brain retains a copy of the signals sent to muscles to initiate movements (efference copy) and so can compare the desired movement with the information that comes back from the sensors about the actual movement. In general, the semicircular canals and position sense are best for high-frequency motions, while the otolith organs and eyes sense low-frequency motions. The integrative centers can adapt to a variety of unusual sensory situations (reversing prisms, labyrinthectomy, slow rotating room).

an overall sense of the body's location and movement. These sensors also provide feedback to the brain on whether a particular movement or action was executed as planned.

As anyone who has learned to ride a bicycle or ice skate knows, the balance system is very adaptable. With training and practice, a shaky and fall-prone skater can learn to glide smoothly across the ice. Also, deficits within the balance system can be compensated. In fact, the capacity for adaptation is extraordinary because some individuals can function well despite major deficits in one of the core sensory systems (vision, position sense, acceleration sense). For example, many of those who lack a functioning vestibular system (e.g., patients with bilateral labyrinthine dysfunction) can walk, run, and maintain their balance. Graybiel [6] studied two individuals who had lost all vestibular function early in life but retained their hearing. They were not aware that they had a vestibular deficit, and neither were their families or physicians. In fact, these individuals functioned at an above average level in sports [6]. Vestibular testing, however, easily revealed the deficit and demonstrated that these individuals had difficulties when visual cues were inadequate.

Deficits in proprioception also can be compensated. The book *Pride and a Daily Marathon* by Jonathan Cole tells the remarkable story of a young man, Ian Waterman, who lost almost all of his proprioception [7]. Waterman found that all position sense below his neck was gone after a flulike illness. Initially, he was totally incapacitated, and his neurologist gave him a poor prognosis. But over time he trained himself to walk and stand by using his other senses. For him, each day became a marathon because of the effort required to think through each movement he made. Eventually, he defied expectations and returned to work—where his co-workers remained unaware of his major neurological deficit.

These examples from patients with major sensory deficits illustrate the adaptive powers of the brain and nervous system. The integrative centers within the brain can change the weighting given to particular senses or compensate for the complete lack of a particular sense. This capability is important in spaceflight because the otolith organs, which typically receive continuous input from gravity on Earth, lose that stimulus in weightlessness.

The Otolith Organs and Semicircular Canals

Figure 6-2 shows the semicircular canals and otolith organs. The three semicircular canals are at right angles to each other and respond to changes in head position. The canals sense acute accelerations and respond well to high-frequency motion. This is important for one of their major roles: the stabilization of vision during motion. The semicircular canals are not involved in sensing gravity, however, and so might not be expected to differ in their function in weightlessness. The otolith organs, in contrast, do sense gravity. The otolith organs consist of two parts, the utricle and saccule. These both include a mass of stones that rest upon a sensory epithelium (macula) with hairlike projections. The utricle senses motion in the horizontal plane (left/right, forward/back and their combinations), while the saccule is oriented vertically and senses combinations of up/down and forward/back motions [6, 8]. Both the saccule and utricle are important for sensing body tilt and the direction of the gravitational vector. They respond best to low-frequency motions. The input these sensors receive changes dramatically in weightlessness when the chronic acceleration due to gravity is removed.

Vision

Vision is a powerful sense. Many vestibular deficits only become apparent with the eyes closed. With the eyes open, the information from vision often dominates so that aberrant vestibular information is ignored. This dominance is important because even a normally functioning vestibular system can provide erroneous information. For example, when pilots fly in low-visibility conditions, vestibular illusions can be produced [9]. In certain situations, pilots can get the sensation that they are turning when, in fact, they arc flying straight and level (or that they are pitching when they are in fact accelerating forward). To prevent accidents, pilots learn to rely on their instruments and their vision when flying in clouds or fog and learn to ignore compelling but unreliable vestibular information.

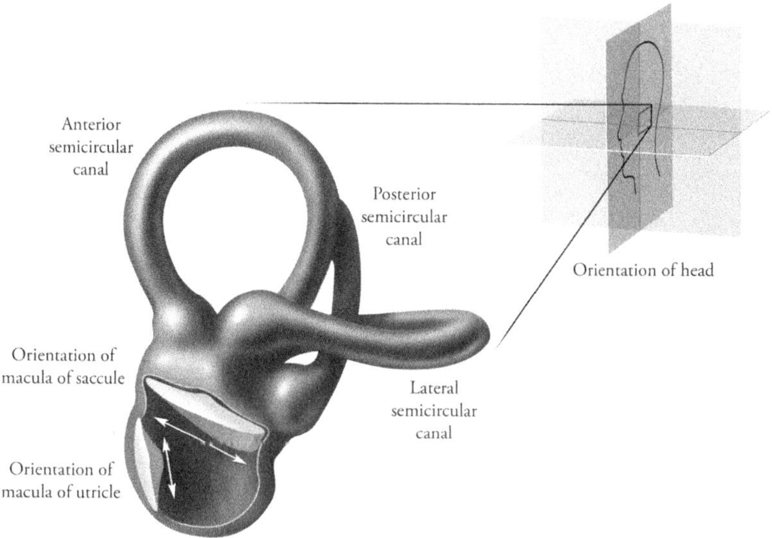

Figure 6-2. The vestibular labyrinth. The labyrinth is the important sensor for the vestibular system. The otolith organs (saccule and utricle) sense the static pull of gravity and provide information on translational movements and linear accelerations. The otolith organs are also an important part of the ocular counter rolling reflex (when the head tilts, the eyes rotate in the opposite direction to keep the visual image upright). The semicircular canals are accelerometers. Angular accelerations of the head will stimulate the canals. If the angular accelerations are prolonged, as happens when rotating in a chair, then the output from the semicircular canals will decline. The semicircular canals are important in vestibular–ocular reflexes because they keep a visual image focused on the retina in spite of body movement. (©Kestrel Illustrations, LLC.)

Vision is also an important sense for perceiving motion. When a driver is sitting at a stoplight and the car alongside moves, the driver of the stopped car often feels as if her or his car is moving and may press the brake. Similarly, if a passenger on a train sees the train on the next track move, this can cause the passenger on the stationary train to feel as if he or she is moving. In these cases, stimuli to the visual system overcome the information from the inner ear and position sense to produce the sensation of motion. This visually induced sense of motion is called vection. Visual cues become even more important in individuals with vestibular dysfunction (bilateral labyrinthine defective subjects) [10]. Because these individuals must rely on vision almost exclusively for their sense of orientation, visual illusions of motion can be very powerful for them. In weightlessness, where the information from the otoliths is suspect, it might be expected that visual cues might start to carry more weight with the astronauts.

Proprioception

Proprioception includes the sensors in the muscles, tendons, joints, and skin that provide information on movement and location. Proprioception is often called the hidden sense because it is not under conscious control and is taken for granted in everyday activities. As the case of Ian Waterman shows, the absence of position sense is disabling [7]. The control of movement and the maintenance of a stable posture require continuous attention and effort if the proprioceptive system is not functioning. Patients with tabes dorsalis or diabetic neuropathy demonstrate the problems that the lack of proprioception can cause.

Proprioception can be improved through training. Proprioception enhancement is a physical therapy technique used in some patients with balance disorders [11]. Muscle receptors can adapt to increased use to improve athletic performance over time [12]. In weightlessness, proprioception remains important for determining the position of the limbs and providing feedback on movements. The loading of the soles of the feet, however, is missing unless the astronaut uses bungees or some other method to provide contact with the spacecraft surface. It is likely that the gain and sensitivity of some position and pressure sensors may change to adapt to the situation in space.

The Major Vestibular Reflexes

Some of the functions the vestibular system performs require rapid responses. For example, to keep an image focused on the retina while the body is moving requires a rapid-responding system that can sense head movements and swiftly adjust the eye to keep the image stable. Similarly, someone who is falling needs to respond rapidly without thinking. These fast responses are accomplished through reflexes.

Vestibular–ocular reflexes

Vestibular–ocular reflexes keep a visual image stable on the retina despite body movement [8]. For example, a rightward movement of the head is associated with a leftward movement of the eyes, which keeps the retinal image in place. Both the semicircular canals and the otoliths are involved in vestibular–ocular reflexes. The semicircular canals are primarily involved in rapid side-to-side and up-and-down pitching movements. The otoliths are important for ocular counter-rolling, where tilting the head to one side causes torsion of the eyes in the opposite direction. If these reflexes are not working or if they are changed substantially, the image on the retina can slip during body motion, leading to a blurring of the image and difficulty in tracking visual targets [5]. The simplest vestibular–ocular reflex involves a three-neuron arc, where the signal originates in the vestibule, connects through an interneuron in the midbrain, and then travels to the extraocular muscles through one of the nuclei for cranial nerve III, IV, or VI.

The vestibular–ocular reflexes are very adaptable. For example, if a subject wears reversing prisms, head movements cause the image on the retina to move opposite to the expected direction. If the person continues to wear the reversing prisms, over time the gain of the reflex will decrease (the amount of eye movement for a given amount of head movement is reduced) and then even switch direction [8]. When the reversing prisms are removed, the system will adapt back. In other words, the system has the capacity to adapt to a marked change in visual input. While the adaptation is taking place, however, performance on demanding visual tracking tasks is reduced.

Vestibular-ocular reflexes are not the only means available to maintain a stable visual image. In general, the vestibular-ocular reflexes are optimal for stabilizing vision during externally imposed, sudden movements. If the body movements are slower or if they are predictable, other sensory information can be used to track the targets. The dynamic visual acuity (visual acuity while moving) of patients with a vestibulopathy can be quite good despite the damage to the vestibular–ocular reflexes [13].

Vestibular–ocular reflexes can be assessed in a variety of ways. Rotational testing, using a rotating chair while monitoring eye movements, is the most commonly used technique to assess canal-based reflexes [14]. Ocular counter-rolling can be measured using off-axis rotation (which produces a laterally directed G force) or by tilting the head and measuring eye movements [15]. From an operational standpoint, the most important measures of the vestibular-ocular reflexes are whether an individual's visual acuity is decreased with head motions and whether an individual has difficulty reading instruments while flying during reentry.

Vestibular spinal and vestibular colic reflexes

The vestibular spinal reflexes originate in the vestibule and connect to the postural muscles through nerves in the spinal cord. These reflexes are important for maintaining posture and preventing falls [8]. The reflexes also depend on the otoliths and so might be expected to show some changes while in weightlessness. The vestibular system also has reflex connections to the neck muscles (vestibular colic reflexes). In this way, the neck muscles can stabilize the head as needed to maintain tracking ability.

The function of the vestibulospinal connections can be assessed using posturography [14]. With posturography the patient stands on a platform that can remain motionless or can move with the patient's postural sway. By moving the platform in a variety of ways with eyes open and eyes closed, the contribution of vision, proprioception, and vestibular inputs to postural stability can be assessed. The different testing conditions are shown in figure 6-3. Alterations mainly in vestibulospinal reflexes after spaceflight would be manifest by poor performance in tests 5 and 6 on the dynamic posturography device.

Integration of the Senses and Adaptive Ability

Precisely where in the brain the senses needed for balance are integrated is not known. But studies in labyrinthectomy patients have shown that the vestibular nuclei, cerebellum, and the connections they make throughout the brain play a critical role [6, 8, 16]. This system can adapt to extreme changes in sensory input such as from labyrinthectomy, reversing prisms, and chronic exposure to a slowly rotating room.

Although the vestibular system has great adaptive capability, there are certain clinical situations where adaptation does not occur or occurs slowly. As mentioned in chapter 9, some individuals develop chronic motion sickness in a new motion

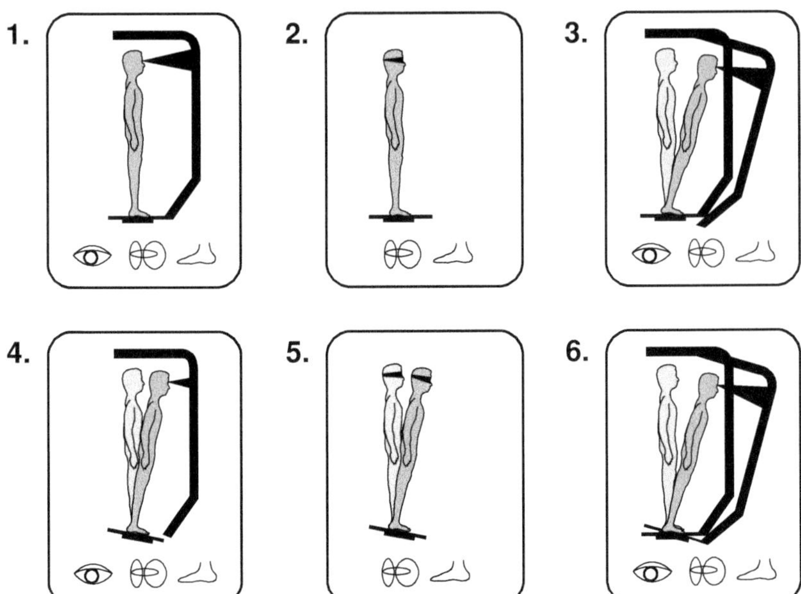

Figure 6-3. The sensory organization test from the Equitest posturography system. In this test the subject stands on a platform and faces an enclosure. The testing system can isolate different components of the postural system by removing the information from one of the senses. For example, in condition 5, the subject has his eyes closed, which removes visual input, and the platform moves along with his body sway, which removes proprioceptive input (i.e., if the platform moves to the right as the body moves to the right, the ankle proprioreceptors will not be stimulated). As a result, in condition 5, the subject must depend on vestibular signals alone to maintain a stable posture. Condition 6 also forces the subject to rely on vestibular information by referencing both proprioception and vision to body sway. (Sensory Organization Test, used with permission of NeuroCom International, Inc.)

environment [17]. Although most individuals placed in a novel motion environment will adapt and have their symptoms abate, some will not adapt or will take significant time to lose their symptoms. It is not clear whether this represents a physiological or psychological problem.

Another instance where adaptation can be prolonged is after a sea voyage. Rarely, some individuals will have a syndrome where feelings of swinging, swaying, unsteadiness, and disequilibrium persist after returning to dry land [18–20]. In some cases, the symptoms can last for months. This syndrome, called *mal de débarquement*, tends to affect mainly middle-aged women after sea voyages. The reasons for *mal de débarquement* are not clear but may represent a diminished ability to adapt to new environments [21]. No similar syndrome of prolonged adaptation has been described after spaceflight, but the number of people who have been in space is small.

Graybiel [6] used an adaptive capacity test to measure individual differences in the ability to acquire adaptation to a rotating environment. Graybiel used the slow rotation room, a continuously rotating room on a turntable. The subjects made head movements in the rotating environment. The speed of the rotating room was increased once they could make the head movements without producing vestibular symptoms. The subjects showed a variety of responses. Some could adapt rapidly, some required a large number of head movements, and some could not adapt at all on the schedule they were given and stopped the test [6].

In general, most people can adapt readily to a new motion environment. The examples of chronic motion sickness, *mal de débarquement*, and adaptation to the slow rotation room show that adaptive capabilities vary markedly. This may be a significant factor for spaceflight, as long-duration missions and missions to other planets require crew members to adapt rapidly to differing gravitational environments.

Changes in the Balance System Produced by Spaceflight

The main effect of spaceflight on the vestibular system is the unloading of the otolith organs. On Earth, the gravity vector acts upon the macula in the saccule and utricle as the body moves. In space, however, the static stimulus to the otolith organs is the same no matter what the body orientation. The otolith organs still respond to dynamic stimuli, since side-to-side, front-to-back, and up-and-down motions will stimulate the otoliths due to the inertia of the otoliths on the macula. But the continuous, steady input from gravity is gone. Proprioception is still used in space, although the magnitude of the stimuli may be reduced. For example, walking and running provide major inputs to the soles of the feet. In space, these inputs can only be produced by walking, running, or jumping while using bungees to bring the body into contact with a surface in the spacecraft.

Evidence from weightlessness exposure shows there is considerable plasticity within the balance control system. Immediate early gene activity is markedly increased in the vestibular nuclei and cerebellum of rats exposed to weightlessness, suggesting that the architecture and functions of the parts of the nervous system controlling balance and gaze may be modified early on in a spaceflight [16, 22]. Although the changes occur rapidly, some time is needed to complete the adaptation. For example, in the case of motion sickness, symptoms are usually gone within 3 days, and this may reflect the time required for the adaptation. During the adaptation, it is likely that cues from the otolith organs are given less weight, and new weighting is given to visual and proprioceptive cues. This adaptation seems to work well because most astronauts are able to function effectively in space. Problems arise, however, once gravity is sensed again during reentry. The system must adapt back, but this takes time. While the readaptation is taking place, there may be difficulties in areas that are influenced by the vestibular system, such as gaze control, posture, and locomotion. Also, while in weightlessness, the characteristics of the gravity sensors themselves (the otolith organs) may have changed.

Plasticity in the Otolith Organs and Changes in Sensitivity

Studies done on rats on several Spacelab missions show that the otolith organs may respond to weightlessness by increasing the number of synapses made within the macula. Ross et al. [23] showed an increase in synapses in the type II hair cells in the utricular macula in rats exposed to weightlessness. One interpretation of these findings is that, with the relative absence of stimulation in weightlessness, the otoliths increase their sensitivity to compensate for the lack of gravitational stimulation [24].

Other data suggest that the sensitivity of the vestibular end organs may be increased. The output from the utricle was measured using a wafer electrode implanted into the eighth cranial nerve of toadfish that flew in space [25]. Nerve fibers carrying information from the utricular otolith traveled though the wafer, allowing a direct recording of the signals. The measurements showed a hypersensitivity to side-to-side movements after spaceflight [25]. One possible reason for this could be plasticity in the utricle that increased the sensitivity of this gravity sensor.

Studies on humans also suggest changes in the sensitivity of the otolith organs. Before and after three Spacelab missions lasting from 9 to 14 days, several crew members rode on a sled that moved on rails. While random translations moved the sled off center, the crew member used a joystick to bring the sled back to its starting position. This task tested the ability of the crew members to sense translations of the sled. Surprisingly, after spaceflight, most of the crew members improved in their ability to do this task in the head-to-foot and side-to-side directions. Over time they returned to their baseline performance [26, 27]. One intriguing, but not proven, explanation for these findings is that the sensitivity of their otolith organs was increased after spaceflight, which allowed for improved detection of translational movements.

To date, most of the studies performed in space have been done on a limited number of subjects. It is not known if the findings from animal studies can be directly applied to humans. Nevertheless, the bulk of evidence suggests that the otolith organs have the ability to make new connections in response to weightlessness. This may change their output when they are again stimulated by the static force of gravity on Earth.

Altered Ability to Sense Tilt

The otolith organs are important in the determination of body tilt, particularly with the eyes closed. Because the sensitivity of the otoliths may change in space, and the weight given to otolithic inputs may be altered in space, this may be reflected in alterations in the ability to sense body tilt after spaceflight. Several studies have shown that tilt perception is changed after spaceflight.

Merfeld [2] studied two subjects the day of landing after a 14-day spaceflight [2]. In this experiment, the crew members were seated in a modified Link aviation trainer. During the experiment, the trainer was tilted by variable amounts in the roll plane (around an axis directed front-to-back through the nose). Changes in the amount and direction of the tilt were presented at random. The crew members had to control the trainer when the tilt occurred and bring the trainer back to the straight and level position. When tested after the flight with their eyes open, the crew members showed no changes compared to preflight (a visual cue was presented that corresponded to the tilt). With their eyes closed, however, the crew members exhibited a marked reduction in their ability to control the tilt.

Other studies have shown that with the eyes closed, crew members sense tilt inaccurately after spaceflight. Clement and colleagues [3] studied four subjects on landing day after the 16-day *Neurolab* mission [3]. The crew members were seated in a chair that was tilted in the roll direction by variable amounts in a random pattern. The crew members provided their best estimate of the degree of tilt. On average, they overestimated the degree of tilt by 15°.

Changes in the sensation of tilt are also seen when crew members are tested in a rotating chair after flight. Sideways (lateral) accelerations stimulate the otolith organs (particularly the utricle) and can be interpreted as a body tilt. Sideways accelerations can be produced by riding in a rotating chair, where the chair is displaced from the axis of rotation (like a chair on a carousel). The speed of rotation determines how much of a sideways acceleration is produced. Normally, sitting blindfolded in a rotating chair in Earth's gravity where the rotation produces 1 *G* of sideways acceleration, a person feels as if he or she is tilted sideways by about 45° (figure 6-4). This is because the downward 1 *G* acceleration sums with the sideways 1 *G* acceleration to produce a resultant tilt of 45°. Before and after the 16-day *Neurolab* mission, crew members rode in an off-axis rotating chair. When rotating at a sideways acceleration of 1 *G* before the flight, they experienced the expected degree of sideways tilt. When tested the day after landing, they experienced significantly more tilt (figure 6-4). The tilt estimations were back to normal after 7 days [3].

Taken together, the data from spaceflight show that vestibular sensations are unreliable on landing day. If visual cues are removed, crew members make significant errors in their estimation of tilt, and lateral accelerations will produce much larger sensations of tilt than was the case before flight. These data are significant because Shuttle commanders and pilots are called upon to fly the Shuttle during the final portion of the landing. If they experience sudden roll tilts due to wind gusts or lateral accelerations due to tight turns, they need to be aware that the vestibular sensations of tilt they experience will be inaccurate.

Changes in Posture

Dynamic posturography (figure 6-3) allows visual, proprioceptive, and vestibular cues to be isolated in different test sequences so that the main contribution to postural instability can be identified. Postural stability has been studied on a series of 34 crew members on landing day after Shuttle flights using posturography [4]. This testing clearly shows that when the crew members must rely on vestibular information to maintain posture, they show the greatest deficits in postural stability and may even fall (table 6-1). If the crew members can use visual and proprioceptive inputs, they do much better (although they still have greater instability than before flight). Overall, however, the data point to changes in the vestibular system as being the most important contributor to postural stability after a spaceflight. This testing does not isolate where in the vestibular system the problem resides (i.e., the otolith organs, central integration, etc.).

Changes in Gaze Control

When the space-adapted vestibular system again receives otolith inputs upon reentering a gravitational field, this can produce problems with gaze control. Occasionally, some crew members report oscillopsia (instability of the visual scene during move-

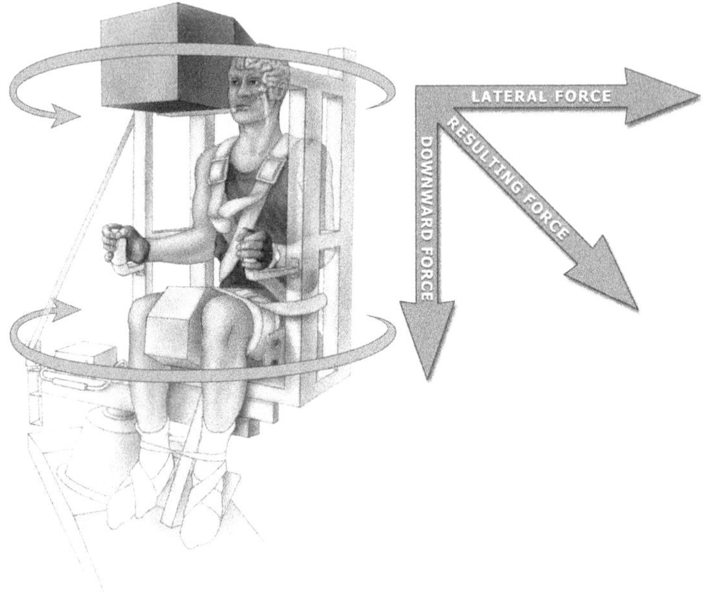

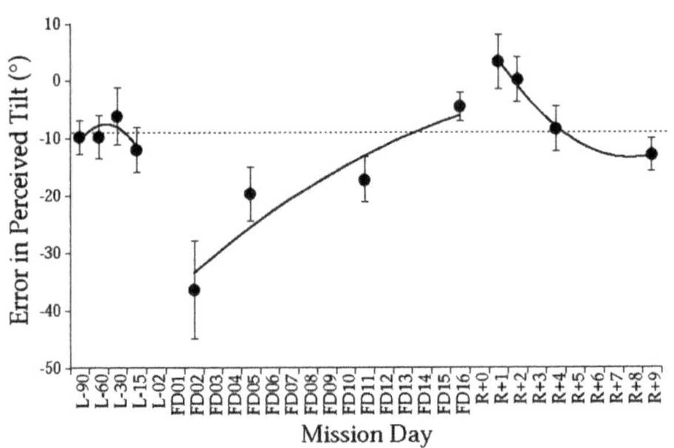

Figure 6-4. Off-axis vertical rotation. (Top panel) An astronaut riding in the off-axis vertical rotator used on the Neurolab Spacelab mission [35]. The lateral centripetal force combines with the force of gravity to give the subject the sense of body tilt. (©Kestrel Illustrations, LLC.) (Bottom panel) The tilt sensations in the chair for the four Neurolab subjects. During the first couple of days after landing, the perception of tilt was different from what it had been before the flight. (Courtesy of NASA.)

Table 6-1. Results of dynamic posturography testing, using the sensory organization test, on 34 astronauts after Shuttle flights.

Test	Visual cues	Somatosensory cues	Preflight sway		Postflight sway	
			Degrees	SEM	Degrees	SEM
1	Normal	Normal	0.76	±0.1	1.03	±0.1
2	Absent	Normal	1.37	±0.1	1.72	±0.1
3	Sway-ref	Normal	1.00	±0.1	1.60	±0.1
4	Normal	Sway-ref	1.36	±0.1	2.30	±0.2
5	Absent	Sway-ref	3.12	±0.2	5.09	±0.3
6	Sway-ref	Sway-ref	3.00	±0.2	6.12	±0.4

There was a significant increase in postural sway in all test conditions, but the increased sway threatened stability in only conditions 5 and 6. In those conditions, useful visual and somatosensory cues are removed, requiring the astronaut to rely on vestibular cues alone. Adapted from Paloski et al. [4].

ment) after a spaceflight, which is an indication that the vestibular ocular reflexes are not adequately compensating for head and body movements [28]. Sensitive tests of target-tracking ability also show that gaze control is impaired after spaceflight [29, 30]. It can take more time to acquire a target after spaceflight, and the eye movements may not be as smooth or coordinated as they were before flight. These findings are significant because the ability to read instrument panels clearly when the spacecraft is moving is an important operational skill. At present, however, it is difficult to assess the operational impact of the gaze-control deficiencies that have been observed in research studies.

Changes in Sensory Integration

Because the otoliths and pressure receptors will not provide as much useful information while floating in space as they do on Earth, the centers that integrate sensory information may start to place more weight on visual cues. Experiments done in space suggest that this is the case.

On the *SLS 1* and *SLS 2* Spacelab missions, Young and colleagues [31] used a rotating drum to produce visual sensations of motion. The drum was cylindrical, with a random pattern of dots painted within it. One end of the drum was open, and the other end was closed and attached to a rotating shaft. A subject looking into the open end of the rotating drum developed a sense of self-rotation. On Earth, the sense of rotation was usually mild and produced a slight head tilt opposite to the direction of rotation. The experiment team hypothesized that crew members in orbit would experience much stronger senses of self-motion (vection). Their hypothesis was confirmed. In space, some of the crew members even had the feeling that they were rotating completely around while watching the rotating drum [32].

These findings were confirmed on the *Neurolab* mission. Oman et al. [33] showed that crew members were more susceptible to visually induced sensations of self-motion while in weightlessness. These sensations could be significantly reduced if the crew members were given proprioceptive inputs by putting them into contact with the spacecraft floor using constant force springs. After landing the sensations of vection were the same as before flight.

Effect of Spaceflight on Walking, Running, and Egress

While in orbit, astronauts are not aware of the adaptive changes taking place within the balance system. These become manifest after landing, when the crew members once again are walking upright in a gravitational field. As has been described above, after flight, crew members will have more postural sway and will need to rely on vision for accurate information on body tilt. After spaceflights, crew members walk with a wide-based gait and make turns slowly. Turning or tilting the head rapidly can produce vestibular illusions of pitching or tilting excessively [1]. The changes in the balance system, combined with weakness in postural muscles and some degree of orthostatic intolerance, mean that walking and running may be compromised. By moving slowly, minimizing head movements, and relying on visual cues, most crew members can walk, climb stairs, and navigate successfully after flight. Most would have difficulties with running around obstacles in the launch and entry suit (the heavy protective suit worn during reentry) and would find ambulating in the dark challenging.

At present it is hard to estimate how well any given crew would perform in an emergency after landing. Most studies of the neurovestibular effects of spaceflight have been done in controlled laboratory conditions that do not replicate the conditions that a crew member might face in an emergency. After landing, crew members are wearing heavy, hot launch and entry suits. These suits are cumbersome to move around in and present a major heat stress to the crew member if they cannot be connected to an external cooling source. The combination of these operational factors with the physiological changes would determine how successful or unsuccessful an emergency egress would be.

Approaches to Balance System Changes after Spaceflight

To date, there have not been any reports of crew members experiencing long-term problems with their vestibular systems as a consequence of spaceflight (this includes the 438-day trip by Russian physician Valery Polyakov, the longest spaceflight to date) [34]. All crew members have successfully readapted to Earth's gravity after returning from orbit. The duration of this recovery, however, varies among crew members. Recovery time seems to increase with mission duration, but the variability between crew members, missions, and the use of countermeasures make this hard to prove conclusively. Nevertheless, all the evidence suggests that crew members will eventually adapt to different gravitational environments. Therefore, for operations the critical period is the time while the adaptation is taking place, since during this time the crew members' performance may not be optimal.

Time

The simplest approach to the changes in the balance system is to organize the flight so that the astronauts have time to adapt to the new environment before they have to function at a high level. Most studies have shown rapid improvements in balance and posture within hours of landing [4]. If crew members landing on the Moon or Mars are allowed time to begin the adaptation process, they can then venture out of the spacecraft once they feel stable. This approach, however, does not accommodate emergencies, which would demand that crew members function immediately after landing.

Centrifugation/Artificial Gravity

It might seem reasonable to try to prevent or minimize changes in the balance system by using centrifugation or artificial gravity. Using artificial gravity, however, raises several questions. Is artificial gravity necessary for the entire exposure to weightlessness, or can it be introduced intermittently or shortly before landing? If it can be used intermittently, how much is needed? What is the most effective prescription (G level, duration) for using artificial gravity? Unfortunately, the answers to these questions are not known precisely, but there has been research into different options. The main trade-off with the use of artificial gravity is the mass and complexity of the spacecraft. If the entire spacecraft can rotate to provide artificial gravity, this could prevent many of the adverse affects of spaceflight. Unfortunately, this also greatly increases the size, mass, and complexity of the spacecraft. If a smaller, short-arm centrifuge can be used, this decreases mass and complexity but offers only an intermittent countermeasure. This intermittent countermeasure could be used frequently before landing to reintroduce gravitational stimuli to the labyrinth and position sensors.

Intermittent centrifugation

On the *Neurolab* mission, off-axis rotation was used to stimulate the vestibular system (figure 6-4) [35]. The rotator could provide both side-to-side and head-to-foot accelerations and was used at least 5 times during the 16-day flight by the crew members in the study. Before, during, and after the flight, investigators measured ocular counter-rolling. In contrast to previous studies that measured ocular counter-rolling after spaceflight, no change was noted in the ocular counter-rolling responses in this study [15]. One hypothesis for these findings was that the centrifugation the crew members experienced in flight prevented the changes in ocular counter-rolling that had been seen in previous spaceflights. The crew members did show all the usual signs of the adaptation to spaceflight (postural instability after landing, inaccuracies in tilt perception), suggesting that in-flight centrifugation did not have a dramatic effect on these other symptoms. Nonetheless, centrifugation might be helpful and should be further studied.

The centrifuge described above was optimized for stimulating the inner ear; it would be a useful device for studying intermittent centrifugation to mitigate postflight vestibular problems. This design, however, would not be useful as a means to provide skeletal loading to prevent bone loss. Other short-arm centrifuges have been proposed [36, 37]. These involve the astronaut generating the rotating force by cycling. With these designs, however, the astronaut's head is very close to the axis of rotation, so that the inner ear would experience very little centripetal force. The ideal short-axis rotator would need to be flexible enough so that intermittent artificial gravity could be used for both musculoskeletal and vestibular applications.

One other potential concern with intermittent centrifugation is that the crew members would need to be dually adapted [38]. This means that they would need to be able to transfer between the rotating and nonrotating environments without experiencing readaptation symptoms. If motion sickness symptoms occurred every time the crews changed from weightlessness to a rotating centrifuge and back, intermittent centrifugation would be unacceptable. If the crew members developed dual adaptation (as occurred in some subjects in rotating room experiments), this would permit crews to move easily and repeatedly between gravito-inertial environments.

Considerable basic research is needed to establish if intermittent centrifugation is effective and whether crew members can become dually adapted. If it were effective, the appropriate G levels and durations of exposure would need to be established. Overall, intermittent centrifugation would be a very attractive countermeasure because potentially it would be useful for several different physiological changes (cardiovascular, musculoskeletal, vestibular).

Continuous artificial gravity

If an entire spacecraft could be rotated (similar to the torus-shaped rotating space station design featured in the movie *2001: A Space Odyssey*), then the crew members could live continuously in a gravitational environment. For example, a station with the diameter of 1 km rotating at 1 rpm would produce 1 G at the rim and would be well tolerated by the inhabitants [38]. This could be done either by spinning a torus-shaped station or perhaps by attaching a cylindrical spacecraft to a tether that has a counterweight on the other end. The spacecraft and counterweight could rotate while connected to the tether (like two cans connected to different ends of a string) to produce the 1-G force. Both of these approaches, however, carry significant engineering costs, although the tether approach could potentially be economical [39].

If the radius of the spinning station could be reduced, however, this would reduce the cost and complexity of the station. Because the gravity level produced by the rotation would depend on the radius of the spacecraft and the rotation rate, the smaller the radius used, the higher the rotation rate would need to be to produce the desired gravity level [38]. Working in a rotating craft, however, would present challenges to the vestibular system. The semicircular canals do not sense continuous rotation. Therefore, crew members standing still in the rotating craft parallel to the rotation axis would not feel as if they were rotating. If, however, the crew members made head movements out of the plane of rotation, this would stimulate the canals and create a strong illusions of movement in unexpected directions (i.e., cross-coupled accelerations). Over time, individuals can habituate to these illusions, but the crew would probably experience motion sickness until adaptation occurred.

Early studies performed in the slow rotation room showed that 3–4 rpm is typically the highest continuous rotation to which people can readily adapt [6]. More recent studies, however, have shown that if the terminal velocity is approached using a series of gradual steps and if many body movements are made at each velocity, then full adaptation of head, arm, and leg movements to higher rotation rates is possible. Rotation rates as high as 7.5–10 rpm may be reached [40]. This would allow for a fairly short-radius spacecraft. One 10-rpm design from the Massachusetts Institute of Technology has a radius of 4 m and produces 0.5 G at the rim [38].

Proprioceptive Aids

If crew members cannot preadapt while in space to the gravity they will experience back on Earth (or on the Moon or Mars), then they may need assistance. If they have to be mobile shortly after landing and are experiencing unsteadiness, astronauts could employ the usual proprioceptive aids used in rehabilitation. Handholds, walking sticks, and other simple measures could increase stability by compensating for vestibular defects with increased proprioceptive input [41].

Enhanced proprioceptive inputs can also be used to help with spatial disorientation during flying. Rupert et al. [42] developed a vest that provides cues on spatial orientation by using a matrix of mechanical tactile stimulators (tactors) applied on the torso and limbs to convey orientation cues (e.g., the gravity vector) to the skin. When using the vest, pilots were able to fly complex maneuvers with no instruments or outside visual references (i.e., when blindfolded) with less than 20 minutes of training. They also could recover easily from unusual aircraft attitudes solely by using the tactile cues from the vest. This technology might be useful if piloting is required when reentering a gravitational field because it could help overcome erroneous vestibular information. Wall and colleagues [43] also developed a tactor-based system that can be used with patients with vestibular disorders. In this application, the tactors provide information about body tilt. Several other vestibular prostheses are also in development [44].

Recommendations Based on Current Knowledge

Immediately after landing, astronauts may not be aware of how much bone or muscle they lost while in space, but they can easily feel the changes in their balance systems. The loss in postural stability after spaceflight results mainly from an inability to use vestibular information appropriately. Recently returned crew members do not estimate tilt accurately and do not perform as well on visual tracking tasks as they did before flight. Nevertheless, most crew members can walk and navigate successfully as long as they can rely on visual cues. The changes in balance improve rapidly on landing day and resolve more gradually during the first week or so back on Earth. Here are some recommendations on how to deal with the changes:

1. Avoid "seat of the pants" flying after spaceflight. Perhaps the most controversial issue raised by the changes in the vestibular system after spaceflight is whether astronauts should fly manually during reentry. Currently, the Shuttle commander flies the orbiter manually during the final approach and landing. The remainder of the reentry is automated. Two factors have to be considered to determine whether manual control during reentry is an advantage: one is the nature of the flying task and the second is what benefits manual flying may offer over an automated approach. If astronauts were planning to fly air combat missions in a fighter jet immediately after landing, this would clearly be unwise. This kind of flying requires rapid head movements, high G forces, and

quick target acquisition. After spaceflight, the changes in the vestibular system combined with the loss of some orthostatic tolerance would make this kind of high-performance flying difficult. In contrast, a Shuttle landing mainly requires visual monitoring, does not demand excessive head movements, and ordinarily does not involve "seat of the pants" flying. The astronauts are well versed in the landing task because they have practiced it many times. They also are familiar with vestibular illusions because they have encountered them in their flying careers. These factors probably explain why Shuttle commanders have brought the Space Shuttle back successfully for over 100 safe landings.

The larger question, however, is whether the manual control of the Shuttle offers benefits over an automated approach. When the Shuttle was first built, autoland capabilities were in their infancy, but such systems now land commercial aircraft every day. It seems likely that over time, a properly tested automated system would outperform human operators in a task like landing the Shuttle. As spaceflights become longer, it is reasonable to transition from pilot control to pilot monitoring of a landing.

2. Examine artificial gravity as a countermeasure. Countermeasures for spaceflight-induced changes in the balance system may not be absolutely necessary for travel to the Moon or Mars, but they would reduce risks during emergencies after landing. The difficulty of providing a countermeasure needs to be balanced against the need for crew members to be in optimal condition when they arrive at another planet. Currently, crew members returning to Earth from low-earth orbit are not in optimal condition; on Earth, this would only be a problem in some emergency landing situations, and this is a risk that has been accepted. On Mars, if crew members could stay within the spacecraft while they gain some adaptation to Martian gravity, they would then be able to function effectively on the surface. Artificial gravity is a countermeasure that could potentially help preadapt crews before entering a new gravitational environment. As artificial gravity could be useful for other systems (musculoskeletal, cardiovascular) in addition to the vestibular system, it should be studied aggressively as a potential countermeasure to preadapt crews to new gravitational environments.

3. Provide proprioceptive aids as needed. No one likes to fall, but crew members should especially avoid falling while wearing a spacesuit. The fall could damage the suit, and the crew member may find it difficult to get back up unaided (depending on the suit design). Crews should be provided with simple proprioceptive aids (such as walking sticks, etc.) to use to if they need to be ambulatory on another planet immediately after landing. Systems that use tactors also may be useful if the crew members are experiencing unsteadiness over a long period of time, or if they need to walk and navigate before they have fully adapted to the new environment.

4. Select individuals with an adaptable vestibular system. Before sending an astronaut on a long-duration flight, it would be reasonable to ensure that the crew member can adapt to weightlessness and readapt to Earth's gravity on a short-duration flight first.

References

1. Reschke, M.F., et al., Posture, locomotion, spatial orientation, and motion sickness as a function of space flight. Brain Research Reviews, 1998. 28(1–2): 102–17.

2. Merfeld, D.M., Effect of spaceflight on ability to sense and control roll tilt: human neurovestibular studies on *SLS-2*. Journal of Applied Physiology, 1996. 81(1): 50–57.
3. Clement, G., et al., Perception of the spatial vertical during centrifugation and static tilt, in *The Neurolab Spacelab Mission: Neuroscience Research in Space*, J.C. Buckey and J.L. Homick, eds. 2003, NASA, Houston, TX, pp. 5–10.
4. Paloski, W.H., et al., Recovery of postural equilibrium control following space flight, in *Extended Duration Orbiter Medical Project*, C.F. Sawin, G.R. Taylor, and W.L. Smith, eds. 1999, NASA-Johnson Space Center, Houston, TX, pp. 5.4,1–5.4,16.
5. Furman, J.M., and S.P. Cass, Vestibular anatomy and physiology, in *Vestibular Disorders: A Case-Study Approach*. 2003, Oxford University Press, New York, pp. 3–15.
6. Graybiel, A., The vestibular system, in *Bioastronautics Data Book*, J.F. Parker and V.R. West, eds. 1973, NASA, Washington, DC, pp. 533–609.
7. Cole, J., *Pride and a Daily Marathon*. 1995, MIT Press, Boston, MA.
8. Goldberg, J.M., and C. Fernandez, The vestibular system, in *Handbook of Physiology*, section 1, *The Nervous System*, J.M. Brookhart and V.B. Mountcastle, eds. 1984, American Physiological Society, Bethesda, MD, pp. 977–1022.
9. Young, I.R., Perception of the body in space: mechanisms, in *Handbook of Physiology*, section 1, *The Nervous System*, J.M. Brookhart and V.B. Mountcastle, eds. 1984, American Physiological Society, Bethesda, MD, pp. 1023–86.
10. Johnson, W.H., F.A. Sunahara, and J.P. Landolt, Importance of the vestibular system in visually induced nausea and self-vection. Journal of Vestibular Research, 1999. 9(2): 83–87.
11. Chong, R.K., et al., Source of improvement in balance control after a training program for ankle proprioception. Perceptual and Motor Skills, 2001. 92(1): 265–72.
12. Hutton, R.S., and S.W. Atwater, Acute and chronic adaptations of muscle proprioceptors in response to increased use. Sports Medicine, 1992. 14(6): 406–21.
13. Demer, J.L., et al., New tests of vestibular function. Annals of the New York Academy of Sciences., 2001. 942: 428–45.
14. Furman, J.M. and S.P. Cass, Vestibular laboratory testing, in *Vestibular Disorders: A Case-Study Approach*. 2003, Oxford University Press, New York, pp. 30–40.
15. Moore, S., et al., Ocular counter-rolling during centrifugation and static tilt, in *The Neurolab Spacelab Mission: Neuroscience Research in Space*, J.C. Buckey and J.L. Homick, eds. 2003, NASA, Houston, TX. pp. 11–17.
16. Holstein, G.R., and G.P. Martinelli, The effect of spaceflight on the ultrastructure of the cerebellum, in *The Neurolab Spacelab Mission: Neuroscience Research in Space*, J.C. Buckey and J.L. Homick, eds. 2003, NASA, Houston, TX, pp. 19–26.
17. Reason, J.T., and J.J. Brand, *Motion Sickness*. 1975, Academic Press, London.
18. Brown, J.J., and R.W. Baloh, Persistent mal de debarquement syndrome: a motion-induced subjective disorder of balance. American Journal of Otolaryngology, 1987. 8(4): 219–22.
19. Gordon, C.R., et al., Clinical features of mal de debarquement: adaptation and habituation to sea conditions. Journal of Vestibular Research, 1995. 5(5): 363–69.
20. Hain, T.C., P.A. Hanna, and M.A. Rheinberger, Mal de debarquement. Archives of Otolaryngology – Head & Neck Surgery, 1999. 125(6): 615–20.
21. Furman, J.M., and S.P. Cass, Mal de Debarquement Syndrome, in *Vestibular Disorders: A Case-Study Approach*. 2003, Oxford University Press, New York, pp. 219–21.
22. Pompeiano, O., Gene expression in the rat brain during spaceflight, in *The Neurolab Spacelab Mission: Neuroscience Research in Space*, J.C. Buckey and J.L. Homick, eds. 2003, NASA, Houston, TX, pp. 27–38.
23. Ross, M.D., A spaceflight study of synaptic plasticity in adult rat vestibular maculas. Acta Oto-Laryngologica Supplement, 1994. 516: 1–14.
24. Ross, M.D., and J. Varelas, Ribbon synaptic plasticity in gravity sensors of rats flown on Neurolab, in *The Neurolab Spacelab Mission: Neuroscience Research in Space*, J.C. Buckey and J.L. Homick, eds. 2003, NASA, Houston, TX, pp. 39–44.
25. Boyle, R., et al., Neural readaptation to earth's gravity following return from space, in *The Neurolab Spacelab Mission: Neuroscience Research in Space*, J.C. Buckey and J.L. Homick, eds. 2003, NASA, Houston, TX, pp. 45–50.

26. Arrott, A.P., L.R. Young, and D.M. Merfeld, Perception of linear acceleration in weightlessness. Physiologist, 1991. 34(1 Suppl): S40–43.
27. Merfeld, D.M., K.A. Polutchko, and K. Schultz, Perceptual responses to linear acceleration after spaceflight: human neurovestibular studies on SLS-2. Journal of Applied Physiology, 1996. 81(1): 58–68.
28. Layne, C.S., P.V. McDonald, and J.J. Bloomberg, Neuromuscular activation patterns during treadmill walking after space flight. Experimental Brain Research, 1997. 113(1): 104–16.
29. Reschke, M.R., et al., Visual-vestibular integration as a function of adaptation to space flight and return to Earth, in *Extended Duration Orbiter Medical Project*, C.F. Sawin, G.R. Taylor, and W.L. Smith, eds. 1999, NASA-Johnson Space Center, Houston, TX, pp. 5.3,1–5.3,41.
30. Kornilova, L.N., A tracking function of human eye in microgravity and during readaptation to earth's gravity. Aviakosmicheskaia i Ekologicheskaia Meditsina, 2001. 35(6): 30–38.
31. Young, L.R., M. Shelhamer, and S. Modestino, M.I.T./Canadian vestibular experiments on the Spacelab-1 mission: 2. Visual vestibular tilt interaction in weightlessness. Experimental Brain Research, 1986. 64(2): 299–307.
32. Young, L.R., et al., Spatial orientation and posture during and following weightlessness: human experiments on Spacelab Life Sciences 1. Journal of Vestibular Research, 1993. 3(3): 231–39.
33. Oman, C.M., et al., The role of visual cues in microgravity spatial orientation, in *The Neurolab Spacelab Mission: Neuroscience Research in Space*, J.C. Buckey and J.L. Homick, eds. 2003, NASA, Houston, TX, pp. 69–82.
34. Grigoriev, A.I., et al., Main medical results of extended flights on space station Mir in 1986–1990. Acta Astronautica, 1993. 29(8): 581–85.
35. Cohen, B., et al., Adaptation to linear acceleration in space (ATLAS) experiments: equipment and procedures, in *The Neurolab Spacelab Mission: Neuroscience Research in Space*, J.C. Buckey and J.L. Homick, eds. 2003, NASA, Houston, TX, pp. 279–84.
36. Kreitenberg, A., et al., The "Space Cycle" self powered human centrifuge: a proposed countermeasure for prolonged human spaceflight. Aviation, Space, and Environmental Medicine, 1998. 69(1): 66–72.
37. Greenleaf, J.E., et al., Cycle-powered short radius (1.9M) centrifuge: exercise vs. passive acceleration. Journal of Gravitational Physiology, 1996. 3(2): 61–62.
38. Young, L.R., Artificial gravity considerations for a Mars exploration mission. Annals of the New York Academy of Sciences, 1999. 871: 367–78.
39. Zubrin, R., The Mars direct plan. Scientific American, 2000. 282(3): 52–55.
40. Lackner, J.R., and P. DiZio, Artificial gravity as a countermeasure in long-duration space flight. Journal of Neuroscience Research, 2000. 62(2): 169–76.
41. Furman, J.M., and S.P. Cass, Vestibular rehabilitation, in *Vestibular Disorders: A Case-Study Approach*. 2003, Oxford University Press, New York. pp. 47–53.
42. Rupert, A.H., Tactile situation awareness system: proprioceptive prostheses for sensory deficiencies. Aviation, Space, and Environmental Medicine, 2000. 71(9 Suppl): A92–99.
43. Wall, C.R., et al., Balance prosthesis based on micromechanical sensors using vibrotactile feedback of tilt. IEEE Transactions on Biomedical Engineering, 2001. 48(10): 1153–61.
44. Wall, C.R., et al., Vestibular prostheses: the engineering and biomedical issues. Journal of Vestibular Research: Equilibrium and Orientation, 2002. 12(2–3): 95–113.

7

Cardiovascular Changes: Atrophy, Arrhythmias, and Orthostatic Intolerance

Introduction

On May 16, 1963, Gordon Cooper splashed down after the longest (34 hours) and last Mercury flight. Thirty-five minutes later, his *Faith 7* capsule was hoisted onto the deck of the USS *Kearsarge*. His blood pressure and pulse were monitored as he left the capsule and stood up. While standing on the deck, his heart rate climbed to 188 beats/minute. The postflight report noted: "After standing on the deck for approximately 1 minute, the pilot began to look pale and, although his face was already wet, new beads of perspiration appeared on his forehead. He swayed slightly and reported symptoms of impending loss of consciousness including lightheadedness, dimming of vision, and tingling of his feet and legs" [1, p. 317].

Cooper was experiencing postflight orthostatic intolerance. Although this episode was not the first in the U.S. space program (it had been seen on the previous flight), it was the most significant case at the time. Subsequently, orthostatic intolerance would emerge as a regular feature of weightlessness exposure. Other cardiovascular effects, such as rhythm disturbances and cardiac atrophy, also would be documented in space. In addition, exposure to weightlessness would be shown to reduce aerobic capacity.

To prevent cardiovascular events in flight and enable crew members to function effectively after landing, the cardiovascular effects of spaceflight need to be understood and managed. This chapter reviews cardiovascular changes that can occur in space and describes measures that can be taken to ensure they do not interfere with the mission.

Cardiovascular Physiology Relevant to Spaceflight

The cardiovascular system adapts and functions well in space. Adaptation to weightlessness, however, produces changes in blood volume, aerobic capacity, and cardiac mass that can present problems after a flight. Also, just as on Earth, certain stressors and events can increase the incidence of cardiac arrhythmias.

Blood Volume Regulation

Blood volume affects orthostatic tolerance. In general, high blood volume increases and low blood volume decreases orthostatic tolerance—all other physiological factors being equal. Blood volume (composed of plasma volume and erythrocyte volume) can be chronically increased or chronically decreased depending on environmental or physical demands on the body. Several factors, such as exercise, heat acclimatization, hypoxia, salt intake, and administration of exogenous factors (erythropoietin), can change blood volume.

Effect of exercise

Aerobic training increases blood volume [2]. The increase in blood volume includes both an increase in plasma volume and erythrocyte volume, although the hematocrit either remains at pretraining levels or falls slightly. Untrained subjects who begin a training program increase their orthostatic tolerance [3], and this may be due to the increase in blood volume. With continued intense aerobic training, however, adaptive changes in cardiac compliance, vascular conductance, and baroreflex control occur so that orthostatic tolerance in highly aerobically fit individuals can be compromised despite an elevated blood volume [4, 5]. Three hours or more of aerobic exercise a day may lead to reduced, rather than improved, orthostatic tolerance [3].

Heat acclimatization

Most, but not all, studies on acclimatization to hot environments suggest that this process increases blood volume (mainly by increasing plasma volume) [6, 7]. The mechanism for this change appears to be increased sodium and water retention along with an increase in total plasma protein content [7]. Exposure to heat while exercising appears to be a more potent stimulus for acclimatization than heat exposure at rest.

Whether the changes in blood volume produced by heat acclimatization improve resting orthostatic tolerance is not clear, but there is a benefit after exercise. In a small study, Keren et al. [8] did not detect a difference in baseline orthostatic tolerance between heat-acclimatized and nonacclimatized subjects. Heat-acclimatized subjects, however, can exercise longer in the heat before they become dehydrated, and dehydration definitely leads to orthostatic intolerance. Whether the heat stress is applied during exercise or at rest is important. Subjects who rested in a hot environment daily for 8 days did not improve their orthostatic tolerance after an exercise stress, compared to those who performed exercise in heat over the 8 days [9].

The ability to tolerate exercise in the heat for longer periods with less physiological stress is significant for space operations. Space suits often have poor ventilation and cooling, so that working in a suit can quickly raise body temperature.

Erythropoietin

Erythropoietin is a potent regulator of red blood cell mass [10]. It is secreted mainly by the kidney, and a renal oxygen sensor that is not fully understood controls its release. A reduction in hematocrit, such as might occur with anemia, will increase erythropoietin production. The administration of exogenous erythropoietin stimulates red cell production and increases hemoglobin concentration and hematocrit [11]. Consequently, total blood volume is also increased. One effect of the extra hemoglobin and blood volume is improved aerobic performance during exercise [10,

12]. In addition, erythropoietin can be used successfully to treat patients who have autonomic failure and orthostatic hypotension [13, 14].

Recombinant human erythropoietin is available for clinical use (Epoetin Alfa) and is administered either intravenously or subcutaneously. Second-generation erythropoiesis proteins are also available (Darbepoetin Alfa) that can be administered weekly. Erythropoietin can produce adverse effects during long-term use. Hypertension, hyperviscosity of the blood, and thrombosis have been encountered in clinical use [10]. Some cases of sudden death have been attributed to illicit use of exogenous erythropoietin in athletes. In these cases, thromboembolism was believed to have resulted from a high pre-exercise hemoglobin level rising even higher due to exercise-induced dehydration.

Hypoxia

A reduction in oxygen delivery due to altitude exposure will increase endogenous erythropoietin production and hematocrit [15]. This process is part of the acclimatization to altitude. Once an individual returns to sea level, the acclimatization is gradually lost as erythropoietin production declines, red cell mass falls, and hematocrit is restored to normal values. Athletes have used altitude or hypoxia exposure to improve athletic performance. One approach is to "live high and train low" to increase red blood cell mass and improve performance [16]. In this training scenario, the athlete sleeps at altitude. The altitude exposure begins a process of altitude acclimatization that includes erythropoietin secretion.

Effect of central fluid shifts

Stretch receptors in the central circulation sense changes in central blood volume. These cardiopulmonary receptors keep central blood volume at a set point. Evidence suggests that the set point is the upright (sitting or standing) central blood volume [17, 18]. As humans spend most of their waking lives upright, this set point seems reasonable.

As a consequence of this set point, increases in central blood volume produced by water immersion, supine bed rest, or head-down tilt bed rest stimulate adaptive mechanisms to reduce blood volume. Stretch receptors in the atria and pulmonary arteries sense the increased central blood volume and decrease renal sympathetic nerve activity, plasma renin activity and aldosterone secretion [19]. Also, atrial stretch releases atrial natriuretic peptide, which causes salt and water loss. The net effect of these changes is a reduction in plasma volume, which moves central blood volume back toward the standing/sitting level.

The reduction in plasma volume in response to these headward fluid shifts leads to an increased hematocrit. The bulk of the evidence suggests that this increased hematocrit inhibits erythropoietin secretion, which subsequently reduces red cell mass [19, 20]. Overall, the net effect of prolonged bed rest, head-down tilt exposure, or water immersion is a reduced red blood cell mass, plasma volume, and total blood volume. This reduced blood volume is an important contributor to the orthostatic intolerance that occurs after these interventions.

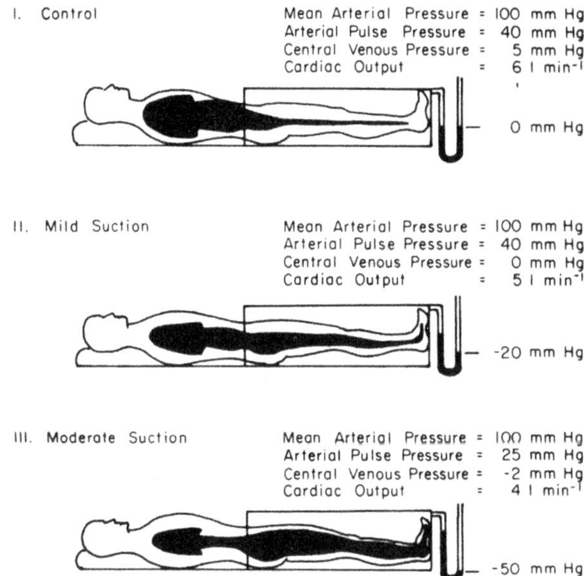

Figure 7-1. Lower body suction or lower body negative pressure (LBNP) provides a way to move fluid from the upper to lower body and thereby stimulate both the low pressure (cardiopulmonary) and high pressure (aortic and carotid) baroreceptors. Low levels of LBNP (−20 mmHg or less) stimulate only the low-pressure baroreceptors, while higher levels stimulate both. A level of −40 to −50 mmHg LBNP is approximately the same hemodynamic stress as standing [27]. LBNP can also be used to provide an orthostatic stress to stimulate the release of fluid regulatory hormone (e.g., renin, aldosterone). Figure reprinted from Rowell [27], with permission of Oxford University Press.

Some of the volume loss can be prevented by interventions that reduce central blood volume periodically. A short period of standing each day (4 hours) during bed rest can prevent bed-rest–induced orthostatic intolerance [21]. Similarly, the use of lower body negative pressure (or lower body suction) can move fluid from the chest to the lower body and stimulate fluid retention. With lower body negative pressure (LBNP), the hemodynamic effects of standing can be produced while a subject lies supine. Figure 7-1 illustrates the use of LBNP. LBNP is ideal for use in space because it provides a nongravity-dependent way to shift fluid from the upper to lower body.

Fluid ingestion

Salt and water intake can affect plasma volume. Heer and colleagues [22] studied six normal subjects over three consecutive 8-day periods. During each period they ingested a different amount of sodium chloride per day (5 g, 10 g, 15 g). Heer et al. found a strong direct relationship between the amount of sodium ingested and plasma volume. Damgaard and colleagues [23] showed that cardiovascular variables are also affected. They studied 14 subjects on both a low (1.6 g/24 hours) and high sodium intake (5.6 g/24 hours). Plasma volume, stroke volume, and cardiac output were all increased on the high sodium intake. In clinical practice, salt loading is used to treat some patients with orthostatic intolerance [24]. Taken together, most studies show that ingestion of salt containing liquids can increase plasma volume. The magnitude and duration of the increase depends on several factors (e.g., heat stress, exercise, posture).

When fluid ingestion is used to rehydrate dehydrated subjects, the composition of the fluid is important. Salt tablets with water have been used to rehydrate athletes after significant exertion. The right proportion of salt tablets and water can provide an isotonic solution when mixed properly, but a cluster of salt tablets in the gut can create very high local salt concentrations. This can produce vomiting. In general, premixed rehydration fluids that contain some carbohydrate (6% carbohydrate/electrolyte solution) are the best absorbed and tolerated [6].

Blood Pressure Control

The autonomic nervous system regulates blood pressure. High-pressure baroreceptors in the carotid arteries, aorta, and ventricles, along with low-pressure baroreceptors in the pulmonary arteries and atria, sense pressure within the system. The brain integrates the information from the sensors and sends out signals via the parasympathetic and sympathetic nervous system to regulate blood pressure. Blood pressure can be thought of as the product of cardiac output and peripheral resistance. Increasing or decreasing heart rate alters pressure by changing cardiac output; dilating or constricting resistance vessels changes pressure by increasing or decreasing peripheral resistance. The system shows continuous slight oscillations as the pressure sensors and brain work jointly to adjust blood pressure continuously. Figure 7-2 is a block diagram of the system, and figure 7-3 shows how the system works during the adjustment to the upright posture.

Lying down increases central blood volume, stroke volume, and cardiac output. This increases blood pressure and leads to reflex heart slowing and blood vessel dilation. Standing decreases central blood volume, stroke volume, and cardiac output. In this situation, the cardiovascular control system increases heart rate and constricts vessels to maintain blood pressure (figure 7-3). Orthostatic intolerance refers to those situations when either the cardiovascular response is inadequate or the drop in stroke volume is too large, and blood pressure is not maintained.

Factors affecting the reduction in stroke volume with standing

When a person moves from the supine to the upright posture, approximately 500 ml of blood leaves the thorax and moves to the lower body. This drops filling pressure to the heart and reduces stroke volume. Several factors can influence the decrease in stroke volume. First, if total blood volume is reduced (e.g., due to blood loss or dehydration), supine stroke volume will also be reduced. In the standing position, even if the amount of blood that moves from the upper to lower body is the same as it was

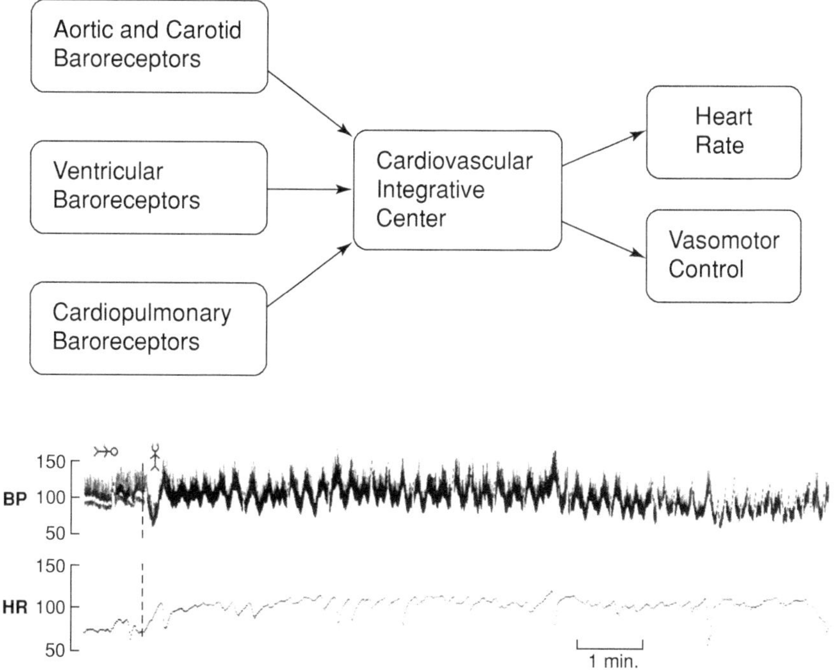

Figure 7-2. Sensors in the aorta, carotids, ventricles, atria, and pulmonary arteries send information on pressure to the cardiovascular control center in the brain. The results are changes in heart rate and vasomotor activity to regulate blood pressure. The bottom panel shows a blood pressure (top) and heart rate (bottom) tracing during a stand test (the dashed line indicates when the subject moved from the supine to the standing position). The top tracing shows that the blood pressure control system oscillates as heart rate and peripheral resistance are continuously adjusted to regulate blood pressure.

before the loss of blood volume, stroke volume will reach a lower level. As a result, the stress to the blood pressure control system will be greater.

Second, the relationship between the volume of blood leaving the chest and the decrease in stroke volume that results is not linear. A compliance curve describes the relationship between the filling pressure to the heart and stroke volume. The nature of this relationship can change [25]. Some conditions may move the supine operating point on the pressure volume relationship from a relatively flat portion to a steep portion. If the operating point should move to the steep portion of the pressure volume relationship, the drop in filling pressure caused by blood leaving the thorax will increase, causing a much greater reduction in stroke volume. This mechanism may be a reason for the orthostatic intolerance seen after bed rest [26] or in athletes [25]. After prolonged bed rest, there may be cardiac atrophy, resulting in a stiffer ventricle. When subjects stand after prolonged bed rest, they may experience a greater drop in stroke volume than might be expected from the degree of blood volume loss. This may be due to changes in cardiac compliance and movement to a steeper portion of

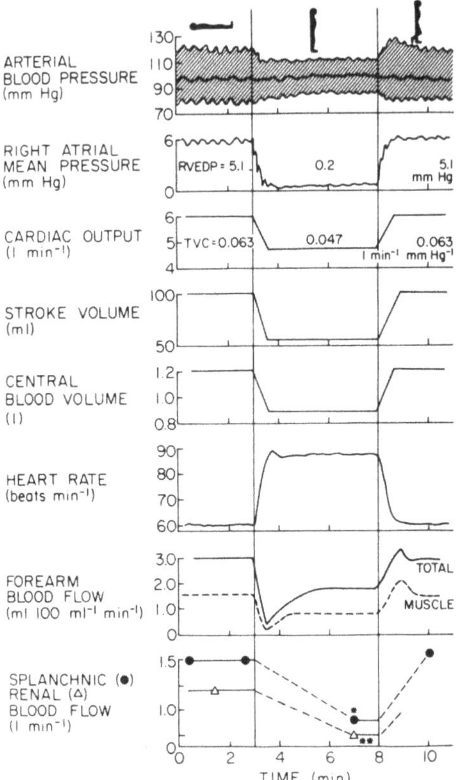

Figure 7-3. The hemodynamic response to standing. Blood pressure can be thought of as the triple product of heart rate, stroke volume, and peripheral resistance. With standing, blood leaves the thorax and moves to the lower body. Right atrial pressure declines, and stroke volume falls. Cardiac output also drops, leading to unloading of the arterial baroreceptors. The response to the fall in blood pressure is an increase in heart rate and increased vasoconstriction in several vascular beds (as indicated by the drop in gut and forearm blood flow). Overall, in the standing position cardiac output is allowed to fall from supine values, but blood pressure is well maintained. Reprinted from Rowell [27], with permission of Oxford University Press.

the pressure volume curve. Studies have shown that, after bed rest, if supine stroke volume is restored to pre-bed-rest levels by the infusion of fluid, the drop in stroke volume still exceeds pre-bed-rest levels [18].

Third, the compliance of the lower body may change. The amount of blood that moves from the chest to the lower body depends in part on the compliance of the lower body [27]. If, for example, a subject loses muscle mass and tone in the legs, it is possible that the legs will become more compliant and distend to a greater degree when a subject stands up. In this case, more blood will move from the upper to lower body with standing, and the reduction in stroke volume will be greater [28].

Fourth, it is not possible to completely separate the stimulus (the drop in stroke volume) from the response (the increase in heart rate and vasoconstriction). The abil-

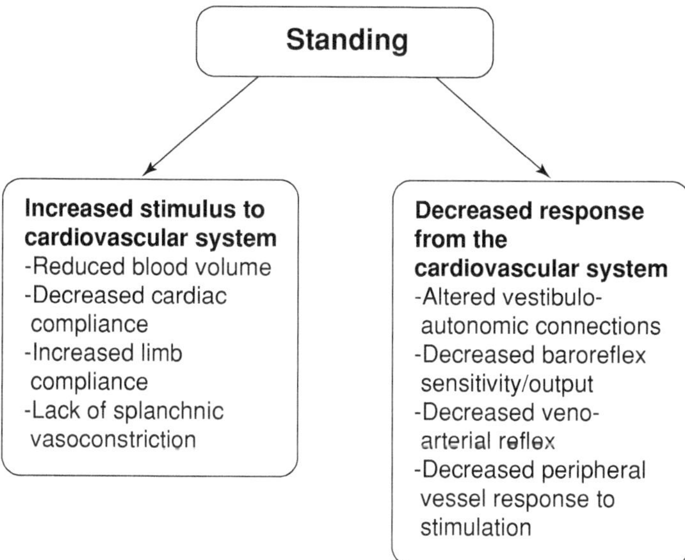

Figure 7-4. Possible reasons for orthostatic intolerance after spaceflight. A greater stimulus to the cardiovascular system, a reduced response, or a combination of the two could lead to fainting with standing. This assessment assumes that environmental factors are the same during testing before and after the flight. Blood volume is reduced, cardiac mass appears to decrease, and baroreflex regulation may be altered after spaceflight.

ity to vasoconstrict can have a major effect on how blood moving from the chest to the lower body affects cardiac filling pressure. Vasoconstriction in splanchnic areas during standing provides an important way to increase blood on the venous side of the circulation. Intense arterial vasoconstrictions in the gut and other regions allow the venous blood in the splanchnic veins to move to the central venous reservoir. This helps minimize the reduction in stroke volume seen with standing [27]. Finally, venoconstriction can help minimize the postural shift in blood volume.

Because multiple factors can influence the drop in stroke volume with standing, it can be difficult to isolate a single cause for a significant decrease in stroke volume. Figure 7-4 summarizes some of the factors that can affect stroke volume with standing.

Heart rate response to orthostatic stress

Blood pressure can be thought of as the triple product of heart rate, stroke volume, and peripheral resistance (i.e., heart rate × stroke volume × total peripheral resistance = blood pressure). The baroreflex-mediated increase in heart rate that occurs with standing increases cardiac output (heart rate × stroke volume = cardiac output) and helps restore blood pressure. This increase is mediated both by vagal withdrawal and sympathetic stimulation. During exercise, increases in heart rate dramatically increase cardiac output. Exercising muscle enhances the return of blood to the heart, and the heart fills rapidly and fully. Even though the amount of time available for filling in diastole decreases as heart rate increases, the increased filling pressure and rapid blood flow during exercise compensate for this.

Using heart rate to increase blood pressure during standing, however, has limitations. As heart rate increases during standing, cardiac output also increases initially. Yet filling pressures decrease with standing. As heart rate increases further, therefore, the reduced filling time at a low filling pressure can decrease stroke volume, so that a faster heart rate no longer increases cardiac output. This explains why tachyarrhythmias can cause orthostatic intolerance.

Ventricular baroreceptors can complicate the response to standing. Stimulation of the ventricular baroreceptors decreases heart rate and dilates blood vessels (the Bezold-Jarisch reflex) [29]. Typically, this is a response to high intraventricular pressure due to high blood pressure or excessive demands on the heart, but, ventricular baroreceptors can be stimulated when the ventricle is nearly empty and the ventricular walls contract forcefully against themselves. In this situation, the drive from the ventricular baroreceptors to reduce heart rate and dilate blood vessels can cause a faint. The stimulation of ventricular baroreceptors is thought to be the one mechanism for vaso-vagal (or vasodepessor) syncope. With vasodepressor syncope, sympathetic output to blood vessels is paradoxically decreased, and heart rate drops just before fainting.

Factors affecting the vasoconstrictor response

The constriction of blood vessels is a powerful force to control blood pressure. Gut, kidney, and muscle blood flow are all reduced with standing as vasoconstriction takes place. The greater the orthostatic stress, the greater the degree of vasoconstriction. Ultimately, however, this mechanism has limits. Vasoconstriction can redistribute flow so that the brain continues to be perfused, but in the process, overall flow through the cardiovascular system is reduced. This can only go so far before filling pressure is compromised. An individual will faint when standing once the blood flow to the brain is reduced below the point where consciousness can be maintained.

Some factors affect how much constriction will be produced for a given amount of nervous outflow from the sympathetic nervous system. Unloading of the peripheral arteries and arterioles, as occurs in bed rest when hydrostatic gradients are removed, can lead to atrophy of smooth muscle in the vessel wall and perhaps diminish responsiveness to sympathetic stimulation [30].

Autoregulation of brain blood flow

One additional factor that can influence orthostatic tolerance is the autoregulation of brain blood flow [27]. Blood flow can be maintained in the face of falling blood pressure by dilating blood vessels within the brain. Similarly, autoregulation can prevent excessive flow when blood pressure is high by constricting blood vessels. This autoregulatory function operates over a wide range of pressures. If, however, the autoregulatory range were reset, brain blood flow might not be as well regulated. Individuals could experience fainting at blood pressure levels that might previously have been tolerated before the autoregulatory range was reset or altered [31].

Summary

Figure 7-4 summarizes the possible causes for orthostatic intolerance. Changes in blood volume, cardiac compliance, limb compliance, heart rate regulation, vasoconstrictor regulation, and cerebral autoregulation are all potential contributors to postflight orthostatic intolerance. Often the cause can be multifactorial, with elements of each of the factors contributing to the total problem.

Cardiac Mass

Just as with skeletal muscle, heart muscle mass can change depending on the loading. Patients with high blood pressure experience heart hypertrophy; individuals who begin an exercise program increase their cardiac mass. Conversely, treating high blood pressure allows cardiac hypertrophy to regress, and heart mass decreases after stopping a training program [32].

Exercise increases cardiac mass, bed rest decreases it

The increase in cardiac mass with training depends on the type of training. Weightlifters and strength training athletes increase cardiac wall thickness, but the chamber size remains unchanged. Endurance training increases both chamber size and wall thickness. After training is stopped, cardiac size reduces back toward age-adjusted norms [32].

Perhonen and colleagues [33] demonstrated that 6 weeks of bed rest decreased cardiac mass by an average of 8% [33]. The limit of cardiac atrophy may be inferred from studies on patients with lesions high in the spinal cord. These patients, who experience years of very restricted activity, show a reduction in cardiac mass of 25–35% compared to nonathletic normal controls [34]. This degree of atrophy may represent a plateau.

Decrease in cardiac mass may affect orthostatic tolerance

One physiologic consequence of cardiac atrophy has been well outlined by Levine et al. [26] and Perhonen et al. [35]. Bed rest deconditioning can produce a smaller and less distensible left ventricle. In addition, the relationship between cardiac filling pressure and stroke volume moves to a steeper portion of the pressure volume curve. The net result is more profound orthostatic intolerance than would be expected from the reduction in plasma volume found after bed rest [35].

Cardiac Rhythm Disturbances

Cardiac arrhythmias sometimes indicate serious cardiac disease. Patients with coronary artery disease are more likely to have ectopy during exercise compared to normal subjects [36], but normal individuals can have cardiac rhythm disturbances in a variety of situations, with no pathology present. Continuous ambulatory electrocardiogram monitoring of healthy volunteers revealed that only 13% showed normal sinus rhythm throughout [37]. Crew members on spaceflights may be exposed to a variety of cardiovascular stresses, such as exercise or an electrolyte imbalance. Therefore, when an arrhythmia occurs in flight, it can be difficult to determine whether weightlessness or some other factor is the cause.

Effect of exercise

Almost any type of cardiac arrhythmia can be produced with exercise; the challenge is deciding when pathology is present [36]. In a study of patients who were undergoing cardiac catheterization, exercise-induced ventricular arrhythmias were observed in 25% of normal patients. In a study of 345 men and women who exercised regularly, ventricular ectopy was found in 18% of subjects less than 30 years old, and in >50% of those older than 50 [38]. In general, however, exercise-induced ventricular arrhythmias are more common in patients with coronary artery disease than in healthy subjects. The presence of coronary artery disease should be suspected when the ventricular arrhythmias are provoked by minimal exercise with less than 70% of predicted maximal heart rate [36].

Effect of electrolyte imbalance

Alterations in serum electrolytes, particularly potassium and calcium, but also magnesium and sodium, can lead to cardiac arrhythmias and electrocardiographic abnormalities. Typically these are seen in hospitalized patients, in whom electrolyte levels are significantly beyond the normal range. Nevertheless, dehydration and exercise in the heat can lead to significant changes in electrolyte levels and may predispose to arrhythmias [36].

Effect of stress

Stress and emotional upset can increase the tendency toward cardiac arrhythmias. Stamler and colleagues [39] performed ambulatory cardiac monitoring on interns on call in a hospital and noted a correlation between ectopy and stress. Similarly, Insulander and colleagues [40] demonstrated that mental stress could affect electrophysiologic variables. As spaceflight can produce stress, particularly during emergencies or at demanding flight phases, stress needs to be considered as a factor when arrhythmias are seen.

Effect of Spaceflight on the Cardiovascular System

When discussing the cardiovascular effects of spaceflight, the reference body position on Earth used for comparison is important. In this chapter, the upright position is used for reference. Gauer and Thron [17] outlined the rationale for using the upright position for comparisons in 1965. Because humans spend approximately two-thirds of their time upright, adaptive mechanisms work to restore upright hemodynamics during perturbations such as bed rest, head-down tilt, water immersion, and spaceflight. In the sections below, changes observed in space are referenced to a standing baseline, unless otherwise noted.

Another important consideration is that spaceflight investigations are "field science." Although investigators try to control confounding variables, research in space occurs in an operational environment. Crew members may need to take medications or follow a countermeasure program. Adherence to the countermeasure program may vary depending on the demands of the flight. Motion sickness symptoms may interfere with fluid balance and dietary intake. On longer flights, psychosocial stresses and in-flight emergencies can affect sleep, diet, and physiologic variables. Because of these constraints, results from spaceflight must be interpreted cautiously. Although a physiological event or change may be detected in space, that does not necessarily mean the change is due to weightlessness. Other factors, such as medications, psychological stress, or the countermeasure program, could also be contributing causes.

Entry into Weightlessness: Physiological Effects

Shortly after reaching orbit, leg volume decreases and the face becomes puffy [41]. Leg volume decreases by approximately 1 l per leg [42], and forehead tissue thickness increases by approximately 7% compared to a preflight supine control [43]. Pulmonary capillary blood volume increases by 25% [44], and intraocular pressure increases by 92% [45]. These changes reflect a significant fluid shift from the lower to upper body.

The fluid shift initially increases heart size and stroke volume. Early in flight (the first 24–48 hours), cardiac dimensions are increased [46–48]. Cardiac output and stroke volume are both elevated [44]. Over time, however, heart rate, stroke volume, and cardiac output return approximately to preflight sitting levels [49]. Blood pressure may be slightly reduced from usual preflight upright levels [50].

Several findings suggest that the early adaptation to weightlessness may differ significantly from ground-based simulations using bed rest or water immersion. First, the magnitude of the fluid shift may be quantitatively greater than what is seen with moving from upright to supine. Thornton et al. [51] showed that weightlessness simulations using bed rest or water immersion produced 50% or less of the fluid loss seen in the leg with weightlessness. Second, such a significant fluid shift might be expected to raise central venous pressure, but this does not occur [46, 52]. Third, ground-based simulations produce diuresis, but diuresis early in flight has not been seen in astronauts [53–56]. Data on urine output on the first flight day exists for 15 crew members; only 1 showed an increase above preflight levels.

Because the leg volume, tissue thickness, and stroke volume data all show a significant headward fluid shift initially, the central venous pressure and urinary output data appear contradictory. Two main hypotheses exist to explain the central venous pressure findings from spaceflight. The changes in central venous pressure and cardiac output might be explained by a reduction in intrathoracic pressure or the loss of gravitational compression of the cardiac tissues [52]. The lack of diuresis may be due to an interaction with the neurovestibular changes early in flight. Motion sickness produces marked elevations in antidiuretic hormone levels and is associated with a reduction in urine output [57]. As motion sickness symptoms are extremely common early in spaceflight, it is likely that these neurovestibular influences interact with the cardiovascular changes taking place to reduce urine output. Ultimately, however, after the initial fluid shift, stroke volume and cardiac output move toward preflight sitting values.

Reduction in Blood Volume

Although urine output does not appear to increase on the first day in orbit, plasma volume drops rapidly. Spaceflight data suggest that part of the reduction of plasma volume may be due to the movement of albumin-containing fluids from the intravascular to the extravascular space [58]. On the first day of flight, plasma volume is reduced by 17%. This rapid plasma volume loss increases hematocrit, which likely serves as a stimulus to reduce erythropoietin secretion. Erythropoietin levels fall in space [58]. The net result is a reduction in red blood cell mass. The combined plasma volume and red cell mass losses add up to an approximate 11% total blood volume reduction. This new level of blood volume appears to provide a central blood volume similar to what is seen on Earth in the upright position.

Cardiac Atrophy

Once in space, some of the daily volume and pressure stimuli that the heart receives are removed. On Earth, periodically lying down presents intermittent volume loads to the heart, but this does not occur in space. Also, the overall level of activity is reduced so that the integrated pressure load that the heart encounters during the day may be less (although this will depend on the countermeasure program). Blood pressure on average may be slightly reduced from usual $1\text{-}G$ levels [50]. Measurements made after flight suggest an 8–10% decrease in cardiac muscle mass occurs during spaceflight [33].

Cardiac Rhythm Disturbances

Serious cardiac arrhythmias have been documented in space. On *Apollo 15*, one crew member experienced ventricular bigeminy [59]. Ventricular ectopy was reported on *Skylab* during both extravehicular activity and exercise tests [60, 61]. On *Mir*, a crew member experienced a 14-beat run of ventricular tachycardia [62]. It has been difficult to discern whether the ventricular ectopy that occurs in space is related to weightlessness or whether it is a consequence of other flight factors. For example, the *Apollo 15* crew member had been involved in a stressful extravehicular activity on the Moon and was severely dehydrated. The ectopy on *Skylab* was reported to have occurred during times of significant stress. On *Mir*, commander Vasily Tsibliyev developed a significant heart rhythm abnormality [63]. This occurred after the collision of a *Progress* capsule with *Mir* during the time Tsibliyev was in command. He was under stress and not sleeping well. Also, it is likely that despite screening, some astronauts fly with occult coronary artery disease. They may experience ectopy during stress because of their underlying cardiac disease and not because of any independent effect of weightlessness.

Whether weightlessness increases the tendency to have ventricular arrhythmias is not known. Holter monitors done during extravehicular activities do not show a greater rate of ectopy [64], and Holter records recorded during short-duration flights showed a decrease in ectopy [50]. It is hard to assess whether the amount of significant ectopy that has occurred in space is excessive considering the potential for electrolyte imbalance, stress, and sleep loss in space operations.

Changes in Aerobic Capacity

Although exposure to weightlessness does reduce blood volume, measurements on short-duration spaceflights show that maximal oxygen uptake is well-maintained in flight [65]. On longer missions, if crew members did not exercise, they would lose aerobic capacity—just as is the case on Earth. Exercise in space, however, can increase maximal oxygen uptake. During the Skylab program, the *Skylab 4* astronauts were attentive to their in-flight exercise program and increased their maximal oxygen uptake during the flight [66].

After flight, however, maximal oxygen uptake is markedly reduced. After short-duration spaceflights (9–14 days), upright peak oxygen uptake was reduced by 22% on landing day [65]. The cause was a decreased peak stroke volume and cardiac output. Peak heart rate, blood pressure, and systemic arteriovenous oxygen difference were unchanged. Levine and colleagues [65] concluded that systemic peak oxygen uptake is well maintained in the absence of gravity for 9–14 days but is significantly reduced immediately on return to Earth, most likely because of reduced intravascular blood volume, stroke volume, and cardiac output. Aerobic capacity is important after landing because crews may need to do work on the Moon or Mars and may be exposed to significant heat stress in a spacesuit. Also, in case of an emergency or unplanned event, the crew may need to do significant aerobic work.

Postflight Orthostatic Intolerance

Orthostatic intolerance is commonly seen after returning to Earth from space. The precise incidence is difficult to determine because orthostatic intolerance can be defined in many different ways (frank syncope, lightheadedness, a 20 mmHg drop in systolic blood pressure, etc.). Also, orthostatic intolerance detected during a stand test may not be operationally significant because walking and movement help alleviate orthostatic tolerance by activating the muscle pump [67, 68]. A crew member may be able to function upright when active, but may get lightheaded while standing still. Nevertheless, on average, crew members will show higher heart rates and lower blood pressures when standing upright after spaceflight. After short-duration flights, as many as 63% of crew members could not complete a 10-minute stand test that all of them had completely easily before the flight [69].

Some factors do not play a major role in postflight orthostatic intolerance. Increased compliance in the limbs could lead to a greater postural fall in stroke volume, but data on leg compliance do not show increased leg compliance after short-duration spaceflights [70]. Reduced sensitivity of the baroreflexes is a potential contributor, and the heart-rate limb of the carotid baroreflex does show changes in weightlessness [71]. On average, however, astronauts show marked and brisk increases in heart rate during orthostatic stress after spaceflight [31, 69, 72], suggesting that the heart-rate limb of the baroreflex is grossly intact. Testing both during and after spaceflight show that baroreflex mechanisms (both heart rate and sympathetic nerve activity) are intact and respond appropriately to orthostatic stress [31, 71, 73]. Similarly, studies to date have not shown changes in cerebral autoregulation. Cerebral autoregulation was measured before, during, and after spaceflight in four subjects as part of the *Neurolab* mission. No significant changes were seen [31].

Reduced blood volume is likely a major contributor to postflight orthostatic intolerance. Supine stroke volume is reduced after flight, reflecting the reduced blood volume [69]. Although the reduction in blood volume may be a primary factor, efforts to restore blood volume have not been totally effective at preventing orthostatic hypotension. Crew members typically use a fluid-loading countermeasure (8 g of salt plus 1 l of water) before reentry to expand plasma volume and minimize orthostatic intolerance. Overall, this approach improves, but does not eliminate, orthostatic intolerance [69]. The data suggest that other mechanisms may also contribute to postflight orthostatic intolerance.

One potential factor could be limitations on vasoconstriction. In the study by Buckey et al. [69], individuals who could not complete a stand test did not raise their total peripheral resistance above preflight levels [69]. Levine et al. [31] showed that, although sympathetic nerve traffic was greatly increased during upright tilt after spaceflight, this was not accompanied by a greater increase in total peripheral resistance. Taken together, these findings suggest that, in addition to blood volume reduction, the ability to vasoconstrict may change after spaceflight. Possible explanations for this include changes in the veno-arterial reflex [74] or atrophy of smooth muscle in resistance arteries in the periphery [30]. Overall, data collected from space investigations suggest that interventions aimed at increasing blood volume and enhancing vasoconstriction should have the highest likelihood of success.

Approaches to Cardiovascular Changes in Space

Whether returning to Earth or landing on another planet, the cardiovascular system faces a variety of challenges after spaceflight. During reentry, the crew might be exposed to head-to-foot *G* forces (such as the ones experienced during a Shuttle reentry), which introduce a significant orthostatic stress at a vulnerable time. Similarly, the front-to-back *G* forces of a capsule (e.g., *Soyuz*) reentry also stress the cardiovascular system and reduce cardiac filling.

Once the crew arrives on Earth or another planet, they may be wearing space/protective suits and need to perform physical work, either for an emergency or for nominal operations. Most suits have limited cooling and can present a major heat stress to the body (see chapter 5). Also, prolonged work in a suit may require superior endurance. If crew members land with reduced blood volume, poor heat tolerance, and a reduced exercise capacity, they could quickly become dehydrated and fatigued while working. Mission success may depend on the crew members maintaining good physical condition in space so they are ready for landing.

Aerobic Exercise

Maintaining good aerobic capacity while in flight makes sense for space crews. Aerobic training increases maximal oxygen uptake, improves endurance, and offers positive psychological effects. The ability to perform aerobic work may provide an essential edge in emergencies or during demanding work in a space suit. A trained individual is better able to adjust to heat than an untrained one [6]. Aerobic training also increases blood volume, which may be beneficial for orthostatic tolerance, provided the aerobic training is not excessive.

Thirty minutes or more at submaximal levels of exercise (50% of maximum or an approximate increase in heart rate of 60 or more beats/minute) three or four times a week can improve endurance. Large-muscle exercise at high intensity for 3–5 minutes followed by rest or mild exercise for the same duration can develop aerobic power when repeated frequently. Bursts of intense activity can develop muscle, tendon, and ligament strength [32].

Heat Acclimatization

Heat acclimatization improves endurance and work capacity in the heat. Non–heat-acclimated individuals become dehydrated faster during equivalent work and can develop exercise-induced orthostatic intolerance. Because crews working on other planets will be working in space suits, and because most suits have a limited cooling ability, crew members can expect extravehicular activity to often be hot and uncomfortable (see chapter 5).

Studies suggest that 1 hour of exposure to a hot environment daily will result in some acclimatization within a week. To work a significant amount in heat, however, requires a more aggressive acclimatization strategy with 4 hours of heat exposure (34°C) daily for 8–9 days [6].

Strength Training

Strength training is addressed in chapter 4. Strength training can help improve orthostatic tolerance through two main methods. Because cardiac atrophy may be a contributor to orthostatic intolerance after spaceflight, the increased pressure load that strength training presents to the heart may help maintain or increase cardiac mass. Similarly, although leg compliance does not seem to decrease in space, increased lower body muscle tone may help decrease postural pooling and enhance the effectiveness of the skeletal muscle pump.

Artificial Gravity

Continuous 1-G artificial gravity should prevent the cardiovascular adaptation to weightlessness. The lowest level of artificial gravity that could be used that still would have a beneficial effect is not known. Because continuous artificial gravity may not be available, intermittent artificial gravity offers another option. In weightlessness, the loss of hydrostatic pressure gradients in the leg may reduce pressure loading on the resistance vessels in the lower body. This could lead to smooth-muscle atrophy in these vessels and diminish their ability to increase resistance in response to sympathetic stimulation [30, 75]. If this is true, then intermittent gravitational loading of the vessels with artificial gravity may help minimize orthostatic intolerance.

Table 7-1. Lower body negative pressure training program used in the days before reentry in the Russian space program.

Training session	Pressure levels (mmHg)	Time at each pressure (minutes)
Preliminary		
Day 1	10, 15, 20, 25	5
Day 2	15, 20, 25, 30	5
Day 3	20, 25, 30, 35	5
Day 4	25, 30, 35, 40	5
Day 5	25, 35, 40, 45	5
Day 6	25, 35, 40, 45	5
		When each training session is complete gradually return pressure [over 1 minute] to 0
Final 2 days		
1 cycle	25, 35, 40, 45 mmHg, followed by returning pressure to 0 over 1 minute	5
2 cycle	25, 35, 45, 30 mmHg, followed by returning pressure to 0 over 1 minute	5

The crews are instructed to drink 200 ml of fluid before the test and to contract their leg muscles alternately during the exposure.

Intermittent artificial gravity, depending on how it is applied, could also reduce central blood volume. This reduction in central blood volume, similar to what occurs with standing, might activate the low-pressure baroreceptors and activate fluid retention mechanisms to increase plasma volume. At present, it is not known how much gravitational force is needed and how frequently it should be applied to stimulate fluid retention.

Lower Body Negative Pressure Exposure

Lower body negative pressure (or lower body suction) stresses the cardiovascular system in a way similar to standing (figure 7-1). The lower body is enclosed in an airtight chamber and then exposed to a vacuum. The reduction in pressure on the outside of the legs causes fluid to move from the central circulation to the legs (similar to what occurs when standing on Earth). At low levels of suction (20 mmHg or less), the cardiopulmonary receptors are stimulated without an effect on arterial blood pressure; at higher levels, aortic and carotid baroreceptors are unloaded. A level of –40 to –50 mmHg produces hemodynamic changes similar to standing in 1 G. As LBNP exposure continues, plasma renin activity and other fluid regulatory hormones (e.g., aldosterone) are increased [27, 76].

The rationale for LBNP exposure is that it reduces central blood volume and stimulates fluid retention mechanisms. Although LBNP has an immediate effect on

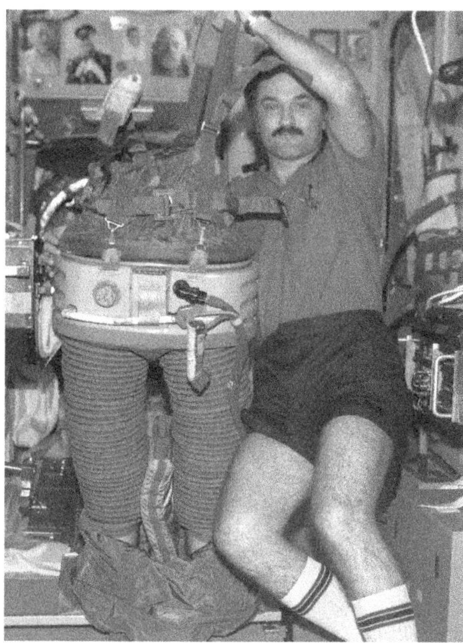

Figure 7-5. Cosmonaut Alexander Kaleri, Russia's Federal Space Agency Expedition 8 flight engineer, poses in the Zvezda Service Module, with the Russian lower body negative pressure (LBNP) or Chibis suit. The suit applies suction to the legs and lower body but still allows for some mobility. (Photo courtesy of NASA.)

stroke volume, blood pressure, and heart rate, the humoral effects can take longer to develop. Bevegard et al. [77] showed that plasma renin activity started to rise after 19 minutes of exposure. Also, higher levels of LBNP (–40 mmHg) are needed to stimulate renin release. Lower levels (–10 to –20 mmHg) do not appear to be as effective for stimulating a hormonal response.

LBNP has been a consistent feature of the Russian countermeasure program for long-duration flights [78]. LBNP is used in the week before reentry to prepare crews for landing. Fluid is ingested before LBNP, and then graded levels of LBNP are applied to the lower body with pauses at each step (table 7-1 outlines the Russian protocol). The "Chibis" LBNP garment allows the crew member to be mobile while the LBNP is applied (figure 7-5). The Russian program also includes orthostatic testing using LBNP (a graded increase in LBNP level while heart rate and blood pressure are monitored) to assess the effectiveness of the countermeasure. LBNP was also evaluated as a countermeasure for use on the Shuttle [50]. In that study, 8 g of salt plus 1 l of water were ingested before a 3.5 hour exposure to LBNP at –30 mmHg. This countermeasure appeared effective, but it was not adopted because of the time commitment. LBNP combined with treadmill exercise has been studied because this combination could provide both an orthostatic stress and a musculoskeletal countermeasure [66, 79, 80]. Unfortunately, the combination of 40 minutes of moderate exercise during –50 mmHg of LBNP did not improve orthostatic tolerance.

Medications and Other Interventions

Several medications and physical approaches also could be useful after spaceflight. Erythropoietin, fludrocortisone, and oral fluid loading are potential ways to increase blood volume. Vasoconstrictors have been used after bed rest studies with good results. G-suits and body cooling are also physical methods to increase total peripheral resistance.

Vasoconstrictors

Because studies have shown that people who cannot complete a stand test after spaceflight do not increase total peripheral resistance beyond preflight levels [69, 72], the augmentation of vasoconstriction with drugs is a reasonable approach. In clinical practice, several different vasoconstrictors have been tried against orthostatic hypotension (ephedrine, phenylephrine, ergotamine). Midodrine, which is an alpha-receptor agonist with few cardiac effects, has been used to treat patients with orthostatic hypotension [81]. Midodrine is primarily a peripheral arterial and venous constrictor. This drug has been shown to be effective after bed rest [82] and has been used after spaceflight in selected astronauts.

The exact mechanism of action for vasoconstrictors after spaceflight is not totally clear. In contrast to patients with autonomic failure, astronauts seem to show brisk responses from the sympathetic nervous system when standing after spaceflight, suggesting that blood vessels are getting adequate sympathetic stimulation [31]. It may be, however, that venoconstriction is impaired or that splanchnic vasoconstriction is inadequate, or that the addition of the drug can elicit vasoconstriction beyond what the sympathetic nervous system can provide. In those cases where the ventricular and arterial baroreflexes are in conflict (due to the Bezold-Jarisch reflex described above), vasoconstrictors might help minimize or prevent the vasodilation that occurs as part of the vasodepressor reaction.

Fludrocortisone

Fludrocortisone is a mineralocorticoid commonly used to treat patients with orthostatic hypotension. It is similar to aldosterone and leads to sodium retention, potassium wasting, and an increase in plasma volume. Fludrocortisone was effective as a countermeasure for the orthostatic intolerance produced by bed rest [83]. Fludrocortisone was also evaluated as part of the Extended Duration Orbiter Medical Program as a potential spaceflight countermeasure. In the setting of spaceflight, fludrocortisone did not seem to improve orthostatic intolerance, and it has not been adopted for use in space [50]. The drug was not administered for as long in the space study, however, as it was in the bed rest study.

Erythropoietin

Erythropoietin is commonly used in patients on renal dialysis and for cancer patients with anemia. In general, erythropoietin is well tolerated, although antibodies can develop to the drug, and some patients have hypersensitivity reactions to some components in the commercially available preparations. Recombinant human erythropoietin can be given as a daily injection (Epogen alfa), although a newer preparation (Darbepoetin alfa) can be given as weekly injections. Gene-activated erythropoietin also has been produced, and encapsulated forms of the drug that can be taken orally are under development.

In patients with anemia, the goal is to raise hematocrit. For spaceflight, the aim would be to increase red cell mass—and therefore blood volume—before reentry. In this setting, however, hematocrit starts out at a normal level, so there could be a hazard in increasing hematocrit beyond the normal range. Fatal events have occurred in athletes who increased their hematocrit and then became dehydrated. If attempted, spaceflight use would need to be limited to the week before reentry and should include careful monitoring.

Hypoxia

Rather than administer erythropoietin, endogenous erythropoietin can be stimulated by exposure to hypoxia. Erythropoietin secretion requires altitudes of approximately 2500 m or greater, and exposures of 6 hours or longer [15]. In athletic training this is done by sleeping (and living) at altitude and training at sea level. Altitude exposure can also be accomplished by sleeping in an altitude chamber. The combination of altitude acclimatization and sea-level training is called the "live high–train low" approach to training and has been shown to be effective [16, 84]. At present this approach has not been studied as a way to prepare astronauts for landing either back on Earth or on another planet.

Autologous transfusion

The donation of blood for reinfusion before an athletic contest is called blood doping. Blood transfusion increases hemoglobin concentration above an individual's normal values and increases maximum oxygen uptake. Studies have shown that reinfusion of 900–1350 mL of blood elevates blood oxygen-carrying capacity and increases maximum oxygen uptake by 4–9%. Improvements were seen 24 hours after the infusion [10]. Reinfusion of autologous blood avoids most of the disease transmission problems associated with transfusion. The main issues for spaceflight use would be storage of the blood and the avoidance of bacterial contamination.

Fluid loading

Drinking salt-containing solutions has been component of both the U.S. and Russian countermeasure program. The U.S. fluid loading countermeasure offers the crews a variety of options, including taking salt tablets (8 g) and water (1 l) or drinking 1 l of electrolyte-containing beverages. Salt tablets have the potential to induce vomiting, so they may not be ideal for spaceflight use. Several different fluid-loading regimes have been evaluated in ground-based testing, and those with a slighter higher salt concentration (1.07 N) are retained longer [85]. Carbohydrate in the rehydration solution (such as is in sports drinks) is probably also important to enhance absorption [6].

G-suits

G-suits are a simple, effective, and time-proven method to improve orthostatic tolerance. The G-suit includes calf, thigh, and abdominal bladders that compress the tissues. The compression increases total peripheral resistance, reduces leg vein compliance, and improves G tolerance [86]. In fighter aviation, a properly fitted and functioning G-suit can increase G tolerance by 2–4 G.

In the Russian space program, a tight-fitting garment called the Kentavr suit is worn before reentry. This garment compresses the lower body like a G-suit and helps provide protection against orthostatic intolerance. The launch entry suit used during Shuttle reentry incorporates a G-suit that can be inflated before and during reentry. The G-suit can be a very potent countermeasure. Most stand test studies are done without the G-suit in place. With the G-suit on, the rate of failure in stand-test studies would be low, but exact data on effectiveness are not available.

Body cooling

Body cooling also provides an alternative method to increase vasoconstriction. Heat stress dilates skin blood vessels and decreases total peripheral resistance. The combination of vasodilation and fluid loss makes heat stress a major threat to a crew after landing. Body cooling, however, works in the opposite direction. The cooling serves to constrict skin blood flow, thereby improving central circulation. Also, cold applied to the hand and/or face can markedly elevate sympathetic nerve activity (e.g., the cold pressor test). Skin cooling has been tested as a method to improve orthostatic tolerance and is effective. In a study by Durand et al. [87], subjects were tested both with and without skin cooling using an LBNP tolerance test. LBNP tolerance increased, as did blood pressure and norepinephrine levels.

Monitoring Cardiovascular Changes

Techniques are needed to keep track of the cardiovascular changes while in space to assess whether the countermeasure program is working and whether the crew is ready for operational demands (e.g., extravehicular activity or reentry).

Holter Monitoring

Twenty-four hour monitoring of heart rate and the electrocardiogram is the best way to assess the overall incidence of cardiac arrhythmias. The results have to be interpreted with caution. Rhythm disturbances of many types can be seen in normal individuals and may have no clinical significance [37, 38, 88–90]. Even arrhythmias that might be considered to be dangerous (such as ventricular tachycardia) can sometimes occur in normal individuals and lead to no long-term adverse consequences [91]. Nevertheless, an increase in the frequency of cardiac arrhythmias, particularly those that occur at low levels of exercise, may be of concern. Also, without solid data on the "normal" level of cardiac ectopy expected in space, flight surgeons may attach more significance to a rhythm disturbance than might be warranted if the incidence of the arrhythmia were known. As a basis for comparison with in-flight recording, there should be adequate data collected before the flight in stressful situations (extra-vehicular activity simulations, exercise).

Ultrasound

The degree of cardiac atrophy that might be expected in space has not been firmly established. By monitoring cardiac mass, a set of baseline data could be collected to assess whether a given astronaut's change in heart mass is significant and represents a level outside of the expected range. Magnetic resonance imaging (MRI) offers the most accurate way to measure cardiac mass [33], but MRI is impractical in space. Cardiac ultrasound offers the most sensible method to track cardiac mass in space crews. Advances in ultrasound technology allow for three-dimensional ultrasound that should provide more accurate and repeatable measurements than are possible with individual measurements of cardiac wall thickness.

Functional Tests

Exercise tests

Aerobic capacity can be readily measured in flight using a bicycle ergometer or treadmill. In-flight fitness assessments have been a regular feature of the Russian long-duration flight program and are performed on the International Space Station [78]. Concern has been expressed about whether a maximal exercise test is prudent in space because this might increase the incidence of arrhythmias. Most good exercise training programs will involve efforts that reach maximal oxygen uptake, so limiting exercise testing to submaximal levels does not seem necessary. Extravehicular activity, exertion on landing day, and heat stress all have the potential to take the cardiovascular system to maximal heart rates and oxygen uptakes. Because the crews must be able to tolerate these stresses, the stress of a maximal oxygen uptake test does not seem excessive or overly demanding. Submaximal assessments of fitness are done on the International Space Station at 30-day intervals [92].

Lower body negative pressure tests

The current Russian countermeasure program includes exposure to LBNP at increasing levels in the week before reentry (table 7-1). In addition, a test of orthostatic function using LBNP is incorporated into the Russian program. Pressure is ramped up to −50 mmHg while heart rate and blood pressure are measured. This approach provides a way to assess how well the countermeasure program is working while the crew member is still in orbit. In this way, the crew member can increase his or her exercise program or alter fluid intake to try to improve orthostatic performance before reentry.

Recommendations Based on Current Knowledge

The most concerning cardiovascular change that occurs in space is decreased orthostatic tolerance. The primary, but probably not only, factor affecting orthostatic tolerance after spaceflight is the reduction in blood volume that occurs in space. Cardiac atrophy, diminished peripheral vascular smooth-muscle tone, and changes in the venoarterial reflex may also contribute. Baroreflex function seems largely intact, but small changes in baroreflex function could be important at times of peak orthostatic stress. The drop in orthostatic tolerance is also accompanied by a reduction in upright exercise capacity because the same basic mechanism (a decrease in upright stroke volume) contributes to both effects. Orthostatic intolerance has the potential to be hazardous during reentry (particularly if the crew members are sitting upright and experiencing increased G loads, such as during a Shuttle reentry) and could complicate performance in an emergency after landing.

If the crew members can reenter supine and gradually assume the upright posture after landing, crews can return with significant orthostatic intolerance. After landing, the gradual introduction of head-to-foot G loads will stimulate fluid retention and initiate the readaptation to gravity. The case could be made that countermeasures to orthostatic intolerance could be minimal during the flight because crew members will be able to adapt to the new environment once they arrive there. This approach has some problems, however.

Front-to-back G loads (G_x) experienced during supine reentry, although less of a stress on the cardiovascular system than head-to-foot (G_z) loads, also reduce cardiac filling. High G_x loads can markedly reduce stroke volume and potentially lead to hypotension [93]. The supine G loads the crew experiences depend on the mission plan, but in off-nominal scenarios these loads can become quite high (10–13 G_x). The margin for safety is increased if the crews have an adequate blood volume.

Also, after landing, the crews cannot be sure that they will have time to gradually adapt to the new environment. An emergency or equipment problem may require them to get to work immediately in extravehicular activity suits or other protective gear. The heat stress and workload that this would present could quickly lead to dehydration and incapacitation if the crews are not acclimated to heat and do not have satisfactory exercise tolerance. In the worst case, significant heat stress and dehydration after landing could lead to rhabdomyolysis and kidney failure.

To help prepare for the potential cardiovascular stresses that might occur after landing and to maintain the crews in space, the following recommendations can be made:

1. Use a G-suit. After landing, either a G-suit or lower body elastic garment should be provided to improve orthostatic tolerance. Heat stress should be avoided, and, when possible, cooling should be used to prevent hypotension.
2. Fluid load before landing. Crews should fluid load before reentry with a carbohydrate-containing electrolyte solution within an hour of when G loading is expected.
3. Maintain aerobic fitness. Crew members should exercise in space to maintain a level of aerobic fitness at or beyond preflight levels. Also, in the weeks before landing, the exercise could be done at an elevated ambient temperature to stimu-

late heat acclimatization. Monthly testing of aerobic fitness should be used to track progress and assess whether the training program should be increased or decreased. ECG monitoring should be performed during the testing. The aerobic training program should include short bursts of high-intensity exercise, interval training, and endurance training.

4. Do strength training. Maintaining lower body muscle strength (see chapter 4) may help with orthostatic tolerance after the flight in addition to its importance for maintaining postural muscles.

5. Test orthostatic responses in space. A baseline test of orthostatic function using LBNP should be performed before flight and early in flight to establish a baseline. In the weeks before reentry, LBNP testing should be performed on a ramped protocol to assess the heart rate and blood pressure responses. Countermeasures can then be applied in space (plasma volume expansion, LBNP training) and orthostatic responses retested before reentry.

6. Consider LBNP as a countermeasure. In addition to its use as a test of orthostatic tolerance, LBNP can be used as a countermeasure. The Chibis LBNP suit used in the Russian program allows the crew members to move around the cabin while the LBNP is applied (figure 7-5). Extended duration LBNP exposure can be used in the weeks before reentry to stimulate plasma renin activity and aldosterone. Studies suggest that exposure should exceed 20 minutes to produce a hormonal response [77]. The optimal duration, frequency, and intensity of LBNP exposure needed to increase central blood volume are not precisely known. By performing in-flight orthostatic testing, the results of countermeasures can be evaluated before landing. The application of LBNP to increase plasma volume should be accompanied by an increase in salt and water intake.

7. Mild hypoxia could be studied to increase blood volume. Hypoxia, either by breathing a low-oxygen mix or by decreasing oxygen tension in the spacecraft, could be studied as a possible intervention to stimulate erythropoietin secretion before reentry and increase blood volume.

8. Study methods to increase blood volume before reentry. Methods such as autologous transfusion and erythropoietin should be evaluated for possible use in space.

9. Use the results of in-flight orthostatic tests to guide therapy. The results of in-flight orthostatic testing, in addition to previous experience with a particular crew member, can be used to decide if a vasoconstrictor should be taken. Midodrine reaches a peak concentration in the blood approximately 1 hour after an oral dose, and so ingestion has to be appropriately timed. Usual doses of midodrine range from 2.5 mg to 10 mg.

10. Holter monitoring should be performed until the issue about whether ectopy is increased in space is settled. Holter monitoring should be performed periodically in space to provide baseline data on the rate of ventricular ectopy.

11. Monitor cardiac atrophy. Cardiac mass should be measured periodically throughout a long-duration flight to establish in flight norms and to assess the effectiveness of countermeasures.

References

1. Catterson, A.D., et al., Aeromedical observations, in *Mercury Project Summary*. 1963, Manned Spacecraft Center, Houston, TX, pp. 299–324.

2. Convertino, V.A., Blood volume: its adaptation to endurance training. Medicine and Science in Sports and Exercise, 1991. 23(12): 1338–48.

3. van Lieshout, J.J., Exercise training and orthostatic intolerance: a paradox? Journal of Physiology, 2003. 551(Pt 2): 401.

4. Ogoh, S., et al., Carotid baroreflex responsiveness to head-up tilt-induced central hypovolaemia: effect of aerobic fitness. Journal of Physiology, 2003. 551(Pt 2): 601–8.

5. Levine, B.D., et al., Physical fitness and cardiovascular regulation: mechanisms of orthostatic intolerance. Journal of Applied Physiology, 1991. 70(1): 112–22.

6. Astrand, P.P., et al., Temperature regulation, in *Textbook of Work Physiology*. 2003, Human Kinetics, Champaign, IL, pp. 31–70.

7. Mack, G.W., and E.R. Nadel, Body fluid balance during heat stress in humans, in *Handbook of Physiology*, section 4, *Environmental Physiology*, M.J. Fregly and C.M. Blatteis, eds. 1996, Oxford University Press, New York, pp. 187–214.

8. Keren, G., et al., Orthostatic responses in heat tolerant and intolerant subjects compared by three different methods. Aviation, Space, and Environmental Medicine, 1980. 51(11): 1205–8.

9. Shvartz, E., N.B. Strydom, and H. Kotze, Orthostatism and heat acclimation. Journal of Applied Physiology, 1975. 39(4): 590–95.

10. Gaudard, A., et al., Drugs for increasing oxygen and their potential use in doping: a review. Sports Medicine, 2003. 33(3): 187–212.

11. Berglund, B., and B. Ekblom, Effect of recombinant human erythropoietin treatment on blood pressure and some haematological parameters in healthy men. Journal of Internal Medicine, 1991. 229(2): 125–30.

12. Shaskey, D.J., and G.A. Green, Sports haematology. Sports Medicine, 2000. 29(1): 27–38.

13. Biaggioni, I., Erythropoietin in autonomic failure, in *Primer on the Autonomic Nervous System*, D. Robertson, P.A. Low, and R.J. Polinsky, eds. 1996, Academic Press, New York, pp. 332–33.

14. Kawakami, K., et al., Successful treatment of severe orthostatic hypotension with erythropoietin. Pacing and Clinical Electrophysiology, 2003. 26(1 Pt 1): 105–7.

15. Ge, R.L., et al., Determinants of erythropoietin release in response to short-term hypobaric hypoxia. Journal of Applied Physiology, 2002. 92(6): 2361–67.

16. Stray-Gundersen, J., B.D. Levine, and C.G. Blomqvist, "Living high and training low" can improve sea level performance in endurance athletes. British Journal of Sports Medicine, 1999. 33(3): 150–51.

17. Gauer, O.H., and H.L. Thron, Postural changes in the circulation, in *Handbook of Physiology. Circulation.*, W.F. Hamilton, ed. 1965, American Physiological Society, Washington, DC, pp. 2409–39.

18. Blomqvist, C.G., Cardiovascular adjustments to gravitational stress, in *Handbook of Physiology*, J.T. Shepherd, ed. 1983, American Physiological Society, Washington, DC, pp. 1025–63.

19. Fortney, S.M., V.S. Schneider, and J.E. Greenleaf, The physiology of bed rest, in *Handbook of Physiology*, section 4, *Environmental Physiology*, M.J. Fregly and C.M. Blatteis, eds. 1996, Oxford University Press, New York, pp. 889–939.

20. Gunga, H.C., et al., Erythropoietin under real and simulated microgravity conditions in humans. Journal of Applied Physiology, 1996. 81(2): 761–73.

21. Vernikos, J., et al., Effect of standing or walking on physiological changes induced by head down bed rest: implications for spaceflight. Aviation, Space, and Environmental Medicine, 1996. 67(11): 1069–79.

22. Heer, M., et al., High dietary sodium chloride consumption may not induce body fluid retention in humans. American Journal of Physiology – Renal Fluid and Electrolyte Physiology, 2000. 278(4): F585–95.

23. Damgaard, M., et al., Effects of sodium intake on cardiovascular variables in humans during posture changes and ambulatory conditions. American Journal of Physiology – Regulatory Integrative and Comparative Physiology, 2002. 283(6): R1404–11.

24. Mtinangi, B.L., and R. Hainsworth, Early effects of oral salt on plasma volume, orthostatic tolerance, and baroreceptor sensitivity in patients with syncope. Clinical Autonomic Research, 1998. 8(4): 231–35.
25. Levine, B.D., and C.G. Blomqvist, Regulation of central blood volume and cardiac filling in endurance athletes: the Frank-Starling mechanism as a determinant of orthostatic tolerance. Medicine and Science in Sports and Exercise, 1993. 25(6): 727–32.
26. Levine, B.D., J.H. Zuckerman, and J.A. Pawelczyk, Cardiac atrophy after bed-rest deconditioning: a nonneural mechanism for orthostatic intolerance. Circulation, 1997. 96(2): 517–25.
27. Rowell, L.B., Adjustments to upright posture and blood loss, in *Human Circulation Regulation During Physical Stress*. 1986, Oxford University Press, New York, pp. 137–73.
28. Streeten, D.H., and T.F. Scullard, Excessive gravitational blood pooling caused by impaired venous tone is the predominant non-cardiac mechanism of orthostatic intolerance. Clinical Science (London), 1996. 90(4): 277–85.
29. Bishop, V.S., A. Malliani, and P.P. Thoren, Cardiac mechanoreceptors, in *Handbook of Physiology*, section 2, *The Cardiovascular System*, J.T. Sheperd and F.M. Abboud, eds. 1983, American Physiological Society, Bethesda, MD, pp. 497–556.
30. Zhang, L.F., Vascular adaptation to microgravity: what have we learned? Journal of Applied Physiology, 2001. 91(6): 2415–30.
31. Levine, B.D., et al., Neural control of the cardiovascular system in space, in *The Neurolab Spacelab Mission: Neuroscience Research in Space*, J.C. Buckey and J.L. Homick, eds. 2003, NASA, Houston, TX, pp. 175–85.
32. Astrand, P., et al., Physical training, in *Textbook of Work Physiology*. 2003, Human Kinetics, Champaign, IL, pp. 313–68.
33. Perhonen, M.A., et al., Cardiac atrophy after bed rest and spaceflight. Journal of Applied Physiology, 2001. 91(2): 645–53.
34. Nash, M.S., et al., Reversal of adaptive left ventricular atrophy following electrically-stimulated exercise training in human tetraplegics. Paraplegia, 1991. 29(9): 590–99.
35. Perhonen, M.A., J.H. Zuckerman, and B.D. Levine, Deterioration of left ventricular chamber performance after bed rest : "cardiovascular deconditioning" or hypovolemia? Circulation, 2001. 103(14): 1851–57.
36. Chung, E.K., Exercise-induced cardiac arrhythmias, in P*rinciples of Cardiac Arrhythmias*. 1989, Williams and Wilkins, Baltimore, MD, pp. 613–37.
37. Stinson, J.C., et al., Use of 24 h ambulatory ECG recordings in the assessment of new chemical entities in healthy volunteers. British Journal of Clinical Pharmacology, 1995. 39(6): 651–56.
38. Ekblom, B., L.H. Hartley, and W.C. Day, Occurrence and reproducibility of exercise-induced ventricular ectopy in normal subjects. American Journal of Cardiology, 1979. 43(1): 35–40.
39. Stamler, J.S., et al., The effect of stress and fatigue on cardiac rhythm in medical interns. Journal of Electrocardiology, 1992. 25(4): 333–38.
40. Insulander, P., et al., Electrophysiologic effects of mental stress in healthy subjects: a comparison with epinephrine infusion. Journal of Electrocardiology, 2003. 36(4): 301–9.
41. Thornton, W.E., T.P. Moore, and S.L. Pool, Fluid shifts in weightlessness. Aviation, Space, and Environmental Medicine, 1987. 58(9 Pt 2): A86–90.
42. Moore, T.P., and W.E. Thornton, Space shuttle inflight and postflight fluid shifts measured by leg volume changes. Aviation, Space, and Environmental Medicine, 1987. 58(9 Pt 2): A91–96.
43. Kirsch, K.A., et al., Fluid shifts into and out of superficial tissues under microgravity and terrestrial conditions. Clinical Investigator, 1993. 71(9): 687–89.
44. Prisk, G.K., et al., Pulmonary diffusing capacity, capillary blood volume, and cardiac output during sustained microgravity. Journal of Applied Physiology, 1993. 75(1): 15–26.
45. Draeger, J., et al., Self-tonometry under microgravity conditions. Aviation, Space, and Environmental Medicine, 1995. 66(6): 568–70.

46. Buckey, J.C., et al., Central venous pressure in space. Journal of Applied Physiology, 1996. 81(1): 19–25.
47. Frey, M.A.B., J.B. Charles, and D.E. Houston, Weightlessness and response to orthostatic stress, in *Circulatory Response to the Upright Posture*, J.J. Smith, ed. 1990, CRC Press, Boca Raton, FL, p. 65–120.
48. Lathers, C.M., et al., Echocardiograms during six hours of bedrest at head-down and head-up tilt and during space flight. Journal of Clinical Pharmacology, 1993. 33(6): 535–43.
49. Verbanck, S., et al., Pulmonary tissue volume, cardiac output, and diffusing capacity in sustained microgravity. Journal of Applied Physiology, 1997. 83(3): 810–16.
50. Charles, J.B., et al., Cardiovascular deconditioning, in *Extended Duration Orbiter Medical Project Final Report*, C.F. Sawin, G.R. Taylor, and W.L. Smith, eds. 1999, NASA, Houston, TX, pp. 1.1–1.19.
51. Thornton, W.E., et al., Changes in leg volume during microgravity simulation. Aviation, Space, and Environmental Medicine, 1992. 63(9): 789–94.
52. Buckey, J.C., Central venous pressure, in *Gravity and the Lung: Lessons from Microgravity*, G.K. Prisk, J.B. West, and M. Paiva, eds. 2001, Marcel Dekker, New York, pp. 225–54
53. Leach, C.S., and W.C. Alexander, Endocrine, Electrolyte and fluid volume changes associated with Apollo missions, in *Biomedical Results from Apollo*, R.S. Johnston, L.F. Dietlein, and C.A. Berry, eds. 1975, NASA, Washington, DC, pp. 163–84.
54. Leach, C.S., and P.P.C. Rambaut, Biochemical responses of the Skylab crewmen: an overview, in *Biomedical Results of Skylab*, R.S. Johnston and L.F. Dietlein, eds. 1977, NASA, Washington, DC, pp. 204–16.
55. Leach, C.S., Fluid control mechanisms in weightlessness. Aviation, Space, and Environmental Medicine, 1987. 58(9 Pt 2): A74–79.
56. Grigoriev, A.I., et al., Metabolic endocrine processes, in *Space Flights in the Soyuz Spacecraft. Biomedical Research*, O.G. Gazenko, L.I. Kakurin, and A.G. Kutznetsov, eds. 1977, NASA, Washington, DC, pp. 307–51.
57. Reason, J.T., and J.J. Brand, Motion sickness. 1975, Academic Press, London.
58. Alfrey, C.P., et al., Control of red blood cell mass in spaceflight. Journal of Applied Physiology, 1996. 81(1): 98–104.
59. Rowe, W.J., The *Apollo 15* space syndrome. Circulation, 1998. 97(1): 119–20.
60. Smith, R.F., et al., Quantitative electrocardiography during extended space flight: the second manned Skylab mission. Aviation, Space, and Environmental Medicine, 1976. 47(4): 353–59.
61. Smith, R.F., et al., Vectorcardiographic changes during extended space flight (M093): observations at rest and during exercise, in *Biomedical Results of Skylab*, R.S. Johnston and L.F. Dietlein, eds. 1977, NASA, Washington, DC, pp. 339–50.
62. Fritsch-Yelle, J.M., et al., An episode of ventricular tachycardia during long-duration spaceflight. American Journal of Cardiology, 1998. 81(11): 1391–92.
63. Foale, C., *Waystation to the Stars: The Story of Mir, Michael and Me*. 1999, Headline Book Publishing, London.
64. Rossum, A.C., et al., Evaluation of cardiac rhythm disturbances during extravehicular activity. American Journal of Cardiology, 1997. 79(8): 1153–55.
65. Levine, B.D., et al., Maximal exercise performance after adaptation to microgravity. Journal of Applied Physiology, 1996. 81(2): 686–94.
66. Watenpaugh, D.E., and A.R. Hargens, The cardiovascular system in microgravity, in *Handbook of Physiology*, section 4, *Environmental Physiology*, M.J. Fregly and C.M. Blatteis, eds.. 1996, Oxford University Press, New York, pp. 631–74.
67. Wieling, W., J.J. van Lieshout, and A.M. van Leeuwen, Physical manoeuvres that reduce postural hypotension in autonomic failure. Clinical Autonomic Research, 1993. 3(1): 57–65.
68. Ten Harkel, A.D., J.J. van Lieshout, and W. Wieling, Effects of leg muscle pumping and tensing on orthostatic arterial pressure: a study in normal subjects and patients with autonomic failure. Clinical Science (London), 1994. 87(5): 553–58.

69. Buckey, J.C., et al., Orthostatic intolerance after spaceflight. Journal of Applied Physiology, 1996. 81(1): 7–18.
70. Watenpaugh, D.E., et al., Effects of spaceflight on human calf hemodynamics. Journal of Applied Physiology, 2001. 90(4): 1552–58.
71. Cox, J.F., et al., Influence of microgravity on arterial baroreflex responses triggered by Valsalva's maneuver, in *The Neurolab Spacelab Mission: Neuroscience Research in Space*, J.C. Buckey and J.L. Homick, eds. 2003, NASA, Houston, TX, pp. 187–95.
72. Fritsch-Yelle, J.M., et al., Subnormal norepinephrine release relates to presyncope in astronauts after spaceflight. Journal of Applied Physiology, 1996. 81(5): 2134–41.
73. Ertl, A.C., et al., The human sympathetic nervous system response to spaceflight, in *The Neurolab Spacelab Mission: Neuroscience Research in Space*, J.C. Buckey and J.L. Homick, eds.. 2003, NASA, Houston, TX, pp. 197–202.
74. Shykoff, B.E., et al., Cardiovascular response to submaximal exercise in sustained microgravity. Journal of Applied Physiology, 1996. 81(1): 26–32.
75. Delp, M.D., Myogenic and vasoconstrictor responsiveness of skeletal muscle arterioles is diminished by hindlimb unloading. Journal of Applied Physiology, 1999. 86(4): 1178–84.
76. Convertino, V.A., Lower body negative pressure as a tool for research in aerospace physiology and military medicine. Journal of Gravitational Physiology, 2001. 8(2): 1–14.
77. Bevegard, S., J. Castenfors, and L.E. Lindblad, Effect of changes in blood volume distribution on circulatory variables and plasma renin activity in man. Acta Physiologica Scandinavica, 1977. 99(2): 237–45.
78. Kozlovskaya, I.B., A.I. Grigoriev, and V.I. Stepantzov, Countermeasure of the negative effects of weightlessness on physical systems in long-term space flights. Acta Astronautica, 1995. 36(8–12): 661–68.
79. Smith, S.M., et al., Evaluation of treadmill exercise in a lower body negative pressure chamber as a countermeasure for weightlessness-induced bone loss: a bed rest study with identical twins. Journal of Bone and Mineral Research, 2003. 18(12): 2223–30.
80. Schneider, S.M., et al., Lower-body negative-pressure exercise and bed-rest-mediated orthostatic intolerance. Medicine and Science in Sports and Exercise, 2002. 34(9): 1446–53.
81. Lamarre-Cliche, M., Drug treatment of orthostatic hypotension because of autonomic failure or neurocardiogenic syncope. American Journal of Cardiovascular Drugs, 2002. 2(1): 23–35.
82. Ramsdell, C.D., et al., Midodrine prevents orthostatic intolerance associated with simulated spaceflight. Journal of Applied Physiology, 2001. 90(6): 2245–48.
83. Vernikos, J., and V.A. Convertino, Advantages and disadvantages of fludrocortisone or saline load in preventing post-spaceflight orthostatic hypotension. Acta Astronautica, 1994. 33: 259–66.
84. Levine, B.D., Intermittent hypoxic training: fact and fancy. High Altitude Medicine and Biology, 2002. 3(2): 177–93.
85. Frey, M.A., et al., Blood and urine responses to ingesting fluids of various salt and glucose concentrations. Journal of Clinical Pharmacology, 1991. 31(10): 880–87.
86. Gaffney, F.A., et al., Hemodynamic effects of Medical Anti-Shock Trousers (MAST garment). Journal of Trauma, 1981. 21(11): 931–37.
87. Durand, S., et al., Skin surface cooling improves orthostatic tolerance in normothermic individuals. American Journal of Physiology – Regulatory Integrative and Comparative Physiology, 2004. 286(1): R199–205.
88. Masini, V., M. Rocchi, and M. Santini, Dynamic ECG in normal subjects. Giornale Italiano di Cardiologia, 1980. 10(10): 1267–79.
89. Faris, J.V., et al., Prevalence and reproducibility of exercise-induced ventricular arrhythmias during maximal exercise testing in normal men. American Journal of Cardiology, 1976. 37(4): 617–22.
90. Turner, A.S., et al., The prevalence of disturbance of cardiac rhythm in randomly selected New Zealand adults. New Zealand Medical Journal, 1981. 93(682): 253–55.

91. Kennedy, H.L., et al., Long-term follow-up of asymptomatic healthy subjects with frequent and complex ventricular ectopy. New England Journal of Medicine, 1985. 312(4): 193–97.
92. International Space Station Integrated Medical Group, ISS Medical Checklist, JSC-48522-E1. 2000, NASA, Houston, TX.
93. Lindberg, E.F., et al., Studies of cardiac output and circulatory pressures in human beings during forward acceleration. Aerospace Medicine, 1962. 33: 81–91.

8

Nutrition: Maintaining Body Mass and Preventing Disease

Introduction

After spending 4 months on *Mir* in 1995, Norm Thagard weighed 8 kg (17.5 pounds) less than when he launched. He had lost approximately 11% of his body weight [1]. When a group of 10 climbers attempted to summit Mt. Everest in 1989, they lost an average of 9.6% of their body weight [2]. A group of 66 polar explorers who spent 10 months in Antarctica maintained a stable weight throughout (although they did show an increase in body fat and decrease in muscle mass [3]). Submarine crews often have skinfold thicknesses higher than similarly aged men and must work to keep their weight from increasing [4].

Although residing on a space station resembles living in a submarine or at an Antarctic station, astronauts sometimes lose more weight than mountain climbers ascending Mt. Everest. This highlights the importance of nutritional support for successful spaceflight. Significant undernutrition can weaken muscles and reduce performance; it also increases the risk of infection. Overnutrition can lead to obesity and an increased risk for cardiovascular disease and cancer. Closely monitored nutritional support should be available throughout a space mission to prevent significant weight loss, maintain immune function, and reduce the risk for cardiovascular disease and cancer. This chapter summarizes some of the major nutritional challenges presented by spaceflight and what can be done to meet them.

Nutritional Issues of Concern for Spaceflight

Many crew members lose weight on long-duration space voyages. Overall, body mass drops by approximately 1–4 kg. Table 8-1 summarizes the weight loss data from a variety of American and European spaceflight missions. There are several possible reasons for the weight loss. While adapting to weightlessness, astronauts can lose both blood and muscle volume, which would be reflected in decreased body mass. At times, psychological factors, such as a depressed mood, may reduce appetite. Exercise, although important for the countermeasure program, can reduce body mass if energy intake is inadequate. On any given mission, some or all of these factors can cause weight loss. It is important to note, however, that not all crew members lose weight in space, and some even gain weight.

On a spaceflight, tension exists between providing a healthy diet and meeting the operational needs of the mission. To fight cancer, heart disease, and tooth decay, nutritionists advise against excessive fat and simple sugars. Instead, they recommend a diet rich in fresh fruits, fresh vegetables, whole-grain cereals, and fiber [5, 6]. On a long-duration spaceflight, however, fresh vegetables are hard to supply, and fiber increases the waste that must be processed. In addition, high-calorie foods rich in fat and simple carbohydrates take up the least space per calorie provided. This makes them desirable operationally, even if they are not the best nutritionally. Ultimately, nutrition in space involves finding the right balance between operational needs and crew health.

Calories

Caloric requirements can be calculated for an individual based on equations determined by the World Heath Organization. These equations are listed in table 8-2. For

Table 8-1. Weight loss, fluid intake and energy balance before and during spaceflight on U.S. and European space missions.

Mission	Flight duration (days)	Weight change (kg)	Fluid intake (ml/day)	Energy intake pre-flight (kcal/kg/day)	Energy intake in-flight (kcal/kg/day)	Energy expenditure in-flight (kcal/kg/day)
Apollo 16, 17	11–12	−3.63	3000	36	25	50
Skylab 2	28	−3.27	2911	42	44	45
Skylab 3	59	−3.67	2670	45	43	47
Skylab 4	84	-1.47	2954	46	44	46
Shuttle (mixed)	5–17	−1.49	2153		27	36
Shuttle, *SLS 1/2*	9–12	−1.8	2700	39	34	34
Shuttle, *LMS*	17	−2.6	1890	37	24	41
Shuttle, *D-2*	16	−2.8	1800		25	
NASA-*Mir*	90–190	−4.6		35	26	
Euro-*Mir*	29	+0.2	1100	34	24	
Euro-*Mir* 95/97	21–179	−2.25	2900		35	

Table reprinted from Wade et al. [10] with permission from Elsevier.

an average male astronaut over age 30, the daily energy requirement is 2875 kcal, 12 MJ, 41 kcal/kg, or 171 kj/kg/day. Although these equations were determined for use on Earth, total energy utilization in space is similar [7].

In rats, food consumption, body mass, and body mass gain in space are similar to a control group on the ground. Overall, spaceflight does not appear to affect the rats' energy intake or energy balance. This differs from what is seen in human spaceflight. Inadequate food intake is a frequent reason for weight loss in long-duration spaceflight [8, 9]. Often astronauts will take in less food a day than they do on the ground, even though the energy requirements are similar. Also, food intake appears to correlate negatively with the amount of exercise performed in flight. Inactive rats do not reduce their food intake in weightlessness, but active humans do [10]. The reasons for this are not totally clear.

Table 8-2. World Health Organization equations for calculating energy requirements for men and women of different ages [7].

Group	Equation
Men, 18–30 years old	kcal/day = 1.7 (15.3 W + 679)
Men, >30 years old	kcal/day = 1.7 (11.6 W + 879)
Women, 18–30 years old	kcal/day = 1.6 (14.7 W + 496)
Women, >30 years old	kcal/day = 1.6 (8.7 W + 829)

The requirements listed are for moderate activity. An additional 500 kcal/day should be added when energy requirements increase (such as for extravehicular activity or increased exercise countermeasure activity). W is body weight in kilograms. To convert kcal/day to kJ/day, multiply by 4.1868.

Possible reasons for reduced intake in space

One factor influencing reduced food intake in space may be that food generates heat. Studies in the past have shown that food intake is decreased in hot environments, or when heat removal is inefficient [11]. Even though the spacecraft temperature is maintained at a comfortable level, removal of heat from the body could be compromised in space [8]. Convective cooling, which depends on the airflow generated by less dense air rising (i.e., hot air rises, carrying heat way), does not exist in weightlessness. As a result, it is possible that astronauts may have difficulties getting rid of excess heat in space, particularly during exercise. This in turn might affect appetite and food intake, as food intake could be expected to decline if heat dissipation were compromised [12].

Exercise can also depress food intake [13]. Studies in animals and humans show that acute exercise transiently reduces food intake and induces weight loss. This may be due to blood flow being diverted from the gut during exercise. The diversion of blood flow signals the appetite center that the gut is not ready to digest food [8]. The food deficit created during exercise can be made up during a period of rest, but a crew member following an aggressive daily exercise program may not match intake to the increased metabolic need.

Another possible reason for decreased food intake could be related to another waste product of metabolism—carbon dioxide. Earth's atmosphere contains about 0.03% CO_2, whereas a spacecraft has levels around 0.3%. Therefore, the removal of CO_2 is less efficient in the spacecraft. The higher CO_2 levels can cause a problem with acid-base balance because they create a mild respiratory acidosis. This requires a metabolic compensation to keep pH constant. Metabolic acidosis can decrease albumin synthesis and create a negative nitrogen balance [14, 15]. It is not clear, however, whether a mild respiratory acidosis or temporary lactic acidosis (due to exercise) would have these effects [16]. Also, while some studies in rats have suggested that a high atmospheric CO_2 can decrease food intake [17], others have not [18]. Whether elevated CO_2 levels may play a role in reducing food intake in space is not clear. Weight loss has not been a documented problem on submarines, even though they operate at higher ambient CO_2 levels (usually in the range of 0.5–1.5%).

Minimizing Bone Loss

As discussed in chapter 1, in weightlessness calcium is mobilized from weight-bearing bones, leading to decreased parathyroid hormone levels and reduced calcium absorption. In addition, because crew members are in a closed environment with low ambient light levels, they are at risk for vitamin D deficiency. Nutritional interventions can help push the balance in bone turnover toward bone formation and away from excessive bone losses.

Exogenous vitamin D must be supplied in space to prevent deficiency. In conditions of reduced ambient light, an intake of 10 µg/day is recommended [7]. The tolerable upper level for vitamin D intake on Earth is 50 µg/day [19]. Bone loss in space is not a result of inadequate calcium intake, so a very high calcium intake in space is not warranted. Calcium intake should be at a level appropriate to prevent deficiency. For adults on Earth in the astronaut age range, the recommended dietary intake of calcium is 1000 mg/day, with the tolerable upper limit for intake set at 2500 mg/day [19]. Keeping calcium intake a little higher than recommended at 1200 mg/day (with phosphorus limited to no more than 1.5 times that of calcium) has been recommended [7]. Calcium citrate might be a superior oral supplement because it has better oral absorption [20].

One concern about calcium (and vitamin D) supplementation in space (where urinary calcium excretion is likely to be elevated) is whether the dietary calcium will aggravate hypercalciuria and increase the risk of kidney stones. Several lines of evidence suggest this will not be the case [21]. Despite aggressive oral calcium and vitamin D supplementation in patients with postmenopausal osteoporosis, kidney stone incidence has not increased in these patients. One possible reason is that calcium may bind to oxalate in the gut and reduce oxalate absorption. Because oxalate is essential for the formation of calcium oxalate stones, this oxalate binding effect may reduce the amount of oxalate absorbed into the body and therefore into the urine. To maximize the oxalate-binding effect, calcium supplements should be taken with meals.

A high sodium intake increases urinary sodium excretion, which in turn increases urinary calcium excretion. As this increased calcium excretion could aggravate ongoing bone loss, excessive sodium intake should be avoided. Current recommendations for sodium intake are between 1.5 and 3.5 g/day. Typically, however, sodium intake has been high on Shuttle flights (> 4 g/day). Crew members may need to be aware of their sodium intake [22] because it can influence bone loss.

Protein intake is also an important consideration. On one hand, metabolism of a diet high in animal protein creates an acid load that may increase the need for skeletal buffering and increase bone loss [23]. On the other hand, an active exercise countermeasure program requires an adequate protein intake. Current recommendations advise that protein constitute 12–15% of calories consumed with a ratio of animal to plant protein of 60:40 [7].

Other dietary factors also can be important. As mentioned in chapter 1, isoflavones are one class of phytoestrogens that may be effective in preserving bone mass on Earth [24]. Postmenopausal women who consumed a diet high in soy protein (which is a good source of isoflavones) showed a significant increase in lumbar spine bone density [25]. Ipriflavone, a synthetic isoflavone, has also shown effectiveness against postmenopausal osteoporosis [26]. The physiology of postmenopausal bone loss differs substantially from that of immobilization or weightlessness, so it is not known whether isoflavones would be effective in the bone loss caused by immobilization or weightlessness.

Minimizing Radiation Damage

As discussed in chapter 3, radiation produces damage by producing free radicals and other oxidizing species in tissues. Antioxidants help minimize the effects of the damage from the reactive molecules formed. They can work by reacting with the damaged molecules, by chemically repairing them, or by reacting with intermediates before they damage key biological molecules [27]. Several nutritional compounds are antioxidants, such as vitamin A (and related retinoids and carotenoids such as beta-carotene), vitamin C (ascorbic acid), vitamin E (D-alpha-tocopherol), alpha-lipoic acid, niacin (vitamin B_3), thiamin (vitamin B_1), folic acid, selenium, and the amino acid cysteine. Plant-based anthocyanins (found in blueberries and strawberries) have powerful antioxidant properties. Also, compounds such as glutathione, *N*-acetylcysteine, and co-enzyme Q10 are antioxidants [28, 29].

Antioxidants Potentially Useful for Radiation Protection

Most of the evidence for the use of antioxidant compounds for radiation protection has been developed in animal studies. Cysteine is an amino acid containing a sulfur atom in a thiol group. This was the first compound demonstrated to provide rats with protection against ionizing radiation [27]. At doses effective for radiation protection, however, cysteine is toxic [30], so it would not be a useful nutritional intervention. Nonetheless, cysteine is part of a group of compounds called thiols or phosphorothioates that can be administered as radioprotective agents. For example, cysteine is a component of both the tripeptide glutathione and the drug *N*-acetyl-cysteine. Glutathione is an antioxidant and is important in several detoxification reactions [31]. *N*-acetyl-l-cysteine (NAC) is one of the least toxic thiols and has been used in clinical medicine as an antidote for acetaminophen poisoning. Both glutathione and NAC are effective in reducing the development of cancer and are considered promising cancer preventive agents [29]. Alpha-lipoic acid is a thiol-containing compound that has been shown to be useful in preventing oxidative stress as a consequence of radiation exposure, UV radiation, or chronic disease [32–34].

As mentioned earlier, vitamin E, vitamin C, vitamin A (and related retinols and carotenoids), folic acid, niacin (vitamin B_3), and thiamin (vitamin B_1) are all antioxidants [35]. Animal studies have shown the vitamins C and E can reduce DNA damage and increase cell survival after radiation [35]. In humans, increased vitamin C and E intake has been associated with a reduced risk of cancer development [29]. A combination of lipoic acid, vitamin C, and vitamin E protected against lens damage produced by low-level radiation in astronauts [36]. Folic acid supplementation may reduce cancer risk [37]. The optimal dosing for these vitamins to provide radiation protection, however, has not been determined. Also, providing supplementation beyond current ground-based guidelines has not been shown to provide extra benefit. In studies on Earth where antioxidant supplementation is given to adequately nourished populations, little effect on biomarkers of DNA damage could be seen [38], suggesting that DNA damage was not further reduced by the extra antioxidants. In space, where the baseline radiation level is higher, the effect might be different because astronauts may need higher levels of dietary antioxidants to prevent damage. Definitive evidence is lacking, however. Fortunately, these compounds do have the benefit of low toxicity, and they can be taken orally and dosed daily. Suggested recommendations for these vitamins appears in table 8-3.

Certain trace elements are also important for radiation protection and reducing cancer risk. For example, in men, low selenium levels have been linked to cancer. Controlled trials have shown that providing selenium can reduce cancer incidence in men [39]. This may be due to the antioxidant effect of selenium. For astronauts, however, no optimal dose or serum level of selenium has been established. Iron, while important for hemoglobin and other metabolic processes, increases oxidative damage. Excessive iron supplementation should be avoided in space; just enough

Table 8-3. Suggested dietary intakes for various antioxidant vitamins and trace elements [7].

Vitamin	Recommendation	Trace element	Recommendation
Vitamin A	1000 µg retinol equivalents/day	Iron	10 µg/day
Vitamin B$_1$ (thiamin)	1.5 mg/day	Manganese	2–5 mg/day
Vitamin B$_3$ (niacin)	1 mg niacin/day	Zinc	15 mg/day
Vitamin C	100 mg/day	Copper	1.5–3 mg/day
Vitamin E	20 mg/day	Chromium	100–200 µg/day
Folate	400 µg/day	Selenium	70 µg/day
		Iodine	150 µg/day
		Fluoride	4 mg/day

iron to prevent deficiency should be provided. Iron stores increase in space, so little additional iron should be needed, particularly in men [40]. Recommendations for trace elements are also provided in table 8-3.

Beta-carotene is an antioxidant that is part of a larger family of retinoids and carotenoids, a family that includes compounds such as all trans-retinoic acid, lutein, and lycopene. Beta-carotene has been shown to reduce radiation-induced apoptosis in cells [41], suggesting that its antioxidant properties would be helpful in radiation protection and perhaps in preventing cancer. Beta-carotene has been studied in large-scale intervention trails to determine if it could be useful in cancer prevention. Contrary to expectations, beta-carotene supplementation in patients at risk for lung cancer increased cancer incidence [42]. These data suggest that the relationship among antioxidants, radiation protection, and cancer prevention may be complex. Other nutritional antioxidants also have been shown to protect against radiation. Blueberry and strawberry extract, which contains anthocyanins thought to be antioxidants, minimized the effects of heavy-ion irradiation in rats [29].

Although there are data from cell culture and animal studies, there are very few epidemiological studies in humans demonstrating how a particular antioxidant (or mixture of antioxidants) might reduce cell damage or cancer risk due to a long-duration, low-level radiation exposure such as would occur on a space mission. The fact that in some cancer prevention studies antioxidant supplementation increased cancer incidence makes it clear that antioxidants are not necessarily benign. Preventing a deficiency of antioxidant compounds during a space mission clearly seems desirable, but not enough data exist to recommend high-dose antioxidant supplementation. Also, because there are so many different antioxidants, considerable work would be required to determine which antioxidants (or mixture of antioxidants) would be best in any given situation.

In the absence of definitive data to guide the use of antioxidants, one approach is to monitor antioxidant status. Markers of oxidant stress can be measured in urine. Various tests exist to measure DNA damage. Levels of different antioxidant compounds can be measured in blood. To do this would require a significant analytic capability onboard the spacecraft, along with supplies to support the testing. Developments in nanotechnology [43] and in mass spectrometry could lead to compact devices that could measure a wide variety of parameters on very small samples.

Preventing Cancer

As mentioned above, dietary antioxidants may be important minimizing long-term radiation damage. Many studies have shown the benefit of dietary antioxidants in reducing DNA damage and reducing cancer risk in populations who are deficient. Other dietary factors also are important in preventing cancer. Significant research shows that diets low in fat and high in fiber, fruits, vegetables, and grain products are associated with reduced risks for many cancers (such as colon, prostate, and breast) [44, 45]. Japanese atomic-bomb survivors who ate green-yellow vegetables and fruits at least two to four times each week had a significantly reduced bladder cancer risk compared with those who ate them once per week or less [46]. The reasons for these results are not completely understood (it is probably a combination of antioxidant and immune-modulating influences), but the epidemiological data clearly show the effect.

Certain nutrients can affect immune system function, and this in turn can influence cancer risk. The relationship between the immune system and cancer is complex. Some dietary factors may work by promoting aspects of immune function, others by suppressing excessive or unwanted immune responses. Zinc deficiency, for example, depresses immune function and increases cancer risk. There is no evidence, however, that once the deficiency is corrected, additional zinc supplementation will decrease cancer risk further [47].

Other dietary factors seem to reduce cytokines that promote inflammation. A growing body of evidence suggests that inflammation and proinflammatory cytokines can promote cancer [48]. Plant polyphenols are one group of compounds that seem to have anti-inflammatory properties. Epigallocatechin-3-gallate is the major component of green tea, and it appears to have important anti-inflammatory and cancer chemopreventive properties [48]. Some dietary agents may work by inhibiting proteolysis. Proteolysis has been shown to increase during the progression of some cancers, suggesting that inhibiting proteolysis might help prevent cancer [49]. The soybean-derived Bowman-Birk inhibitor (a protease inhibitor that inhibits proteolysis) has been shown to be effective in preventing the development of cancer after irradiation [29]. In animal studies, this inhibitor has prevented the initiation, promotion, and progression of carcinogenesis [29]. Another nutritional compound that may have cancer protective properties is flaxseed. Flaxseed is a source of lignans, which is one of the group of phytoestrogens believed to prevent some cancers [39].

Another important class of nutrients related to carcinogenesis is the polyunsaturated fatty acids (PUFAs). Two important types of PUFAs are the omega-6 fatty acids (n-6) and the omega-3 fatty acids (n-3). These fatty acids differ slightly chemically. Some vegetable oils (such as safflower, soy, corn, and sunflower) and animal fats (most meat and some fish) contain high levels of omega-6 PUFAs. Other vegetable oils (such as linseed, rapeseed, and walnut oil) and certain fish (salmon, tuna, and herring) contain high levels of omega-3 PUFAs. Studies have shown that diets containing a high ratio of omega-3 PUFAs to omega-6 PUFAs are associated with a low cancer incidence, possibility due to the ability of the omega-3 PUFA to reduce inflammatory responses [48].

The clearest data for cancer prevention, however, are the epidemiological data showing that a high intake of fresh fruits, fresh vegetables, and fiber is protective.

Cruciferous vegetables (broccoli, cauliflower, cabbage, Brussels sprouts, bok choy, kale) are thought to be particularly worthwhile for cancer prevention. The chemical basis for this effect (i.e., the particular chemical compounds that provide the protection) is not clearly established. There are two ways to offer the cancer-preventive effects of this diet to astronauts. One is to provide fresh or frozen food on the spacecraft. The other approach is to find the specific plant compounds that offer cancer protection through research and provide them in supplements onboard. In the long run, the ability to grow food onboard will become important for very long missions. This would offer the most autonomy and flexibility and would be essential for planetary exploration. Growing food on spacecraft, however, is very complex. If food grown onboard became a major source of calories and the growing system broke down, the crew would be at risk for starvation. Stored food with a long shelf life, while less healthy, is more reliable.

Preventing Cardiovascular Disease

Nutrition is an important component of the cardiovascular prevention program. (The prevention of cardiovascular disease is also discussed in chapter 12.) A low cholesterol intake helps keep blood cholesterol levels low, thereby lowering the risk of cardiovascular disease. High-cholesterol foods include dairy fats, egg yolks, and organ meats (such as liver). A diet high in saturated fat also raises blood cholesterol. Foods such as cheese, whole milk, cream, butter, regular ice cream (i.e., high-fat dairy products), fatty meats, processed meats, poultry skin, lard, palm oil, and coconut oil are all high in saturated fat. Unsaturated fats and oils, however, do not raise blood cholesterol. These fats occur in vegetable oils, most nuts, olives, avocados, and fatty fishes (salmon, herring, tuna). Unsaturated oils can be either monounsaturated or polyunsaturated. Olive, canola, sunflower, and peanut oils are some of the oils high in monounsaturated fats. Vegetable oils such as soybean oil, corn oil, and cottonseed oil are polyunsaturated. Many kinds of nuts are also good sources of polyunsaturated fats [5].

Polyunsaturated vegetable oils are used extensively in processed foods. To make the liquid vegetable oils solid, they are treated using a process called hydrogenation. Through this process trans-fatty acids (also known as trans-fats) are produced in the food. This process improves shelf life and flavor stability, but foods high in trans-fatty acids can raise blood cholesterol. In other words, the hydrogenation process removes the health benefit that the unsaturated oils offer. Many hard margarines, shortenings, and packaged foods (such as cookies and snacks) contain partially hydrogenated oils, which are high in trans-fatty acids. Because food for a space mission must have a long shelf life, there may be a trade-off for some foods between shelf life and the content of partially hydrogenated oils.

Although trans-fatty acids are undesirable, other fatty acids may help prevent cardiovascular disease. Omega-3 fatty acids, for example, may be beneficial for cardiovascular health [50]. They may reduce inflammation, which might offer protection against heart disease. Some fish, such as salmon, tuna, and mackerel, contain omega-3 fatty acids. Alpha linoleic acid is an omega-3 fatty acid that comes from vegetable sources. Although the data are not conclusive about omega-3 fatty acids, a reasonable approach is to ensure that these fatty acids are well represented in the diet [51].

Table 8-4. Suggested dietary guidelines for a sensible diet [5].

Food	Recommendation
Fats and oils	Choose vegetable oils rather than solid fats; avoid partially hydrogenated vegetable oils
Meat, poultry, fish, shellfish, eggs, beans, nuts	Choose 2–3 servings of fish, shellfish, lean poultry, lean meat, beans or nuts daily
	Trim fat, take skin off poultry
	Choose beans, peas, lentils often
	Limit intake of processed meats (bacon, sausages, salami, bologna)
	Limit intake of liver and organ meats
	Use egg yolks and whole eggs in moderation
Dairy products	Use fat-free or low-fat milk
Carbohydrates	Choose whole-grain products
Miscellaneous	Foods in high in omega-3 fatty acids may be beneficial
	Flaxseed may be a useful nutrient (high in omega-3 fatty acids and lignans)

Several epidemiological studies have shown that a "Mediterranean diet" might be the best approach to cardiovascular disease prevention. This diet is mostly vegetarian and is rich in olive oil, omega-3 fatty acids, fiber, B-group vitamins, and various antioxidants. Hu and Willett [50] concluded that a diet in which (1) nonhydrogenated, unsaturated fats are the predominant form of fat, (2) whole grains are the major source of carbohydrate, and (3) fruits and vegetables are well represented, offers considerable protection from cardiovascular disease. Table 8-4 summarizes the components of a sensible diet to promote cardiovascular health. These recommendations also support the goal of cancer prevention.

On a long-duration space mission, the need to prevent cardiovascular disease needs to be put into context. Coronary artery disease has not been a frequent problem in the space program to date, but weight loss has. If a crew member is losing weight, the primary goal is to uncover the cause (excessive exercise, stress, etc.) and increase caloric intake. Whether the foods used to maintain or increase weight in this setting are the best for cardiovascular health would be a secondary consideration. If the crew members are gaining weight and increasing their body fat, then close attention should be paid to the guidelines outlined in table 8-4. Also, dietary intake should be individualized and adjusted to the needs of particular crew members.

Preventing Dental Problems

Diet and nutrition play an important role in preventing dental caries (cavities) and periodontal disease. Dental plaque contains bacteria. Sugars and other fermentable carbohydrates in the diet are metabolized by the plaque bacteria into acids. When the pH of plaque drops below a critical value (approximately 5.5), enamel dissolution can occur. This enamel dissolution starts dental caries. A strong relationship exists between the consumption of fermentable carbohydrates (dietary sugars and starches) and caries [6]. The two primary bacteria involved in forming caries are *Streptococ-*

cus mutans and *Lactobacillus casei*. If an inflammatory response begins in the gums against bacterial products in the plaque, this is called periodontal disease.

Table 8-5 summarizes the cavity-forming potential of various carbohydrates and other foods. Sugars that stay in the mouth for a long time (e.g., hard candies, mints, a sipped sugary beverage) are worse than those with a short contact time. Similarly, those starches with a long retention time in the mouth (e.g., cream sandwich cookies, potato chips) are of more concern than those that are cleared quickly. Foods that are high in calcium, phosphate, and protein may favor remineralization of the tooth. Processed cheese, for example, has been shown to prevent cavities. Highly acidic foods, however, may promote tooth decay, particularly if they stay in the mouth for a long time. Large doses of chewable vitamin C can decrease pH and promote tooth erosion [52].

Several factors can modify the risk of dental caries. An acidic salivary pH or a low salivary flow increases caries risk. Fluoride can decrease the pH needed to dissolve enamel and so is protective. Antibacterial rinses (usually containing chlorhexidine) can reduce the population of streptococci and other bacteria and so protect against caries. A diet low in fermentable sugars, but high in calcium-containing foods (e.g., cheese), can help prevent cavities. Keeping the teeth clean and free of plaque is obviously an important step in avoiding tooth decay. Some studies have suggested that polyphenols (such as the tannins in cocoa, coffee, tea, and fruit juices) may reduce the potential of other foods to produce cavities. Also, sugar-free gums that contain polyols like sorbitol, xylitol, and mannitol may stimulate salivary flow and increase the clearance of sugars from the teeth. Xylitol in particular has been shown to be helpful in increasing salivary flow, increasing pH, and enhancing tooth mineralization. Chewing these gums after meals may assist in cleaning the teeth and preventing cavities [52].

The main dietary recommendations to control dental cavities are to (1) choose a balanced diet rich in whole grains, fruit, and vegetables (and low in simple sugars), (2) eat a combination of foods (i.e., combine diary products with fermentable carbohydrates) to modulate the effect of the carbohydrates, (3) rinse the mouth with water and chew sugarless gum after the consumption of fermentable carbohydrates, and (4) drink, rather than sip, sweetened and acidic beverages.

Food: Operations versus Health

From an operational standpoint, the best food system is one that is simple, compact, uses little power, and generates minimal waste. Dense, high-fat, high-calorie foods that have a long shelf life and take up as little space as possible offer significant advantages. From the health and morale perspective, however, this approach can create problems. On a long-duration mission, food could become very important. Not only is good nutrition essential for health, but good food is also critical for morale. On Earth the special foods associated with holidays create anticipation, and food is often a critical component of celebrations and events. On a long-duration mission, it is likely that crew members will look forward to mealtimes. If these meals become monotonous and unsatisfying, morale could suffer. As noted in chapter 2, a good psychological environment is absolutely critical for successful long-duration spaceflights.

Table 8-5. Caries-promoting activity of various food sources.

Category	Chemical structure	Examples	Caries-promoting potential	Food sources
Sugars	Monosaccharide	Glucose, dextrose, fructose	Yes	Most foods, fruit, honey
		High-fructose corn syrup	Yes	Soft drinks
		Galactose	No	Milk
	Disaccharide	Sucrose	Yes	Fruit, vegetables, table sugar
		Turbinado, molasses	Yes	
		Lactose	Yes	Milk
		Maltose	Yes	Beer
Other carbohydrates	Polysaccharide	Starch	Yes	Potatoes, grains, rice, legumes, bananas, cornstarch
	Fiber	Cellulose, pectin, gums, beta-glucans, fructans	No	Grains, fruits, vegetables
	Polyol-monosaccharide	Sorbitol, mannitol, xylitol, erythritol	No	Fruit, seaweed, exudates of plants or trees
	Polyol-disaccharide	Lactitol, isomalt, maltitol	No	Derived from lactose, maltose or starch
	Polyol-polysaccharide	Hydrogenated starch, hydrolysates, or malitol syrup	No	Derived from monosaccarides
High-intensity sweeteners	Saccharin	Sweet and Low	No	
	Aspartame	Nutrasweet, Equal	No	
	Aceulfame-K	Sunett	No	
	Sucralose	Splenda	No	
Fat replacers made from carbohydrates		Carrageenan, cellulose gel/gum, corn syrup solids, dextrin, maltodextrin, guar gum, hydrolyzed corn starch, modified food starch, pectin, polydextrose, sugar beet fiber, xanthan gum	Unknown	Baked goods, cheese, chewing gum, salad dressing, candy, frozen desserts, pudding, sauces, sour cream, yogurt, meat-based products

Reproduced from Touger-Decker and van Loveren [52], with permission of the *American Journal of Clinical Nutrition*.

On short duration missions, a food system based on freeze-dried or thermostabilized foods clearly makes sense. These foods are compact and easy to prepare. For longer missions, however, three questions will have to be addressed: (1) should the crew have the ability to "cook," (2) how much food should be grown onboard, and (3) can frozen food be provided? Packaged, prepared foods are convenient, but they severely limit creativity. If, instead, the crew had access to certain basic ingredients (such as flour, sugar, beans, rice, spices, etc.), they could prepare a large variety of different meals. This would present a series of engineering challenges because the rudiments of a "space kitchen" would be needed, along with ways to solve issues such as the containment of particles, clean-up, and food safety. On a long-duration mission, however, this ability might be critical for maintaining morale and good performance. Crews on interplanetary flights would likely have time to devote to food preparation. A hybrid system that provides certain basic foods in packages with a long shelf life, but that also allows for different combinations and preparation methods, could help reduce monotony.

Growing food onboard presents another major engineering challenge. A considerable body of research shows that various foods could be grown in small areas onboard a spacecraft [53]. This would allow for fresh vegetables on the spacecraft and provide some autonomy for the crew. This approach adds risk, however, because a complex food production system could fail in a variety of ways, leaving the crew short of food. In contrast, stored, packaged food is unlikely to spoil or degrade. A system that combines stored food with food grown onboard might be the best approach. In this scenario, the crew would have the ability to grow certain crops onboard that could supplement the diet. The bulk of their food, however, would be in a form that could last for the duration of the mission when stored at room temperature.

Submarines have large freezers onboard that allow for a varied and interesting diet. In fact, submarine crews often gain weight. Freezers are typically reliable, but submarines can surface and be resupplied when problems arise. On a long-duration, interplanetary flight, the loss of freezer could be catastrophic if a significant portion of the crew's food were stored there. Nevertheless, even though providing capabilities to prepare, grow, and freeze food would increase cost, weight, and engineering complexity, it might be a wise investment in the long run.

Miscellaneous Dietary Factors

Water systems on spacecraft must include water sterilization. Iodine has been used for this purpose in some water systems. Although this is an effective approach, it is possible for excessive iodine to get into the water supply [22]. Rarely, excessive iodine intake can lead to increased thyroid hormone production. In general, it will be important to monitor iodine and other compounds that are placed into the water supply.

Vitamin K, while important for various clotting factors, also plays a role in bone metabolism. It is needed for the carboxylation of osteocalcin, an important hormone for bone formation. McCormick [7] recommends a vitamin K intake of 80 μg/day of vitamin K_1 (phylloquinone) in space.

Recommendations Based on Current Knowledge

For short-duration spaceflights, the main concern is providing a balanced diet, with adequate calories, in the most efficient way possible. For longer space missions, the details of nutrition become more important. Antioxidant status and the provision of micronutrients need careful attention. Also, as missions lengthen from months to years, the crew members' diets should not increase their risk of cancer or cardiovascular disease. Finally, the psychological importance of food should not be underestimated. Food is a critical part of many social rituals on Earth. In an isolated and confined environment, a limited selection of foods, and restricted ways to prepare those foods, could add to psychological stress.

Also, nutrition needs to be individualized. A crew member losing weight should not be worried about the fat content of the food. On the other hand, a crew member gaining weight should be. A crew member deficient in zinc, selenium, or antioxidant vitamins may be running a greater risk of cancer. Maintaining crews effectively will require a varied food system that includes the ability to monitor nutritional parameters. To provide adequate nutrition, the following recommendations can be made:

1. Body mass needs to be monitored closely and interventions made if a crew member is losing weight. The reasons for weight loss can vary, but it is essential that the loss be detected early, diagnosed properly, and correctly quickly.

2. Crew members need adequate amounts of calcium and vitamin D to prevent deficiency. In addition, crew members should avoid excessive sodium and animal protein, which can aggravate bone loss.

3. Because radiation levels are higher in space than on Earth, it is essential that crew members do not become deficient in antioxidant compounds. Whether to supplement beyond the usual recommended daily allowances has not been clearly established, but antioxidant vitamins should be provided as part of a vitamin supplement. The ability to grow some vegetables onboard also would offer a natural source of various antioxidant compounds.

4. The diet should provide adequate zinc and selenium to prevent deficiency because inadequate amounts of these micronutrients increases cancer risk. Not enough data exist to recommend any particular set of plant-derived chemicals that would definitively reduce cancer risk. If some food were grown onboard, however, it would add variety to the diet and would likely have health benefits.

5. Cardiovascular health is important because cardiovascular disease is the leading cause of death in the astronaut age range. If crew members are maintaining a stable weight, they should avoid saturated fat and hydrogenated oils. Foods that are high in omega-3 fatty acids should be provided. Adequate fiber should be provided either through the diet or with a supplement.

6. Because there will be individual differences in nutritional needs, and because the optimal nutritional intake for each crew member cannot be known precisely, the ability to monitor various biochemical factors could be important. The onboard capability to measure markers of oxidative stress in urine and blood might provide some data that could be used to guide dietary recommendations. Ferritin levels would give a sense of whether extra iron is needed. The ability to measure zinc and selenium levels would confirm that the crew members are not deficient. Markers of protein breakdown can be detected in the urine. Compact

mass spectrometers that can measure a wide variety of nutritionally important molecules are available and could provide important feedback on the nutritional program.

7. Crew members will need to pay attention to dental health. The teeth should be kept free of plaque. Drinking tea can help prevent caries, and chewing xylitol-containing gum can help clean the teeth after sugar-containing meals.

8. For the future, a versatile food system that can accommodate food preparation (i.e., making meals from basic ingredients) and that has some component of grown and frozen food would offer the best balance between operational, nutritional, and psychological needs.

References

1. Ash, J., Rocket man, in Florida State University's Research in Review, 1997. http://www.research.fsu.edu/researchr/fallwinter97/features/rocketman.html.
2. Reynolds, R.D., et al., Intakes of high fat and high carbohydrate foods by humans increased with exposure to increasing altitude during an expedition to Mt. Everest. Journal of Nutrition, 1998. 128(1): 50–55.
3. Belkin, V., and D. Karasik, Anthropometric characteristics of men in Antarctica. International Journal of Circumpolar Health, 1999. 58(3): 152–69.
4. Tappan, D.V., et al., Cardiovascular risk factors in submariners. Undersea Biomedical Research, 1979. 6 (Suppl): S201–15.
5. USDA, *Dietary Guidelines for Americans*. 2000, USDA:U.S. Department of Agriculture, Washington, D.C.
6. Touger-Decker, R., and C.C. Mobley, Position of the American Dietetic Association: Oral health and nutrition. Journal of the American Dietetic Association, 2003. 103(5): 615–25.
7. McCormick, D.B., Nutritional recommendations for spaceflight, in *Nutrition in Spaceflight and Weightlessness Models*, H.W. Lane and D.A. Schoeller, eds. 2000, CRC Press, Boca Raton, FL, pp. 253–74.
8. Stein, T.P., The relationship between dietary intake, exercise, energy balance and the space craft environment. Pflügers Archives, 2000. 441(2–3 Suppl): R21–31.
9. Stein, T.P., Nutrition and muscle loss in humans during spaceflight. Advances in Space Biology and Medicine, 1999. 7: 49–97.
10. Wade, C.E., et al., Body mass, energy intake, and water consumption of rats and humans during space flight. Nutrition, 2002. 18(10): 829–36.
11. Hamilton, C.L., Food and temperature, in *Handbook of Physiology*, section 6, *Alimentary Canal*, C.F. Code, ed. 1967, American Physiological Society, Washington, DC, pp. 303–18.
12. Marriott, B.M., ed. *Nutritional Needs in Hot Environments: Applications for Military Personnel in Field Operations*. 1993, National Academies Press, Washington, DC. 392.
13. LeMagnen, J., Regulation of body energy balance and body weight, in *Neurobiology of Feeding and Nutrition*. 1992, Academic Press, San Diego, CA, pp. 258–90.
14. Mitch, W.E., and A.L. Goldberg, Mechanisms of muscle wasting. The role of the ubiquitin proteasome pathway. New England Journal of Medicine, 1996. 335(25): 1897–905.
15. Mitch, W.E., and S.R. Price, Mechanisms activating proteolysis to cause muscle atrophy in catabolic conditions. Journal of Renal Nutrition, 2003. 13(2): 149–52.
16. Ballmer, P.E., and R. Imoberdorf, Influence of acidosis on protein metabolism. Nutrition, 1995. 11(5): 462–68; [discussion p. 470].
17. Viswanathan, K.R., M.S. Swamy, and N.N. Prasad, Effect of increased level of CO_2 exposure in a closed environment on calcium and phosphorus balance in rats. Indian Journal of Experimental Biology, 1989. 27(2): 151–55.
18. Wade, C.E., et al., Rat growth, body composition, and renal function during 30 days increased ambient CO_2 exposure. Aviation, Space, and Environmental Medicine, 2000. 71(6): 599–609.

19. Weaver, C.M., A. LeBlanc, and S.M. Smith, Calcium and related nutrients in bone metabolism, in *Nutrition in Spaceflight and Weightlessness Models*, H.W. Lane and D.A. Schoeller, eds. 2000, CRC Press, Boca Raton, FL, pp. 179–96.
20. Heller, H.J., et al., Pharmacokinetics of calcium absorption from two commercial calcium supplements. Journal of Clinical Pharmacology, 1999. 39(11): 1151–54.
21. Heller, H.J., The role of calcium in the prevention of kidney stones. Journal of the American College of Nutrition, 1999. 18(5 Suppl): 373S-78S.
22. Volpe, S.L., J.C. King, and S.P. Coburn, Micronutrients: trace elements and B vitamins, in *Nutrition in Spaceflight and Weightlessness Models*, H.W. Lane and D.A. Schoeller, eds. 2000, CRC Press, Boca Raton, FL, pp. 213–32.
23. Sebastian, A., et al., Improved mineral balance and skeletal metabolism in postmenopausal women treated with potassium bicarbonate. New England Journal of Medicine, 1994. 330(25): 1776–81.
24. Tham, D.M., C.D. Gardner, and W.L. Haskell, Clinical review 97: Potential health benefits of dietary phytoestrogens: a review of the clinical, epidemiological, and mechanistic evidence. Journal of Clinical Endocrinology and Metabolism, 1998. 83(7): 2223–35.
25. Potter, S.M., et al., Soy protein and isoflavones: their effects on blood lipids and bone density in postmenopausal women. American Journal of Clinical Nutrition, 1998. 68(6 Suppl): 1375S-79S.
26. Gennari, C., et al., Effect of ipriflavone—a synthetic derivative of natural isoflavones—on bone mass loss in the early years after menopause. Menopause, 1998. 5(1): 9–15.
27. Weiss, J.F., and M.R. Landauer, Radioprotection by antioxidants. Annals of the New York Academy of Sciences, 2000. 899: 44–60.
28. Kennedy, A.R., Prevention of carcinogenesis by protease inhibitors. Cancer Research, 1994. 54(7 Suppl): 1999s-2005s.
29. Kennedy, A.R., and P. Todd, Biological countermeasures in space radiation health. Gravity and Space Biology Bulletin, 2003. 16(2): 37–44.
30. Roberts, J.C., et al., Thiazolidine prodrugs of cysteamine and cysteine as radioprotective agents. Radiation Research, 1995. 143(2): 203–13.
31. Hospers, G.A., E.A. Eisenhauer, and E.G. de Vries, The sulfhydryl containing compounds WR-2721 and glutathione as radio- and chemoprotective agents. A review, indications for use and prospects. British Journal of Cancer, 1999. 80(5–6): 629–38.
32. Cudkowicz, G., and J. Franceschini, alpha-Lipoic acid and chemical protection against ionizing radiation. Archives Internationales de Pharmacodynamie et de Therapie, 1959. 122: 312–17.
33. Packer, L., E.H. Witt, and H.J. Tritschler, alpha-Lipoic acid as a biological antioxidant. Free Radical Biology and Medicine, 1995. 19(2): 227–50.
34. Beitner, H., Randomized, placebo-controlled, double blind study on the clinical efficacy of a cream containing 5% alpha-lipoic acid related to photoageing of facial skin. British Journal of Dermatology, 2003. 149(4): 841–49.
35. Pence, B.C., and T.C. Yang, Antioxidants: radiation and stress, in *Nutrition in Spaceflight and Weightlessness Models*, H.W. Lane and D.A. Schoeller, eds. 2000, CRC Press, Boca Raton, FL, pp. 233–51.
36. Bantseev, V., et al., Antioxidants and cataract: (cataract induction in space environment and application to terrestrial aging cataract). Biochemistry and Molecular Biology International, 1997. 42(6): 1189–97.
37. Strohle, A., M. Wolters, and A. Hahn, Folic acid and colorectal cancer prevention: molecular mechanisms and epidemiological evidence. International Journal of Oncology, 2005. 26(6): p. 1449–64.
38. Choi, S.W., et al., Vitamins C and E: acute interactive effects on biomarkers of antioxidant defence and oxidative stress. Mutation Research, 2004. 551(1–2): 109–17.
39. Collins, A.R., and L.R. Ferguson, Nutrition and carcinogenesis. Mutation Research, 2004. 551(1–2): 1–8.
40. Smith, S.M., Red blood cell and iron metabolism during space flight. Nutrition, 2002. 18(10): 864–66.

41. Ortmann, E.K., et al., Effect of antioxidant vitamins on radiation-induced apoptosis in cells of a human lymphoblastic cell line. Radiation Research, 2004. 161(1): 48–55.
42. Omenn, G.S., Chemoprevention of lung cancer: the rise and demise of beta-carotene. Annual Review of Public Health, 1998. 19: 73–99.
43. Xie, J., et al., Surface micromachined electrostatically actuated micro peristaltic pump. Lab on a Chip, 2004. 4(5): 495–501.
44. Key, T.J., et al., Diet, nutrition and the prevention of cancer. Public Health and Nutrition, 2004. 7(1A): 187–200.
45. Thomson, C.A., et al., Nutrition and diet in the development of gastrointestinal cancer. Current Oncology Reports, 2003. 5(3): 192–202.
46. Turner, N.D., et al., Opportunities for nutritional amelioration of radiation-induced cellular damage. Nutrition, 2002. 18(10): 904–12.
47. Ferguson, L.R., M. Philpott, and N. Karunasinghe, Dietary cancer and prevention using antimutagens. Toxicology, 2004. 198(1–3): 147–59.
48. Philpott, M., and L.R. Ferguson, Immunonutrition and cancer. Mutation Research, 2004. 551(1–2): 29–42.
49. Lippman, S.M., and L.M. Matrisian, Protease inhibitors in oral carcinogenesis and chemoprevention. Clinical Cancer Research, 2000. 6(12): 4599–603.
50. Carrero, J.J., et al., Cardiovascular effects of milk enriched with omega-3 polyunsaturated fatty acids, oleic acid, folic acid, and vitamins E and B6 in volunteers with mild hyperlipidemia. Nutrition, 2004. 20(6): 521–27.
51. Hu, F.B., and W.C. Willett, Optimal diets for prevention of coronary heart disease. Jama-Journal of the American Medical Association, 2002. 288(20): 2569–78.
52. Touger-Decker, R., and C. van Loveren, Sugars and dental caries. American Journal of Clinical Nutrition, 2003. 78(4): 881S-892S.
53. Wheeler, R.M., Bioregenerative life support and nutritional implications for planetary exploration, in *Nutrition in Spaceflight and Weightlessness Models*, H.W. Lane and D.A. Schoeller, eds. 2000, CRC Press, Boca Raton, FL, pp. 41–67.

9

Motion Sickness in Space: Prevention and Treatment

Introduction

Motion sickness can spoil a weekend outing or turn a pleasure cruise into an ordeal, but in most cases it is just a nuisance. In some settings, however, motion sickness can be hazardous. During World War II, troops involved in amphibious landings in high seas and other naval operations suffered an alarmingly high rate of incapacitation due to motion sickness. This spurred intensive research into motion sickness remedies. But even effective remedies can create problems. On D-Day, soldiers on landing boats were issued scopolamine to prevent motion sickness. Some took twice the recommended dose, leading an observer in one area to note, "by noon all the Army personnel were in a drugged stupor" [1, p. 18].

"Motion sickness" refers to the constellation of symptoms (e.g., nausea, vomiting, drowsiness, cold sweat) that can occur in novel motion environments. In space, motion sickness can lead to significant problems. A severely motion sick crew member may not function effectively. Also, vomiting can come on rapidly and unexpectedly. If vomiting occurred in an extravehicular activity (EVA) suit, the results could be disastrous. EVAs are not scheduled during the first 3 days of a flight just to avoid problems with motion sickness [2]. Most motion sickness remedies are sedatives. Too much motion sickness medication can lead to drowsiness, errors, and poor judgment—all undesirable in an operational setting.

On a space mission, motion sickness is usually not a chronic problem. It occurs mainly when the crews are adapting to a new gravitational environment. The first few days of a space mission and the first few days after landing are the peak periods for motion sickness. Motion sickness occurring outside of those times is rare. Nevertheless, the initial entry in orbit and the period after landing are important operationally. To maintain effectiveness, crew members may want to use a medication to prevent or minimize motion sickness. This chapter reviews the physiology of motion sickness and offers strategies to prevent it while minimizing side effects.

The Physiology of Motion Sickness

Many factors can influence motion sickness. While passengers in an airplane or car may become sick, the operator rarely does, even though the operator is in the same motion environment. In flight training, it is well known that giving the student control of the plane may eliminate motion sickness symptoms. These everyday observations show that motion alone often is not sufficient to produce sickness. Other factors can modify susceptibility.

Most people can adapt to motion and thereby lose their motion sickness symptoms, but some people develop a conditioned response. For example, individuals who have become motion sick in a particular setting, say, in an airplane, sometimes get motion sickness symptoms just by climbing back into the airplane and experiencing the smell of the cockpit. Once conditioned, they become motion sick before experiencing any motion. This demonstrates the powerful role that cognitive and psychological factors can play in motion sickness.

Although motion sickness is typically associated with boats, cars, planes, and other devices that move, it is also possible to get motion sickness symptoms without any real motion taking place. Virtual reality simulators, IMAX movies, and other stimuli that produce the sensation of motion by visual stimulation can also produce motion sickness.

The situations that produce motion sickness, and the symptoms that result, are well described, but the actual mechanisms underlying motion sickness remain poorly understood. The best theory to explain motion sickness, the sensory conflict theory, is useful, but it does not provide a mechanistic understanding [3]. Also, several different neurotransmitters and pathways are involved in motion sickness, which makes drug therapy a challenge. Nevertheless, there are some basic facts about motion sickness that are useful for choosing prevention strategies.

Origin of Motion Sickness

In the late nineteenth century, several observers noted the important role the vestibular system plays in motion sickness. Researchers found that deaf-mutes rarely became seasick and that it was difficult to make them sick in a rotating chair [1]. As the processes that produce deafness frequently impair vestibular function as well, the conclusion was that a normal vestibular system was essential for motion sickness. Subsequent studies have shown that individuals without a functioning vestibular apparatus are immune to most forms of motion sickness [4, 5]. There are some situations, however, where even people without a functioning vestibular system can get motion sickness symptoms [6].

Because the vestibular system appeared important to motion sickness, researchers speculated that excessive stimulation of the vestibular apparatus was essential for motion sickness [1]. This seemed to be a reasonable hypothesis for motion sickness due to high seas or rotating chairs, but it did not explain other sickness-producing situations. Motion sickness due to visual stimulation (e.g., due to a fixed-base simulator or virtual reality device) involves no vestibular stimulation. In these cases, postulating vestibular stimulation as the cause of the sickness would make no sense.

The most plausible theory that encompasses almost all forms of motion sickness and that has stood the test of time is the sensory conflict or sensory rearrangement theory [1, 3]. In this theory, it is not the stimulation of any particular sensory modality that leads to motion sickness, but instead it is a conflict between senses or between a sense and past experience that leads to the sickness. The fundamental idea behind this theory is that during most activities on Earth, the sensations from vision, the vestibular system, and proprioception agree and are consistent with past experience. For example, when standing upright the otoliths indicate the head is vertical, the eyes show a vertical scene, and the feet sense pressure. If, however, one of the senses does not agree, this can produce conflicts between senses and perhaps with what the nervous system expects from past experience. Motion sickness symptoms result from this conflict.

This theory can be applied to weightlessness. In space, when a crew member is floating through the cabin, the otolith organs no longer provide a meaningful signal about up or down. The visual scene, however, clearly shows the changing orientation. With head movements, the semicircular canals and vision provide accurate information, but these signals are no longer associated with the usual otolithic information about tilt or pitch. Conflicts arise between different senses (vision and vestibular information), different components of the vestibular system (the otoliths and semi-circular canals), and between these senses and past experience. The conflicts provide a stimulus for adaptation to occur, so that over time the sensations are normalized. Neurophysiologic and anatomic studies on animals have shown that during the first few days of a space mission, significant biochemical changes take place in the vestibular system—consistent with an adaptive process [7–9]. As a result of this adaptation, the crew member no longer experiences symptoms of motion sickness and can function effectively in space.

The adaptation, however, may come at a price. Because the nervous system has learned to accept the new combination of sensory inputs as normal, returning to Earth can prompt a new set of conflicts. This has proven to be the case. After returning to 1 G, crews notice the differences in their balance systems and again become suscep-tible to motion sickness [2]. This time, however, they are Earth sick; because their vestibular systems have adapted to space, the sensory inputs on Earth now seem to be in conflict.

The sensory conflict theory has proven to be a useful way to think about motion sickness. It offers an explanation for why the same basic constellation of symptoms (pallor, salivation, headache, drowsiness, nausea) can occur with so many different stimuli (ocean waves, airplanes, cars, virtual reality simulations, space vehicles, etc.). Unfortunately, the theory does not show where in the brain the conflict takes place or exactly how a conflict is converted into a set of motion sickness symptoms. The adaptive value and evolutionary development of these symptoms remain a mystery.

Signs and Symptoms of Motion Sickness

Nausea is the most commonly reported symptom of motion sickness. Vomiting is the most disturbing symptom and is something crews would like to avoid on a space mission. The two most commonly reported signs of motion sickness—pallor and cold sweating—often accompany the nausea and vomiting. Other symptoms that can occur include headache and dizziness. Salivation, yawning, belching, and flatulence are also frequent signs [1].

Since nausea and vomiting are the most notable and unpleasant symptoms of motion sickness, they receive the most attention in motion sickness studies. There are secondary symptoms of motion sickness, however, that can be significant opera-tionally. A symptom complex involving drowsiness, lack of initiative, lethargy, and apathy has been termed the "sopite syndrome," from the Latin *sopitus*, "to put to sleep" [10]. This constellation of symptoms can occur either with or without the nausea and vomiting of motion sickness and can last after the nausea and vomiting have subsided.

The sopite syndrome is important for several reasons. Because the symptoms can occur even when nausea and vomiting are not present, a crew member suffering from sopite may be thought not to have motion sickness, when in fact the person's performance may be degraded due to lethargy. Also, if a drug relieves nausea and vomiting but does not affect the secondary sopite symptoms, it may not necessarily improve an affected crew member's performance [11]. Finally, medications that pro-duce sedation but do not improve the sopite symptoms might only add to the existing drowsiness.

Motion Sickness and Performance

The most disabling form of motion sickness is chronic motion sickness. In individuals with chronic motion sickeness, the usual adaptive process does not take place, and they can remain sick the entire time they are exposed to the new environment. Based on experience in World War II, this has been estimated to occur in less than 5% of susceptible individuals [1]. These individuals can lose significant amounts of weight and become ineffective at their jobs. Although a rare condition, this kind of motion sickness would be totally incompatible with long-duration spaceflight.

For most people motion sickness is an intermittent problem. Nevertheless, as anyone who has experienced motion sickness knows, the symptoms can markedly reduce motivation. This is balanced by the fact that most people can rise to a challenge despite feeling ill. So when looking at the effect of motion sickness on performance, it is important to make a distinction between peak efficiency (i.e., the kind of performance needed in an emergency) and maintenance efficiency (i.e., the performance needed to do routine, but necessary tasks) [1].

Hettinger et al. [12] summarized the available data on human performance in the setting of motion sickness. Although there have been many studies, using a wide variety of techniques that sometimes show conflicting results, some general trends have emerged. Decrements in performance were greatest in complex tasks that required sustained attention and offered the greatest chance for the subjects to control the pace [12]. In other words, the main effect seemed to be on motivation. If the subjects could force themselves to do the task, they could perform it successfully.

Similarly, in situations where an individual is mentally focused, motion sickness can be delayed or prevented. When subjects performing head motions in a slowly rotating room were given problems to solve projected on a screen in front of them, they developed very few symptoms. Another group who did the head movements in the dark, with no task to perform, developed severe symptoms [1]. These data are supported by wartime observations that emergencies can dramatically improve motion sickness symptoms and restore performance in many individuals. Similarly, one of the best treatments for an air-sick student pilot is to give the student the controls.

The effects of motion sickness on performance can be summarized as follows. In some people, the symptoms are so severe so that they become incapacitated. In most individuals, however, the symptoms cause a decrease in motivation, but satisfactory performance is still possible with effort. In many people, motion sickness symptoms may be resolved or greatly improved in a demanding emergency.

Emetic Pathways

Several regions in the brain seem to be important in the generation of motion sickness. Figure 9-1 is a block diagram of the main areas of interest within the brain for motion sickness. Information from the inner ear is integrated within the vestibular nuclei [3]. The cerebellum is thought to be the site of the "mismatch comparator"— the part of the brain that compares the actual sensory input to what is expected [13]. The limbic system also is thought to be important in the generation of the mismatch signal [13]. The emetic center is the final common pathway for a variety of emetic stimuli (e.g.,

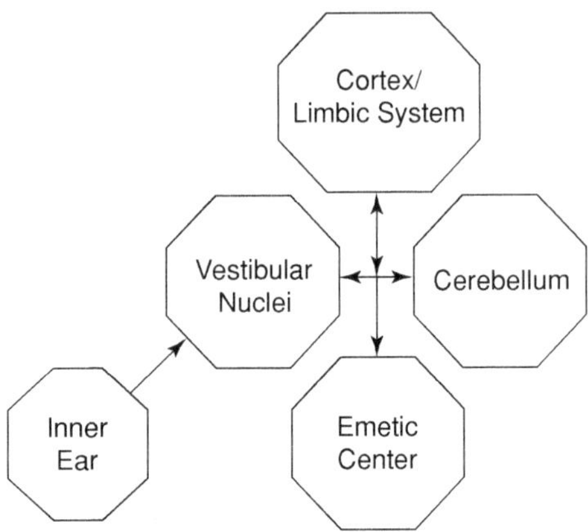

Figure 9-1. The primary brain areas involved with motion sickness. Information from the inner ear is integrated in the vestibular nuclei. Certain parts of the cerebellum seem to be particularly important in motion sickness. The best evidence suggests that the conflict signal (or mismatch signal) is generated by the limbic system and cerebellum. Antihistamines and anticholinergics may work by affecting pathways within the vestibular nuclei and cerebellum [3, 13].

toxins, gagging, gastrointestinal irritation), not just motion sickness. This "center" is not necessarily a single anatomical location, but rather a convenient way to refer to the circuitry within the brainstem that is important for coordinating vomiting [3].

A wide variety of receptors and neurotransmitters are used throughout the pathways shown in figure 9-1. Neurons in the vestibular nucleus, for example, can be influenced by acetylcholine, glutamate, glycine, gamma-aminobutyric acid, histamine, norepinephrine, dopamine, serotonin, substance P, somatostatin, adrenocorticotropic hormone, and enkephalin [3]. This suggests that many medications could potentially affect the vestibular nuclei and influence motion sickness. Similarly, with so many pathways and neurotransmitters involved, the mechanism of action of a motion sickness drug is often elusive. Many of the medications that have been successful in motion sickness have diverse actions at multiple receptors. Nevertheless, there are several receptors that are known to play a role in the production or resolution of motion sickness.

5-HT3 receptors

The hydroxytryptamine (serotonin) receptor 3 (5-HT3) has been shown to be important in nausea and vomiting. Antagonists of 5-HT3 receptors can dramatically decrease nausea and vomiting in many situations. Unfortunately, the 5-HT3 antagonists (e.g., ondansetron, granisetron), which are very effective against nausea and vomiting induced by chemotherapy, are not effective in motion sickness [14, 15]. These findings suggest that 5-HT3 receptors are not involved in the neural pathways that bring about motion sickness.

5-HT1a receptors

Serotonin 5-HT1a agonists (buspirone or 8-hydroxy-2-di-*n*-propylamino-tetralin [8-OH-DPAT]) are effective in preventing motion-sickness–induced vomiting in cats [16, 17]. In addition, 8-OH-DPAT was shown not to interfere with the habituation to motion stimuli (see below for a discussion of habituation) [17]. Currently, there are no published human studies on serotonin 5-HT1A agonists for the treatment of motion sickness. The most commonly available 5-HT1a agonist is buspirone. Lucot [18] suggested that because buspirone is only a partial agonist, too high a dose would be needed for clinical efficacy in humans.

Muscarinic receptors

Anticholinergics acting on muscarinic receptors are the most effective drugs for treating motion sickness [19]. There are five subtypes of muscarinic receptors. Scopolamine, the drug with the longest history and best efficacy in treating motion sickness, blocks all five receptors. Studies in cats have shown that the M1 and M2 receptors may not be important for motion sickness. Blockade of the M3 and M5 receptors in humans with the drug zamifenacin is effective against motion sickness [20]. Zamifenacin, however, has not entered routine clinical use because of concerns about liver toxicity. Newer M3 antagonists are under development.

Histaminic receptors

H1 antihistamines are effective in motion sickness, suggesting that H1 histamine receptors play a role. Most of the antihistamines that are effective for motion sickness also cause sedation, making it difficult to assess whether the effect is due to a specific blockade of histamine receptors or due to a nonspecific effect on neuron excitability [3]. Also, some of the antihistamines that are effective in motion sickness have actions at other receptors. Diphenhydramine and its 8-chloro-theophyline salt dimenhydrinate (Dramamine) both have antimuscarinic properties. Flunarazine, another H1 antihistamine, also blocks calcium channels. Promethazine blocks muscarinic receptors and is also a phenothiazine. Because of the mix of receptors affected, the precise mechanism of action of some drugs classified as antihistamines is hard to determine.

Modern, nonsedating antihistamines that do not cross the blood–brain barrier, such as cetirizine and fexofenadine, are not effective against motion sickness [21]. This supports the view that the antihistamines act centrally.

NK-1 receptors

Substance P has been found in brain-stem regions associated with vomiting, which suggests that neurokinin-1 (NK-1) receptor antagonists might be effective antiemetics. Neurokinin-1 antagonists have been studied against the nausea and vomiting produced by cancer chemotherapy and are effective [22, 23]. Similar to the situation with 5-HT3 antagonists, however, these compounds are not effective in motion sickness. A neurokinin-1 antagonist alone, or in combination with 5-HT3 antagonist, was not effective in improving motion sickness symptoms [15].

Adrenergic receptors

Sympathomimetics (e.g., amphetamine, ephedrine) have been shown to be effective against motion sickness in humans [19]. There are inhibitory alpha-2 adrenergic receptors in the vestibular nuclei, and this might be a potential site of action [3]. Alternatively, however, the sympathomimetics' main role may be to counter the drowsiness and lethargy that accompany motion sickness through other adrenergic pathways [1].

Opioid receptors

Interestingly, although opioids initially can produce nausea and vomiting, after that effect subsides, they have a broad antiemetic action [3]. The opioid antagonist naloxone increases the susceptibility to motion sickness in humans [24], supporting the theory that opioids could potentially play a role in treating motion sickness. Whether the manipulation of endorphins can change susceptibility to motion sickness is not known.

Vasopressin

Studies have shown that a 20- to 30-fold increase in vasopressin levels can occur with the onset of nausea [3]. Infusion of vasopressin can produce symptoms of nausea in humans [25], and blockage of arginine vasopressin (AVP) receptors prevents motion sickness in the squirrel monkey [26]. The AVP_1 (V1) receptor seems to be the active receptor in the squirrel monkey, as V1, but not V2 or combined V1-V2, receptor antagonists were effective against motion sickness. In one study in humans, the nausea produced by vasopressin was blocked by the administration of atropine, suggesting that an anticholinergic pathway may be involved in the nausea and vomiting produced by vasopressin [27]. Vasopressin antagonists are under investigation as agents for the treatment of congestive heart failure (they increase urinary output), but no published trials exist showing if they are effective in motion sickness.

P6 pressure

One interesting method for controlling motion sickness is through the use of acupressure points. The mechanism of action of this approach is not known, but the use of electrical stimulation at the wrist of the P6 acupressure point was effective in reducing the symptoms produced by experiencing vection in a rotating drum [28].

Motion Sickness in Space

Motion sickness is very common on spaceflights. Approximately two-thirds of all crew members on Space Shuttle flights have reported suffering motion sickness [2, 29]. In one study, 13% had severe symptoms [29]. Overall, about half the cases are classified as moderate or severe [2]. Reviews of medications taken on Shuttle flights show that drugs for motion sickness account for almost half of all medications taken [30]. To date, all the data collected on motion sickness incidence has been based on self-reports from the crew members. Because experiencing motion sickness may carry a stigma in a demanding operational setting, there may be a tendency to under-report symptoms. Therefore, the actual incidence of motion sickness during the early phase of spaceflight may be higher than the reported 70%.

The reported data on motion sickness incidence are concerned mainly with nausea and vomiting and not with sopite symptoms. Some crew members may never vomit and have minimal nausea, yet they may have drowsiness, lethargy, reduced motivation, and a reduced appetite. All these symptoms may be due to low-level motion sickness.

In the Shuttle program, the most commonly used medication for motion sickness is promethazine, often given intramuscularly. Promethazine has a variety of actions (e.g., anticholinergic, antihistaminic), and it has been shown to be very effective in motion sickness [19]. In clinical use, the drug produces sedation. In studies where cognitive performance is measured using objective tests, promethazine produces significant performance decrements [31–33]. In the Shuttle program, however, the drug has been very successful and the reported incidence of sedation has not been high [34–36].

The rationale for intramuscular promethazine grew out of the observation that medications taken before flight were often ineffective [34] and that orally administered drugs taken once motion sickness was established in flight were not absorbed. Intramuscular promethazine provided prompt, long-lasting relief [35]. Ground-based studies suggest intramuscular promethazine should lead to a high rate of sedative side effects, but a review of self-reports from flights where promethazine was used did not confirm this [36]. There are several possible reasons for this difference. One is that intramuscular promethazine may be absorbed differently in weightlessness and provide an even blood level of the drug while avoiding the peaks in blood levels usually associated with an intramuscular injection. Another reason is that the level of excitement in the crew on the first flight day when the drug usually is administered is enough to overcome sedative effects. A third explanation is that self-reports are unreliable and that, if objective testing were done, the usual side effects of promethazine would be seen. A final explanation is that the drug may relieve sopite symptoms and so improve the crew members' well-being.

At present, however, the success of intramuscular promethazine has established this drug as the standard of care for motion sickness in space. Formal ongoing assessments of its sedative and performance effects in flight have not been done. As a result, it is unknown whether, on average, having promethazine in the body would be an advantage or disadvantage if an emergency should arise that demands full alertness. On the one hand, crew members might be better able to handle stressful situations if their motion sickness were resolved. Certainly, crew members with severe motion sickness would be much more effective if their symptoms were controlled. On the other hand, studies have shown that emergencies can serve to focus the mind and relieve symptoms. The intramuscular promethazine cannot be removed from the body once it is injected, and crew members have to cope with any ill effects that may result.

After the mission, crews can experience motion sickness again. There are no solid data on the incidence of motion sickness after landing. Nevertheless, it can be a significant problem that can require medication and occasionally intravenous fluids.

Countermeasures for Motion Sickness in Space

Although nausea and vomiting are obvious problems from motion sickness, it is often the lack of motivation, drowsiness, and apathy that are the most operationally significant symptoms. As has been shown in the section on performance above, it may be the lack of motivation that leads to most of the performance deficits. Decreasing nausea may restore well-being and improve motivation. Sometimes, however, nausea may be lessened at the expense of significant sedation, which may relieve one symptom (nausea) but only contribute to the lack of motivation.

Preventive Training

The best motion sickness treatment is a successful adaptation to the new environment. In space, symptoms are typically gone after 2–3 days. Because adaptation is so successful in the majority of cases, studies have been done to determine if the crew members could be preadapted to weightlessness, so that they do not experience motion sickness when they arrive [37].

Preadaptation has been a regular component of cosmonaut training. In the final 2 weeks before launch, cosmonauts ride in a rotating chair daily while executing pitching head movements [38]. These head motions cause illusory sensations of body motion (the Coriolis effect) and produce a situation where the signals from the semicircular canals conflict with the information from the otolith. While these motions usually provoke motion sickness, over time the cosmonauts are able to tolerate more of this stressful motion. The hope is that the habituation that is acquired in the chair can then be transferred to space. No study exists that demonstrates that this program is effective, although the reported incidence of motion sickness in the Russian space program is less that that reported in the U.S. program [38]. Nevertheless, motion sickness is still reported in the Russian program, suggesting that the habituation that is acquired on the ground may not transfer fully to spaceflight.

Preflight adaptation has also been studied in the U.S. space program [37, 39], but it has not been routinely adopted before spaceflights. Protective adaptation has been studied in variety of other settings (rotating rooms, visual simulators, rotating chairs). The results of many of the early studies in the 1960s were summarized by Reason and Brand [1] in this way: "The bulk of the research in this area has indicated that adaptation is normally highly specific to the particular stimulus conditions under which it was acquired" (p. 145). For example, subjects who made tilting head movements to the right while rotating could habituate to this stimulus. If they then tried a run by tilting the head to the left, symptoms recurred [1]. Nevertheless, it is possible that preflight adaptation schedules might produce some generalized reduction in vestibular sensitivity [1]. Also, there have been some circumstances where protective adaptation in one setting can transfer successfully to another if the stimuli are similar [38].

One simple measure that crew members can take while in flight is to minimize head movements. Head movements will generate conflicts between the semicircular canals and otoliths, so by minimizing these movements, crew members might be able to keep their symptoms at a manageable level.

Medications

The ideal motion sickness drug would be easy to administer (i.e., not require an injection), produce no sedation, have no cognitive effects, and not interfere with the adaptation to the new environment. It has been hard to find a single drug that meets all of these criteria (although some combination preparations have come close). The experience in the space program has been that medications taken before entering weightlessness have not been satisfactory [34]. Also, after entering weightlessness, motion sickness symptoms can reduce gastric motility, so that medications taken orally show slow and erratic absorption [40]. A variety of different medications can be used for motion sickness, and each one has advantages and disadvantages. Table 9-1 lists several medications that could be used for the treatment of motion sickness in space.

Scopolamine

Scopolamine, also known as hyoscine, is the most effective medication for motion sickness. The drug is potent, requiring only 0.3–1.2 mg for effectiveness. This characteristic is what makes it amenable to transdermal administration. The drug blocks all types of muscarinic receptors and so has a variety of anticholinergic side effects such as dry mouth, blurred vision, urinary retention, loss of visual accommodation, and sedation. The drug can also affect coordination and short-term memory. Historically, scopolamine often was used as a sedative. An early pharmacology text warns: "Hyoscyamine and hyoscine are very powerful alkaloids and should always be used with great caution. . . . Hyoscine is the less dangerous of the two. Both are used as hypnotics to a considerable extent, and especially in hospitals for the insane" [41, p. 709]. At lower doses, however, this effect is less marked. Scopolamine can be combined with a sympathomimetic to combat the sedative side effect [32]. In the classic study by Wood and Graybiel [19], where a variety of motion sickness drugs were compared for effectiveness using a standardized motion sickness producing stimulus, 1.2 mg scopolamine combined with 20 mg amphetamine was the most effective combination. Doses less than 0.9 mg are recommended, however, to avoid side effects [42]. A more common dosage might be 0.3–0.6 mg of scopolamine combined with 5–10 mg of dextroamphetamine.

The most significant problem with scopolamine, however, may be that it interferes with the habituation to the new environment. This has been demonstrated in experiments using the rotating chair [43]. One hypothesis for this finding is that scopoloamine might suppress the conflict that arises between senses, and this conflict or mismatch signal is essential for the subsequent adaptation [18]. A possible result of this suppression is that motion sickness symptoms could return once the medication is removed.

Table 9-1. Characteristics of major motion sickness drugs.

Drug	Dose	Route[a]	Onset	Peak (hours)	Duration (hours)	Notes
Scopolamine	0.25–0.9 mg	PO	30–60 min	1	4–6	Also known as hyoscine
	0.3–0.6 mg	SC, IM, IV	30 min		4	
	1.0 mg patch	Transdermal	4 hours	24	72	
Promethazine	25–50 mg	PO	15–60 min		4–6	May last as long as 12 hours
		PR	20 min		4–6	
		IM	20 min		4–6	
		IV	Immediate		4–6	
Dimenhydrinate	25–50 mg	PO	15–30 min		3–6	The active moiety of
		IM	20–30 min		3–6	dimenhydrinate is
		IV	Immediate		3–6	diphenhydramine
Meclizine	25–50 mg	PO	1 hour		8–24	Long-acting antihistamine
Chlorpheniramine	8–12 mg	PO	15–60 min	2–6	4–8	Small molecule
Dextroamphetamine	5–10 mg	PO	1 hour	2		
Ephedrine	25–50 mg	PO	15–60 min		2–4	
		SC	20–30 min			
		IM	10–20 min			
		IV	Immediate			
Ginger	1,000–2,000 mg	PO	20 min (approx.)			Pharmacokinetics not established. 250 mg PO 4×/day used in nausea and vomiting of pregnancy.
Phenytoin	500–1000 mg loading, 100–200 mg every 12–24 hours maintenance	PO				Drug adjusted to give a blood level of >9–12 µg/ml

The column labeled "Onset" provides the time to the onset of the medication's effect. The "Peak" column lists the time to the peak effect of the drug (taken to correspond with the peak in blood level). The "Duration" column gives data on how long the therapeutic effect lasts. Data on all these factors are not available for all the drugs. Information from multiple sources [46, 56, 57]

[a]PO, per os; SC, subcutaneous; IM, intramuscular; IV, intravenous; PR, per rectum.

Promethazine

Promethazine has proven to be a useful drug for the treatment of motion sickness in space. This drug has a variety of actions, possessing both antihistaminic and anticholinergic properties. In space, 25–50 mg of the drug is given by intramuscular injection or by suppository. When given by intramuscular injection, the onset of action is prompt. Also, the duration of action in space is longer than the expected 4–6 hours. If crew members take an injection before sleep, they can benefit from the sedative effect, while at the same time relieving motion sickness symptoms [2].

The main side effect of promethazine is sedation, and studies have shown that performance on complex tasks is degraded when promethazine is taken [32, 33]. This indicates that some judgment is needed about when to use this medication. In a severely motion-sick crew member, theoretical decrements in performance would not be an issue—the crew member is already functioning at a low level. In contrast, for a crew member with mild symptoms, the risk–benefit trade-off might be different.

Promethazine can be combined with a sympathomimetic to help combat the sedative side effect. In various studies, this approach has provided relief without significant sedation and decrements in performance [32, 33]. While amphetamine has been the sympathomimetic most studied, ephedrine also has been useful and has less abuse potential. In contrast to scopolamine, promethazine does not appear to interfere with habituation [18].

Promethazine can produce dystonic reactions, such as torticollis, after an intramuscular injection. To date, this has not been reported in the space program. Nevertheless, crew members need to be aware that this disturbing symptom can occur.

Dimenhydrinate and diphenhydramine

Dimenhydrinate and diphenhydramine are H1 antihistamines that are frequently used to treat motion sickness. In general, H1 antihistamines have fewer side effects than other drug classes used for motion sickness. Unfortunately, the antihistamines also appear to be less effective [19]. Dimenhydrinate (Dramamine) is very close chemically to diphenhydramine (Benadryl) and, in fact, diphenhydramine is the active moiety in dimenhydrinate. Dimenhydrinate is the most commonly used of the two for motion sickness symptoms. In a comparison study, these drugs were shown to be the most effective of the antihistamines [19], although antihistamines of this pharmacological class (ethanolamine) are also the most sedating of the H1 antihistamines [44]. The usual dose is 25–50 mg given orally. These drugs also can be given as an intramuscular injection, although there are no reports of this being tried during a spaceflight.

The most frequently reported adverse reactions to these antihistamines are drowsiness, fatigue, and dry mouth. The sedative effects vary considerably among individuals. Other less frequent effects include weakness, confusion, dysarthria (slurred speech), fatigue, or headache. Most side effects reported with dimenhydrinate are attributable to its anticholinergic properties [45]. Anticholinergic effects include thickening of bronchial secretions, urinary retention, dry eyes, dilated pupils, and blurred vision [46]. The anticholinergic effect may result in increased intraocular pressure in susceptible patients. H1 antagonists in general may cause adverse gastro-intestinal effects, including constipation, appetite stimulation, anorexia, or abdominal pain. Cardiovascular effects, like sinus tachycardia, palpitations, or arrhythmias also can occur. In general, however, dimenhydrinate and diphenhydramine are well tolerated, and only the sedation is troublesome in operational settings.

Meclizine

Meclizine (Bonine) is an H1 antihistamine with a good track record for treating motion sickness. The usual dose is 25–50 mg given orally [19, 47]. Compared to dimenhydrinate and diphenhydramine, meclizine has a long duration of action, suggesting that it could possibly be taken before a spaceflight to provide some relief once in orbit.

Chlorpheniramine

Chlorpheniramine is a widely available H1 antihistamine that has only recently been studied for effectiveness against motion sickness [48]. Chlorpheniramine at the 8 and 12 mg dose can prolong the amount of stressful motion that subjects can tolerate in a rotating chair [48]. At present, no data exist comparing chlorpheniramine to other established motion sickness remedies. One potential advantage of chlorpheniramine is that it might be amenable to novel routes of administration (e.g. transdermal, intranasal) because it is a potent antihistamine (requiring 4–12 mg for effectiveness, rather than 25–50 mg for most other antihistamines) and a small molecule.

Amphetamine and dextroamphetamine

Amphetamine and dextroamphetamine are stimulants. They are believed to work by stimulating the release of norepinephrine and other biologic amines from central adrenergic receptors. The result of this is increased alertness, a decreased sense of fatigue, mild euphoria, and improved mood [45, 46]. The most commonly used amphetamine is dextroamphetamine, which is the dextrorotatory form of amphetamine. Dextroamphetamine (Dexedrine) has been shown to be effective against motion sickness as a single agent. Amphetamine was comparable in effectiveness to antihistamines in a comparison study [19]. The mechanism may be due to an effect in the vestibular nuclei [3], although it is also possible that the amphetamine's effect on secondary symptoms, such as drowsiness, may be the most important benefit it provides in motion sickness treatment [1]. Typically, dextroamphetamine is combined with another motion sickness medication to provide a combination treatment. The combination of scopolamine and Dexedrine has been a very effective treatment that is still used extensively to prevent motion sickness during parabolic flight on the NASA C-9 aircraft. The dose ranges for the combination medication are 0.3–0.6 mg scopolamine and 5–10 mg dextroamphetamine. The *Apollo 11* crew used the combination of 5 mg amphetamine and 0.3 mg scopolamine for their moon voyage. A dose of 5–10 mg is used when amphetamine is given as a single agent.

The side effects of amphetamines include insomnia and anorexia. In normal doses, amphetamines may increase both systolic and diastolic blood pressure and exacerbate hypertension. Cardiac arrhythmias may occur. Allergic reactions to amphetamines are rare, but occasionally urticaria is seen. Excessive doses can produce angina, anxiety, agitation, blurred vision, delirium, diaphoresis, flushing, hallucinations, hyperthermia, labile blood pressure, pupil dilation, palpitations, paranoia, purposeless movements, psychosis, sinus tachycardia, tachypnea, or tremor [45]. If minor manifestations of these symptoms occur with usual doses, the dose should be reduced or the drug discontinued.

The major factor limiting the use of amphetamines is their potential for abuse. Because motion sickness is typically self-limited in weightlessness, long-duration treatment with Dexedrine would not be needed, which should limit the potential for abuse. Dexedrine is included in the in-flight medical kit [49] and so is available to crews.

Ephedrine

Ephedrine is a mild sympathomimetic that has less potential for abuse than amphetamines. The compound was first obtained from plants of the genus *Ephedra*, and it has been used in Chinese and eastern Indian medicine for many years. Ephedrine releases endogenous norepinephrine from its storage sites and so has an indirect sympathomimetic effect [45]. Ephedrine is available in over-the-counter asthma preparations such as Bronkaid.

Given as a single agent, usually 25–50 mg orally, ephedrine is effective against motion sickness [19]. Typically, however, it is used in combination with other drugs. The addition of ephedrine to either scopolamine or chlorpheniramine has been shown to eliminate or reduce the performance decrements sometimes seen with those drugs [42, 50].

The side effects of ephedrine are similar to amphetamine, but are usually less pronounced. Nervousness, anxiety, fear, agitation, restlessness, weakness, irritability, talkativeness, and insomnia can all occur. The drug can cause difficulty in urination. Rarely, serious central nervous system reactions have been observed, including stroke, transient ischemic attack, and seizures. With higher doses, dizziness, lightheadedness, tremor, and hyperreflexia have been reported. Ephedrine increases cardiac workload and may produce palpitations and sinus tachycardia. Cardiac events with ephedrine have included hypertension, palpitations, tachycardia, arrhythmias, myocardial infarction, and cardiac arrest (or sudden death) [45].

Other Medications: Ginger, Dexamethasone, Phenytoin

Ginger has been studied as a remedy for motion sickness and has been shown to be effective in some studies [25, 51, 52]. The active compound in the ginger root that produces the therapeutic effect is not known, so the pharmacokinetics of ginger are not established. Dexamethasone is often used in combination with 5-HT3 antagonists to treat nausea and vomiting in patients receiving cancer chemotherapy. Dexamethasone does not, however, prevent motion sickness in a rotating chair [53]. Phenytoin at a blood level of 9–15 µg/ml was effective against motion sickness produced in a rotating chair [54, 55]. An appropriate blood level can be established with a loading dose followed by a maintenance dose.

Recommendations Based on Current Knowledge

The incidence of motion sickness is extremely high during the first few days of a space flight and often occurs after landing as well. Severe cases can lead to dehydration and can cause a crew member to be operationally ineffective. As a result, crews turn to medications to help with the symptoms. This approach is reasonable, provided the drugs are used prudently. Also, if artificial gravity countermeasures are used, these also have the potential to produce motion sickness. Some suggestions on how to deal with motion sickness in space are presented below:

1. Chronic motion sickness, though rare, can occur. Most astronauts have been in fields where they have had to adapt to stressful motion environments (e.g., naval or aviation operations), but this is not uniformly the case. Therefore, it may be a wise practice for astronauts to experience spaceflight on a short-duration (1–2 weeks) flight before being scheduled for a longer (more than a month) mission. A case of chronic motion sickness on a long-duration spaceflight would be hard to manage.

2. Preflight adaptation to stressful motion has not been shown definitively to be helpful. The current Russian program that involves daily runs in a rotating chair before flight may be effective, but proving this is difficult. In cases where motion sickness in flight could be hazardous, such as a demanding docking early in the flight, the time investment in preflight adaptation might be worthwhile. In other cases, medications can be used as needed.

3. Pretreatment for motion sickness using medications was tried in the Shuttle program with mixed results [34]. Several crew members who took a scopolamine/dexamphetamine combination before flight still had motion sickness symptoms in flight. Once in flight, the absorption of oral medications is erratic, making oral medications unsuitable for early spaceflight use. This led to the use of intramuscular promethazine. Although sedative side effects in space due to promethazine have been reported to be low [36], it may be reasonable to reserve intramuscular promethazine for crew members with severe symptoms. In those situations a clear benefit could be readily appreciated. Crew members who take promethazine prophylatically, or for mild symptoms, run the risk of side effects when they may be able to perform well without the drug.

4. The addition of dextroamphetamine or ephedrine to promethazine in flight may be reasonable to help combat sedative side effects. These drugs should not be given shortly before sleep. It is not clear if these drugs, taken as a single agent, might be useful for sopite symptoms in those who have not taken a medication like scopolamine or promethazine. These combinations may lead to urinary retention, and the capability to deal with that should be present.

5. If a preflight medication is desired, agents with a long duration of action (such as meclizine) or transdermally administered agents should be considered.

6. Minimizing head movements during the early portion of a spaceflight may help delay or prevent symptoms. Because head motions are likely to produce a conflict between the semicircular canals and otoliths, aggressive head movements may provide a powerful stimulus for motion sickness.

7. If they are going to land on another surface (such as the Moon or Mars) where a gravitational field is present, crew members should be prepared for the recur-

rence of motion sickness symptoms. Limiting head movements after landing makes sense.

References

1. Reason, J.T., and J.J. Brand, *Motion Sickness*. 1975, Academic Press, London.
2. Jennings, R.T., Managing space motion sickness. Journal of Vestibular Research, 1998. 8(1): 67–70.
3. Yates, B.J., A.D. Miller, and J.B. Lucot, Physiological basis and pharmacology of motion sickness: an update. Brain Research Bulletin, 1998. 47(5): 395–406.
4. Kennedy, R.S., et al., Symptomatology under storm conditions in the North Atlantic in control subjects and in persons with bilateral labyrinthine defects. Acta Oto-Laryngologica, 1968. 66(6): 533–40.
5. Cheung, B.S., I.P. Howard, and K.E. Money, Visually-induced sickness in normal and bilaterally labyrinthine-defective subjects. Aviation, Space, and Environmental Medicine, 1991. 62(6): 527–31.
6. Johnson, W.H., F.A. Sunahara, and J.P. Landolt, Importance of the vestibular system in visually induced nausea and self-vection. Journal of Vestibular Research, 1999. 9(2): 83–87.
7. Holstein, G.R. and G.P. Martinelli, The effect of spaceflight on the ultrastructure of the cerebellum, in *The Neurolab Spacelab Mission: Neuroscience Research in Space*, J.C. Buckey and J.L. Homick, eds. 2003, NASA, Houston, TX, pp. 19–26.
8. Pompeiano, O., Gene expression in the rat brain during spaceflight, in *The Neurolab Spacelab Mission: Neuroscience Research in Space*, J.C. Buckey and J.L. Homick, eds. 2003, NASA, Houston, TX, pp. 27–38.
9. Boyle, R., et al., Neural readaptation to earth's gravity following return from space, in *The Neurolab Spacelab Mission: Neuroscience Research in Space*, J.C. Buckey and J.L. Homick, eds. 2003, NASA, Houston, TX, pp. 45–50.
10. Graybiel, A., and J. Knepton, Sopite syndrome: a sometimes sole manifestation of motion sickness. Aviation, Space, and Environmental Medicine, 1976. 47(8): 873–82.
11. Wood, C.D., et al., Therapeutic effects of antimotion sickness medications on the secondary symptoms of motion sickness. Aviation, Space, and Environmental Medicine, 1990. 61(2): 157–61.
12. Hettinger, L.J., R.S. Kennedy, and M.E. McCauley, Motion and human performance, in *Motion and Space Sickness*, G.H. Crampton, ed. 1990, CRC Press, Boca Raton, FL, pp. 411–42.
13. Crampton, G.H., Neurophysiology of motion sickness, in *Motion and Space Sickness*, G.H. Crampton, ed. 1990, CRC Press, Boca Raton, FL, pp. 29–42.
14. Stott, J.R., et al., The effect on motion sickness and oculomotor function of GR 38032F, a 5-HT3-receptor antagonist with anti-emetic properties. British Journal of Clinical Pharmacology, 1989. 27(2): 147–57.
15. Reid, K., et al., Comparison of the neurokinin-1 antagonist GR205171, alone and in combination with the 5-HT3 antagonist ondansetron, hyoscine and placebo in the prevention of motion-induced nausea in man. British Journal of Clinical Pharmacology, 2000. 50(1): 61–64.
16. Lucot, J.B., Effects of serotonin antagonists on motion sickness and its suppression by 8-OH-DPAT in cats. Pharmacology, Biochemistry and Behavior, 1990. 37(2): 283–87.
17. Lucot, J.B., and G.H. Crampton, 8-OH-DPAT does not interfere with habituation to motion-induced emesis in cats. Brain Research Bulletin, 1991. 26(6): 919–21.
18. Lucot, J.B., Pharmacology of motion sickness. Journal of Vestibular Research, 1998. 8(1): 61–66.
19. Wood, C.D., and A. Graybiel, Evaluation of sixteen anti-motion sickness drugs under controlled laboratory conditions. Aerospace Medicine, 1968. 39(12): 1341–44.

20. Golding, J.F., and J.R. Stott, Comparison of the effects of a selective muscarinic receptor antagonist and hyoscine (scopolamine) on motion sickness, skin conductance and heart rate. British Journal of Clinical Pharmacology, 1997. 43(6): 633–37.
21. Cheung, B.S., R. Heskin, and K.D. Hofer, Failure of cetirizine and fexofenadine to prevent motion sickness. Annals of Pharmacotherapy, 2003. 37(2): 173–77.
22. Hesketh, P.J., et al., Randomized phase II study of the neurokinin 1 receptor antagonist CJ-11,974 in the control of cisplatin-induced emesis. Journal of Clinical Oncology, 1999. 17(1): 338–43.
23. Navari, R.M., et al., Reduction of cisplatin-induced emesis by a selective neurokinin-1-receptor antagonist. L-754,030 Antiemetic Trials Group. New England Journal of Medicine, 1999. 340(3): 190–95.
24. Allen, M.E., et al., Naloxone enhances motion sickness: endorphins implicated. Aviation, Space, and Environmental Medicine, 1986. 57(7): 647–53.
25. Lien, H.C., et al., Effects of ginger on motion sickness and gastric slow-wave dysrhythmias induced by circular vection. American Journal of Physiology – Gastrointestinal and Liver Physiology, 2003. 284(3): G481–89.
26. Cheung, B.S., et al., Etiologic significance of arginine vasopressin in motion sickness. Journal of Clinical Pharmacology, 1994. 34(6): 664–70.
27. Kim, M.S., et al., Role of plasma vasopressin as a mediator of nausea and gastric slow wave dysrhythmias in motion sickness. American Journal of Physiology, 1997. 272(4 Pt 1): G853–62.
28. Hu, S., et al., P6 acupressure reduces symptoms of vection-induced motion sickness. Aviation, Space, and Environmental Medicine, 1995. 66(7): 631–34.
29. Davis, J.R., et al., Space motion sickness during 24 flights of the space shuttle. Aviation, Space, and Environmental Medicine, 1988. 59: 1185–89.
30. Putcha, L., et al., Pharmaceutical use by U.S. astronauts on space shuttle missions. Aviation, Space, and Environmental Medicine, 1999. 70(7): 705–8.
31. Wood, C.D., et al., Side effects of antimotion sickness drugs. Aviation, Space, and Environmental Medicine, 1984. 55(2): 113–16.
32. Wood, C.D., et al., Evaluation of antimotion sickness drug side effects on performance. Aviation, Space, and Environmental Medicine, 1985. 56(4): 310–16.
33. Schroeder, D.J., W.E. Collins, and G.W. Elam, Effects of some motion sickness suppressants on static and dynamic tracking performance. Aviation, Space, and Environmental Medicine, 1985. 56(4): 344–50.
34. Davis, J.R., R.T. Jennings, and B.G. Beck, Comparison of treatment strategies for space motion sickness. Acta Astronautica, 1993. 29(8): 587–91.
35. Davis, J.R., et al., Treatment efficacy of intramuscular promethazine for space motion sickness. Aviation, Space, and Environmental Medicine, 1993. 64(3 Pt 1): 230–33.
36. Bagian, J.P., and D.F. Ward, A retrospective study of promethazine and its failure to produce the expected incidence of sedation during space flight. Journal of Clinical Pharmacology, 1994. 34(6): 649–51.
37. Harm, D.L., and D.E. Parker, Preflight adaptation training for spatial orientation and space motion sickness. Journal of Clinical Pharmacology, 1994. 34(6): 618–27.
38. Clement, G., et al., Effects of cosmonaut vestibular training on vestibular function prior to spaceflight. European Journal of Applied Physiology, 2001. 85(6): 539–45.
39. Reschke, M.F., et al., Posture, locomotion, spatial orientation, and motion sickness as a function of space flight. Brain Research Reviews, 1998. 28(1–2): 102–17.
40. Harm, D.L., et al., Changes in gastric myoelectric activity during space flight. Digestive Diseases and Sciences, 2002. 47(8): 1737–45.
41. Wilcox, R.W., Pharmacology and Therapeutics, 7th ed. 1907, P. Blakiston's Son and Company, Philadelphia, PA.
42. Nuotto, E., Psychomotor, physiological and cognitive effects of scopolamine and ephedrine in healthy man. European Journal of Clinical Pharmacology, 1983. 24(5): 603–9.
43. Wood, C.D., et al., Habituation and motion sickness. Journal of Clinical Pharmacology, 1994. 34(6): 628–34.

44. Brown, N.J., and L.J. Roberts, Histamine, bradykinin and their antagonists, in *Goodman and Gilman's The Pharmacological Basis of Therapeutics*, J.G. Hardman and L.E. Limbird, eds. 2001, McGraw-Hill, New York, pp. 645–68.

45. Gold Standard Multimedia, Clinical Pharmacology Online, 2003, http://cponline.hitchcock.org.

46. The United States Pharmacopeial Convention, USP DI Drug Information for the Health Care Professional, 23rd ed., vol. 1. 2003, Thomson Micromedex, Taunton, MA.

47. Dahl, E., et al., Transdermal scopolamine, oral meclizine, and placebo in motion sickness. Clinical Pharmacology and Therapeutics, 1984. 36(1): 116–20.

48. Buckey, J.C., et al., Chlorpheniramine for motion sickness. Journal of Vestibular Research, 2004. 14(1): 53–61.

49. International Space Station Integrated Medical Group, ISS Medical Checklist, JSC-48522-E1. 2000, NASA, Houston, TX.

50. Millar, K., and R.T. Wilkinson, The effects upon vigilance and reaction speed of the addition of ephedrine hydrochloride to chlorpheniramine maleate. European Journal of Clinical Pharmacology, 1981. 20(5): 351–57.

51. Mowrey, D.B., and D.E. Clayson, Motion sickness, ginger, and psychophysics. Lancet, 1982. 1(8273): 655–57.

52. Niebyl, J.R., and T.M. Goodwin, Overview of nausea and vomiting of pregnancy with an emphasis on vitamins and ginger. American Journal of Obstetrics and Gynecology, 2002. 185(5 Suppl): S253–55.

53. Buckey, J.C., and D.L. Alvarenga, Dexamethasone for motion sickness, Final Report. 2002, Office of Naval Research, Arlington, VA.

54. Woodard, D., et al., Phenytoin as a countermeasure for motion sickness in NASA maritime operations. Aviation, Space, and Environmental Medicine, 1993. 64(5): 363–66.

55. Knox, G.W., et al., Phenytoin for motion sickness: clinical evaluation. Laryngoscope, 1994. 104(8 Pt 1): 935–39.

56. Nissen, D., *Mosby's Drug Consult*. 2003, Elsevier Science, St. Louis, MO.

57. McEvoy, G.K., *AHFS Drug Information*. 2003, American Society of Health-System Pharmacists, Bethesda, MD.

10

Gender: Identifying and Managing Relevant Differences

Introduction

Although Mary Cleave was a fully trained and capable astronaut, she could not do extravehicular activity (EVA): There were no EVA suits small enough to fit her [1]. Similarly, although astronaut Wendy Lawrence was ready to go to the *Mir* space station on the *Soyuz* capsule, she wasn't tall enough to fit safely within the capsule [2]. This problem, however, has not been limited to women. NASA astronaut Scott Parazynski also was ready to go to the *Mir* space station, but he was too big for the *Soyuz* capsule. Size and strength sometimes matter in the space program, and gender can influence both.

Views on the appropriate roles for men and women can differ between cultures and crews. On August 20, 1982, a visiting crew arrived at the *Salyut 7* space station. The visiting crew included Svetlana Savitskaya, a female cosmonaut who was a world-class aerobatic pilot and an avid skydiver. The crew onboard had prepared a special reception, and, after a postdocking communication session, they invited the visitors to the table they had prepared. As cosmonaut Valentin Lebedev, who was onboard to greet the visiting crew, described the event: "We gave Sveta a blue floral print apron and told her, 'Look Sveta, even though you are a pilot and a cosmonaut, you are still a woman first. Would you please do us the honor of being our hostess tonight?'" [3, p. 241] It is not clear if this was taken as a joke, an insult, or an honor because there is no record of what cosmonaut Savitskaya thought of the offer. Nevertheless, the episode indicates that, for this crew at least, there may have been some definite ideas about gender and responsibilities onboard.

Even in areas that are not directly linked to reproduction, men and women often differ both physically and mentally. Yet the average difference between men and women for a particular factor (such as height or strength or ability to detect odors) is typically much smaller than the variability among men or women. Space crews are made up of individuals, not groups, however, so during a mission, an individual crew member's capabilities matter much more than his or her gender. Ultimately, asking the question of whether men or women are better suited for spaceflight may be like asking whether all-male families are better than all-female families. Just as families generally work better when they include members of both sexes, successful space crews should be made up of people with complementary skills who can work together. Over time, situations may arise that play to the strengths of one gender or the other, and the differences in skills and outlook can often strengthen the group rather than weaken it.

For spaceflight, the key factor is knowing what gender differences exist and when these differences are important. Problems may arise concerning intimate relationships, physiological differences, or different susceptibilities to diseases, and plans should be established to deal with these issues.

Gender Differences of Concern for Spaceflight

Clearly, men and women differ. Studies have shown that, compared to men, women on average have better verbal abilities, better articulatory skills, and are superior at fine motor tasks [4]. Men score higher on tests of spatial and quantitative analysis.

Relative to men, women have a higher percentage of body fat, smaller muscle mass, lower blood pressure, higher estrogen levels, higher progestin values, and lower levels of androgen [5]. Females have a more aggressive immune response to infectious challenge but are more likely to develop autoimmune diseases [6]. Men and women may differ in drug metabolism and the response to toxins. For example, women are at a 1.2- to 1.7-fold higher risk than men for all major types of lung cancer at every level of exposure to cigarette smoke. MRI studies show that men and women differ in brain organization for language. Females are more sensitive to pain and, on average, have better hearing and olfactory abilities [7].

These mean differences are often not of consequence for a given spaceflight. In certain areas, however, crew members' genders may put them at an advantage or disadvantage for a given task or put them at higher or lower risk for some of the expected effects of spaceflight. For example, while spaceflight for the most part does not demand great strength, EVA can very taxing. Strength and training requirements for EVAs will probably be more difficult, on average, for women than for men. Postmenopausal women may be prone to lose bone very rapidly in space. Studies have shown that women have lower G tolerance and so are more susceptible to orthostatic intolerance than men. Men are at higher risk for kidney stones and, in general, have higher levels of aggression. An appreciation for these differences is essential for proper planning and execution of the mission.

Size and Strength

In general, small size is an advantage for spaceflight. Smaller people require less food and oxygen. They occupy less space. Compared to men, women on average are shorter, have a smaller sitting height, and have shorter arms and legs [8]. A smaller body size, however, is often accompanied by a lower muscle mass, less strength, and a smaller aerobic exercise capacity. Data on height, weight, and maximal oxygen uptake (Vo_{2max}) for 122 male and 35 female astronauts is shown in table 10-1. On average, women have a lower Vo_{2max} than men. The Vo_{2max} for adult women is 2.0 l/minute, compared with 3.5 l/minute for men [9]. When corrected for differences in body weight, the averages are 40 versus 50 ml/kg/min. On a Moon or Mars mission, where there will be extensive operations in an EVA suit, the exercise demands are likely to be significant. Because of the lower oxygen uptake, for a given task the average woman will be working at a higher percentage of her maximal oxygen uptake compared to the average man. This can lead to earlier fatigue and greater heat stress.

Table 10-1. Anthropometric data for 122 male and 35 female American astronauts.

	Height (cm)	Weight (kg)	Peak Vo_2 (l/minute)	Peak Vo_2 (ml/kg/minute)	% Body fat
Women	166 ± 5	60.7 ± 8.5	2.19 ± 0.48	36.5 ± 7.0	20.8 ± 4.5
Men	180 ± 25	88.6 ± 8.8	3.55 ± 0.63	44.2 ± 7.1	17.6 ± 6.6

The tests used to generate the oxygen update data probably underestimate actual Vo_2max by 15% because the tests are stopped at 85% of age-predicted maximal heart rate. Data from Harm et al. [9], with permission from the American Physiological Society.

Table 10-2. Average women's strength expressed as a
percentage of average men's strength by body segment.

Body segment	Average across strength measurements (%)	Range (%)
Total body	63.5	35–86
Upper extremities	55.8	35–79
Lower extremities	71.9	57–86
Trunk	63.8	37–70
Handgrip	55	—

Data taken from Percival and Quinkert [8] and Wardle et al. [10].

Additionally, working in the suit requires strength. Table 10-2 shows the average strength of the upper extremities, lower extremities, and trunk in women expressed as a percentage of the male value. Median upper body strength in a woman is around 50–60% of a man's. On average, handgrip strength in a woman is about 55% of a man's [10]. Men and women are closer in lower body and trunk strength, where women's values are about 60–70% of men's. In the area of flexibility, the situation is reversed. Women in general have greater flexibility and greater range of motion around most joints than do men [8]. These data are shown in table 10-3. In response to training, both women and men increase lean body mass and strength, but the total muscle girth of women remains lower [9].

Ability to perform EVA

Weightless EVA involves extensive upper body work. Good handgrip strength is essential to maintain stability. The arm and shoulder muscles are needed to manipulate tools. Maximal upper body exercise can elicit a peak oxygen uptake about 70% of whole body Vo_{2max}. Also, upper body exercise can produce higher heart rates and a greater blood pressure response than lower body exercise at the same power output [9]. As mentioned in chapter 5, the average metabolic cost in the pressurized American EVA suit is approximately 0.8 l/minute. The Russian Orlan suit, which operates

Table 10-3. Average range of joint motion for the shoulder, elbow, wrist, hip, knee, and ankle separately for men and women.

Joint motion	Men (degrees)	Women (degrees)
Shoulder abduction (rearward)	59.8	61.4
Elbow flexion-extension	142.1	149.9
Wrist flexion-extension	141.4	154.0
Hip flexion (with extended knee)	83.5	86.8
Hip flexion (with bent knee)	117.9	121.0
Ankle flexion-extension	62.6	66.9

Table reproduced from Percival and Quinkert [8], with permission from John Wiley & Sons.

at a higher pressure, increases the metabolic cost by 40%. Harm et al. [9] estimated that, based on these numbers, weightless EVA work in the American suit would represent an approximate 44% aerobic intensity for the average female astronaut compared to a 28% intensity for the average male. On the Moon or Mars, EVA involves both upper and lower body work. Although the upper body demands for surface EVA will likely be less than for weightless EVA, the metabolic cost of working against the suit will be the same for all astronauts regardless of aerobic capacity. Those with lower aerobic capacity will need to work at a higher percentage of their maximum to accomplish the EVA tasks. For both weightless and surface EVA, flexibility is also important. The range of motion, however, may be limited more by the suit than by the ability of the crew member to make the movements.

Setting design criteria

Overall, men are larger and have greater reach and strength than women. Women have greater flexibility. From these findings, some general guidelines for the design of space equipment, tasks, and procedures can be made. The goal should be to accommodate as many people as possible. As outlined by Percival and Quinkert [8], whenever possible, clearances should be set based on the data for larger men. Reach and strength requirements should be established from the data for smaller women. Flexibility requirements should be based on the male data [8]. Ultimately, in the design of a Mars mission, trade-offs will have to be made. It is possible that not every person can be accommodated in a space suit or spacecraft. Objective requirements for strength and aerobic capacity will need to be set for performing EVA in order to accomplish EVAs safely. These requirements will depend on the operating pressure of the suit and the likely tasks to be performed. Once the standards are set, it is probable that some individuals will not be selected for an EVA-intensive mission, like a Mars mission, because they are too big, too small, have poor flexibility, or a low work capacity. Those excluded because of small size or low work capacity are more likely to be women; those excluded because of poor flexibility or large size are more likely to be men.

Postmenopausal Bone Loss

After menopause, women experience an increase in bone loss. Losses can reach 1–5% a year. Although it is not known if the effects of postmenopausal bone loss and weightlessness exposure are additive, the combination is certainly not desirable. Counteracting the bone loss from weightlessness is difficult even without the added effect of estrogen deficiency.

The most logical approach to postmenopausal bone loss is estrogen replacement. Estrogen helps to retain bone mass and reduces the risk of coronary artery disease. Undesirable effects of estrogen include an increased risk of endometrial and breast cancer [11]. Also, women taking estrogen are at higher risk for thrombophlebitis. Several approaches can be taken to minimize these risks. The risk of endometrial cancer can be eliminated by a hysterectomy, but this exposes the astronaut to possible surgical and postsurgical problems. Endometrial ablation may be a less invasive approach that avoids the potential complications of postoperative adhesions and bowel obstruction [12]. To minimize the risk of thrombophlebitis, women who take estrogen also should be screened for conditions that might increase the likelihood of blood clots (such as factor V Leiden, protein C, and protein S deficiency).

Reducing the risk of breast cancer is more problematic. A person screened for cancer just before departure is unlikely to develop clinically significant breast cancer during a long-duration mission. But if a cancer should occur, definitive care for breast cancer would most likely not be available. Therefore, it is important to consider preventive measures to minimize breast cancer risk. Selective estrogen-receptor modulators (SERMs) offer one method for preserving bone mass in postmenopausal women without increasing the risk of breast or endometrial cancer [13, 14]. Clinical data indicate that SERMs (e.g., raloxifene) have estrogenlike effects on bone (increase in bone mineral density) and on lipid (decrease in total and LDL cholesterol levels) metabolism. They act as estrogen antagonists in uterine and breast tissues. In other words, these drugs are specific for the estrogen receptors in bone and have less of an effect on the estrogen receptors in breast tissue. They still carry the risk of thrombophlebitis, which has to be balanced against the positive effects on bone and blood lipids.

Orthostatic Intolerance

Data from short-duration space missions have shown that, on average, women are more likely to be presyncopal than men. Twenty-eight percent of women experienced presyncope during a postflight orthostatic test, while only 7% of men became presyncopal [9]. These data fit with a larger set of data that suggest that women in general have lower orthostatic tolerance. At a given level of orthostatic stress, women compared to men will often have a higher heart rate, lower blood pressure, and greater increase in sympathetic nervous system activation [15]. Evidence suggests that gender differences in cardiovascular regulation lead to these effects [16–18]. The exact mechanisms for the difference are not clear but could include reduced baroreflex sensitivity, diminished vasoconstrictor reserve, or an effect of estrogen on nitric oxide metabolism. Estrogen may increase the production of nitric oxide, which is a vasodilator.

As is the case with many gender differences, this average difference may not reflect what will happen on a given mission. Just because women, on average, have a higher likelihood of orthostatic intolerance does not mean that the women on a particular flight will have difficulties with orthostatic stress. As was discussed in chapter 7, the best approach to orthostatic intolerance is individualized management. Crew members should be treated based on their individual needs, and their gender should not be the primary concern.

One area where the gender difference in orthostatic tolerance could become an issue is in selection for missions. Severe orthostatic intolerance after landing either on Earth or on another planet could be a significant safety issue if the crew had an emergency that required rapid egress. If, on previous space missions, an individual had significant, symptomatic orthostatic intolerance after landing that was not mitigated by the countermeasure program (G-suit, fluid loading, etc.), this would indicate that he or she might not be a good choice for a long-duration Mars mission. It is likely that there would be more women than men in this group.

Decompression Sickness Susceptibility

Some studies have suggested that women are more susceptible than men to decompression sickness (DCS) during hypobaric exposures. More recent studies, however, have not confirmed these findings. Operational experience in Air Force altitude chambers does not show an increased DCS risk for women [19], and altitude exposures at NASA that simulate EVA-suit exposures also have not shown a higher incidence of DCS in women [20]. Webb et al. [21] performed a prospective study with 197 men and 94 women exposed to simulated altitude [21]. The subjects were monitored for precordial venous gas emboli (VGE) and DCS symptoms. No significant differences in DCS incidence were observed between men (49.5%) and women (45.3%). In fact, VGE occurred at significantly higher rates among men than women under the same exposure conditions. In a summary of the published literature, Bove [22] concluded that there is no strong evidence at present that women are at a higher risk for DCS with either altitude exposure or diving.

Some studies have suggested that menses, or changes in hormonal cycles, increase the risk of DCS. Dixon et al. [23] exposed 30 women to a 6-hour hypobaric exposure. There were five cases of DCS, all in women who were menstruating at the time of exposure. Only 32% of those who did not develop DCS were menstruating. In a study by Webb et al. [21], women using hormonal contraception showed significantly greater susceptibility to DCS during the last 2 weeks of the menstrual cycle compared to those not using hormonal contraception. Overall, however, Bove [22] concluded that, based on the available data, menstruating women do not appear to be at a higher risk for DCS [22]. Unfortunately, no definitive study has been done in this area.

Kidney Stone Susceptibility

In the United States, the prevalence of kidney stones (i.e., percentage who have ever had a kidney stone) is as high as 15% of men and 7% of women [24]. The exact reasons for this are not clear, but men do have greater urinary excretion of calcium, oxalate, and uric acid. Higher calcium and uric acid excretion has been seen in male astronauts [9]. Table 10-4 presents data collected on both male and female astronauts before and after Space Shuttle missions. The data show that men have a significantly greater urinary supersaturation of calcium oxalate than women both before and after spaceflight. This indicates a greater risk of kidney stone formation. As is the case with most other gender-related differences, the variability within genders far exceeds the differences between them. Nevertheless, if a urinary calcium monitoring strategy is adopted to prevent nephrolithiasis (see chapters 1 and 12), those individuals in whom urinary supersaturation of calcium oxalate cannot be kept within limits are more likely to be men than women.

Aggressiveness, Competition, and Cooperation

Men are much more likely than women to be in accidents or involved in violent behavior. Historically, male-on-male homicide occurs 30–40 times more frequently than female-on-female homicide [25]. Men kill each other much more frequently

Table 10-4. Urinary chemistry data collected on male and female astronauts before and after Space Shuttle flights.

Parameter	Preflight			Postflight		
	Men	Women	P	Men	Women	P
pH	5.99	6.15	<.05	5.63	6.03	n.s.
Calcium	202.3	117.3	<.001	254.6	152.0	<.001
Oxalate	36.7	40.3	n.s.	36.2	34.3	n.s.
Citrate	706.3	722.0	n.s.	623.0	539.8	n.s.
Uric acid	663.7	521.6	<.001	597.3	455.0	<.001
Calcium oxalate supersaturation	1.72	1.10	<.001	2.42	1.81	<.02

Data collected on 240 men before flight and 239 after flight (except for calcium and oxalate data, which include 239 and 238 respectively). Data collected before and after flight on 37 women. Data from Harm et al., [9] with permission from the American Physiological Society.

than women do, and they kill women much more frequently than women kill men [25]. The majority of these homicides, as well as serious nonlethal assaults, seem to result from male–male competition for social status or from disputes over relationships with women. When men kill women, it is often related to sexual jealousy [25]. Male violence, however, is not inevitable. In a summary on the evolution of human sex differences, Geary [26, p. 139] states: "men clearly have the capacity for intense and sometimes deadly one-on-one, as well as coalition-based competition, although men are not mindlessly driven to physical aggression nor are they biologically destined to physical combat."

In modern life, homicide is rare in professional work groups. To achieve career success, most people know to keep competition within appropriate bounds and to collaborate and cooperate with others. Most successful Antarctic, wilderness, and space exploration expeditions have been all-male groups that worked cooperatively to achieve remarkable goals. In fact, competition with others often formed the motivation for exploration efforts and gave the explorers a common purpose. But the drive for success and the possibility of violence may be two sides of the same coin. A strong competitive drive is essential to move a project forward, provide a crew with a shared goal, and allow participants to tolerate the hardships involved. If, however, rivalry within the group gets out of control, or if sexual jealousy develops, a strong competitive drive has the potential to destroy a mission. Whether male or female, crew members chosen for a space mission will be successful competitors who know how to push toward a goal. They will need training to make sure that they can recognize when their competitive drives are heading down dangerous pathways.

Although sexual jealousy can occur within mixed-gender groups, there are benefits to mixed groups. Teams where both sexes are represented may be more stable than single-sex groups. In a European Space Agency space simulation study composed of three men and one woman isolated in a hyperbaric chamber, the female crew member was viewed as being a peacemaker and was thought to have played an important role in reducing tension in the group [27]. Also, in a mixed group, team members may be "on their best behavior" and act with restraint and courtesy. In a review of group performance at a French polar station, Rosnet et al. [28] compared the living climate in all-male groups compared to mixed gender groups. They found that women in the wintering group improved the general social climate by moderating men's behavior. Nevertheless, problems with sexual jealousy and unwanted sexual advances have occurred in both Antarctica and in space mission simulations [27]. In the study by Rosnet et al., the mix of genders was a stressor because seduction behaviors appeared, resulting in rivalry, frustration, and sexual harassment [28].

Depression and Anxiety

Roughly twice as many women suffer from significant anxiety and depression compared to men [25]. In the United States, about twice as many women as men experience depressive symptoms or seek treatment for depression. Data from isolated and confined groups at Antarctica also suggest that depressive symptoms were more common in women than in men [29]. The reasons for the differences are poorly understood. Sociology studies suggest that women are more likely to experience anxiety and depression over disrupted personal relationships. In addition, compared to men, women show a higher level of parental investment in the care of children. Mother–child relationships can be a major source of anxiety and concern. Selection for a Mars mission, which could involve a 2.5-year separation, would be stressful for any married individual with a family. It is possible, however, that on average it would be more stressful for women than men.

Menstruation

In general, no significant problems have been reported concerning menstruation in space, and the sanitary products that work on Earth also have been successfully used in space [12]. One concern about menstruation in space is the occurrence of retrograde menstruation, which allows blood products and cellular debris to enter the abdomen. This is thought to be the primary cause of endometriosis. On Earth, some degree of retrograde menstruation is normal, but in weightlessness, retrograde menstruation might be increased, and the products that enter the abdomen may not be confined to the pelvis. This might increase the risk of endometriosis and create atypical presentations for the disease. To date there have been no abdominal symptoms, shoulder pain, or reduced menstrual flow in space that would suggest increased retrograde menstruation [12]. Unfortunately, no formal studies of retrograde menstruation have been undertaken. If needed for operational reasons, it is possible to suppress or alter menstruation in space by using oral contraceptives.

Contraception

Numerous animal studies suggest that embryos will develop differently in weightlessness. Young rats that spent 16 days in weightlessness differed in their righting responses compared to rats that developed on Earth [30]. Similarly, there may be differences in brain organization and function that occur as a result of weightlessness exposure during early development [31]. Although the data are not definitive about exactly what might happen to a human embryo conceived in space, they clearly indicate that every effort should be taken to ensure conception does not occur. The high radiation levels on a spacecraft also would present a risk to a developing fetus.

It is likely that a well-trained and properly selected mixed-gender crew would be aware of the hazards presented by sexuality onboard and would act accordingly. Nevertheless, medical planners need to provide the means to prevent pregnancy. Female oral contraceptives (typically ethinyl estradiol combined with a progestin) offer several benefits in addition to birth control [12]. These noncontraceptive benefits include lower rates of endometrial cancer, lower ovarian cancer rates, more regular menses with less flow, less likelihood of developing anemia, less salpingitis, and increased bone density. Possible, but not proven, benefits include less endometriosis, less benign breast disease, less rheumatoid arthritis, protection against atherosclerosis, fewer fibroids, and fewer ovarian cysts [32]. The use of oral contraceptives also would allow the menses to be controlled, if this were important for operational reasons.

Oral contraceptives, however, increase the production of clotting factors and increase the risk of venous thromboembolism. Typically, screening for hypercoagulable states is not performed routinely for individuals starting oral contraceptive unless they have a family history or a previous episode of venous thromboembolism. In the spaceflight setting, however, where in-flight venous thromboembolism would be difficult to manage, screening for hypercoagulable states may make sense. In this case, the crew members should be checked for antithrombin III deficiency, protein C deficiency, protein S deficiency, the factor V Leiden mutation, the prothrombin gene mutation, and the antiphospholipid syndrome [32]. Another potential problem with oral contraceptives is that they may increase the risk of breast cancer. This risk, however, appears to be small.

For males who no longer want children (or have sperm banked), vasectomy is an option. Similarly, surgical sterilization is an option for women who have completed childbearing. Instead of oral contraceptives, women also have the option to choose either implants or injectable contraception. An intrauterine device (IUD) is also an option. IUDs are used extensively throughout the world (they are the most widely used method of reversible contraception), but are less commonly used in the United States. Modern IUDs include hormone-releasing IUDs that release a progestin into the uterus [32]. Current IUDs do not increase the risk of pelvic infection; the risk of infection from IUDs occurs at the time of insertion.

Ultimately, the prevention of pregnancy on a long-duration voyage depends on the professionalism, wisdom, and training of the crew. Although contraception should be provided, the crew will need to realize that no contraceptive method is 100% effective. In the first year of use, failure rates for the combination pill, levonorgestrel IUD, and male sterilization are all about 0.1%.

Clinical Concerns

There are differences in the susceptibility of men and women to different diseases and toxins. Women are more likely to develop autoimmune conditions, but less likely to develop coronary artery disease at a young age. The metabolism of some drugs and toxins also can differ on average between men and women. Women are more likely to develop gall bladder disease. Migraine headaches are more prevalent in women, while cluster headaches are more common in men. In general, however, each person on a mission will need to be treated as an individual; knowledge about an average difference between sexes may not be helpful in individual cases.

Men need to be concerned about diseases of the prostate and testicles, and women must be aware of potential problems with the breast, ovaries, and uterus. It is possible that during a long-duration mission women might develop uterine bleeding from a variety of causes. On Earth this might require a dilation and curettage, but in space, therapies such as gonadotropin-releasing hormone (GnRH) agonists (e.g., leuprolide or goserelin) can be used to treat leiomyomata uteri (fibroids), endometriosis, adeno-myosis, or other forms of dysfunctional bleeding [12]. GnRH agonists have been helpful short-term in reducing menorrhagia and reducing uterine size. These com-pounds offer a way to treat some gynecological conditions in space without resort-ing to surgery. Another concern is the development of cancer in the urogenital tract, reproductive organs, or breasts. If careful screening is done before the mission, it is unlikely that a crew member would develop a significant cancer of the prostate, ova-ries, or breast over the course of a 2.5-year Mars mission.

Single- versus Mixed-Gender Crews

One issue that arises when discussing male–female differences is whether it would be simpler and more efficient to use single-sex crews. Historically, most expeditions have been all male, but in recent years there have been all-female missions to under-water habitats and to Antarctica [27]. A single-sex crew would provide assurance that conception would not occur and would prevent sexual rivalries and jealousies from arising between sexes. Choosing crew members from just one gender, however, would exclude a large number of qualified people and reduce crew diversity. Also, although it is essential to exclude people with diseases or physical characteristics that are incompatible with the mission, excluding all members of a gender, regardless of their qualifications, does not give individuals a chance to prove their worth. Because a long-duration crew can be evaluated before flight for their ability to work together, it may be possible to include an intermediate step in selection where a mixed-gender crew's ability to work together and deal with sexuality can be assessed.

As already mentioned, mixed-gender crews can experience sexual jealousies and rivalries. This has occurred at Antarctica [28]. Also, the drive for sex is a strong and important part of human nature. A recent National Opinion Research Center (NORC) survey reports that married couples in the age range of the astronaut corps have sexual intercourse approximately 60–70 times a year [33]. On one hand, infidelity is common in everyday life and might easily occur when people are in close contact for a long time in a confined environment. On the other hand, people can restrain their sexual urges. The NORC survey mentioned above summarized scientific sur-veys about extramarital relations. Overall, these data indicate that extramarital affairs are less prevalent than popular media accounts would suggest. Approximately 3–4% of currently married people have a sexual partner besides their spouse in a given year. Only about 15–18% of ever-married people have had a sexual partner other than their spouse while married [33]. In other words, in everyday life, even though many people work closely with members of the opposite sex, they still manage to remain faithful. One conclusion is that many married people know not to seek out or encourage other relationships that would damage their marriage. During times of war, many couples remained faithful through long separations. In Naval service, although some married sailors have been involved in extramarital relationships during long separations, many have not. In fact, the Uniform Code of Military Justice makes adultery illegal in the military. Individuals with normal sexual urges have honored vows of religious celibacy. Almost 30% of unmarried individuals (which includes those who may have been widowed or divorced) in the 40–49 age group reported having no sex at all in the previous 12 months in the NORC survey [33]. Often sexual tension can be relieved without intercourse (e.g., masturbation). Studies suggest that about 45% of married women and 85% of married men masturbate regularly [34]. Overall, the data show that while sex is a powerful urge, it is also possible for people to control this urge for a greater good (fidelity) and to abstain from sexual intercourse when necessary. Selecting a mixed-gender crew does not guarantee that there will be problems with sexual jealousy, rivalries, or harassment, although they clearly can occur. Crew members need to know the nature of the commitment they are making, understand the dangers that sex can pose to their group, and develop a strategy for dealing with sexuality.

Recommendations Based on Current Knowledge

For most factors, the average differences between men and women are much smaller than those among men or women. Because maintaining a crew in space is focused on a particular set of individuals, knowledge about general male–female differences may not be useful routinely. Nevertheless, the differences that exist can be important in certain situations. During selection, more women than men will be unable to meet an upper body strength requirement. More men than women are likely to be excluded due to kidney stones. For a mixed-gender crew, plans will need to be established to ensure that pregnancy does not occur during the mission. Some general recommendations to accommodate gender differences are listed below:

1. For crew selection, the fairest approach is to choose the best qualified people, using objective selection criteria. Prospective crew members should be allowed to demonstrate their ability to work together. The mission planners should design the equipment to fit as many people as possible and provide objective requirements that must be met to complete the tasks the mission requires. These requirements may favor one gender over another (for example, a strict requirement for upper body strength), but because the variability within genders is so high, individuals from both genders should be able to meet most reasonable selection criteria.

2. Selecting crews based on qualifications and background is fair, and this will likely lead to crews with a mix of men and women, who may be single or married. These mixed crews will need training on potential problems that can arise in mixed-gender groups and will need to establish rules for behavior (including sexuality) that they should follow closely. Although it is unfair to assume that

a mixed-gender crew will automatically have problems with sexuality, it is also unreasonable to expect that a team of men and women randomly assembled on the basis of their technical qualifications can always be made to function smoothly. For a long-duration mission (e.g., Mars), it may be prudent to include an observation period early in the training flow that includes isolation and confinement to assess how the crew comes together as a team. A group that works closely with the crew members, such as the psychology team, will need the full authority (free from political pressures) to remove crew members or to reconstitute teams if a problem with harassment, flirtatiousness, or jealousy arises.

3. For postmenopausal women, selective estrogen receptor modulators (e.g., raloxifene) offer a way to minimize bone loss due to estrogen deficiency without increasing the risk of endometrial and breast cancer. Additionally, endometrial ablation could be considered before flight. Screening for a propensity to venous thrombosis (e.g., factor V Leiden) should be carried out if estrogen receptor drugs are given.

4. Although there are no firm data proving menstruation increases the risk of DCS, if possible, the scheduling of EVAs should take this into account.

5. Female crew members must not conceive in space. Abstinence from sexual intercourse is the most effective approach to avoid conception. Contraception also should be provided for crew members on mixed-gender crews. Oral contraceptives are attractive because they have noncontraceptive benefits and also allow for the control of menses. If oral contraceptives are used, the crew members should be screened for factors that increase the risk of venous thromboembolism.

References

1. Wright, R., Mary L. Cleave oral history transcript, in *NASA Johnson Space Center Oral History Project*. 2002, NASA, Houston, TX. (http://www.jsc.nasa.gov/history/oral_histories/oral_histories.htm)
2. Campion, E., and K. Herring, Astronaut Lawrence to Remain in United States [press release]. October 24, 1995, NASA, Washington, DC.
3. Lebedev, V., *Diary of a Cosmonaut: 211 Days in Space*, D. Puckett and C.W. Harrison, eds. 1988, PhytoResource Research, College Station, TX.
4. Institute of Medicine Committee on Understanding the Biology of Sex and Gender Differences, Sex affects behavior and perception, in *Exploring the Biological Contributions to Human Health: Does Sex Matter?* T.M. Wizemann and M.L. Pardue, eds. 2001, National Academy Press, Washington, DC, pp. 79–106.
5. Institute of Medicine Committee on Understanding the Biology of Sex and Gender Differences, Sex begins in the womb, in *Exploring the Biological Contributions to Human Health: Does Sex Matter?* T.M. Wizemann and M.L. Pardue, eds. 2001, National Academy Press, Washington, DC, pp. 45–78.
6. Institute of Medicine Committee on Understanding the Biology of Sex and Gender Differences, Introduction, in *Exploring the Biological Contributions to Human Health: Does Sex Matter?* T.M. Wizemann and M.L. Pardue, eds. 2001, National Academy Press, Washington, DC, pp. 13–27.
7. Baker, M.A., Sensory functioning, in *Sex Differences in Human Performance*, M.A. Baker, ed. 1987, John Wiley and Sons, New York, pp. 5–36.
8. Percival, L., and K. Quinkert, Anthropometric factors, in *Sex Differences in Human Performance*, M.A. Baker, ed. 1987, John Wiley and Sons, New York, pp. 121–41.
9. Harm, D.L., et al., Invited review: gender issues related to spaceflight: a NASA perspective. Journal of Applied Physiology, 2001. 91(5): 2374–83.

10. Wardle, M.G., M.R. Gloss, and D.S. Gloss, Response differences, in *Sex Differences in Human Performance*, M.A. Baker, ed. 1987, John Wiley and Sons, New York, pp. 107–20.

11. Clemons, M., and P. Goss, Estrogen and the risk of breast cancer. New England Journal of Medicine, 2001. 344(4): 276–85.

12. Jennings, R.T., and E.S. Baker, Gynecological and reproductive issues for women in space: a review. Obstetrical and Gynecological Survey, 2000. 55(2): 109–16.

13. Sporn, M.B., et al., Role of raloxifene in breast cancer prevention in postmenopausal women: clinical evidence and potential mechanisms of action. Clinical Therapeutics, 2004. 26(6): 830–40.

14. Barrett-Connor, E., et al., Risk-benefit profile for raloxifene: 4-year data from the Multiple Outcomes of Raloxifene Evaluation (MORE) randomized trial. Journal of Bone and Mineral Research, 2004. 19(8): 1270–75.

15. Custaud, M.A., et al., Orthostatic tolerance and spontaneous baroreflex sensitivity in men versus women after 7 days of head-down bed rest. Autonomic Neuroscience, 2002. 100(1–2): 66–76.

16. Gotshall, R.W., P.F. Tsai, and M.A. Frey, Gender-based differences in the cardiovascular response to standing. Aviation, Space, and Environmental Medicine, 1991. 62(9 Pt 1): 855–59.

17. Gotshall, R.W., Gender differences in tolerance to lower body negative pressure. Aviation, Space, and Environmental Medicine, 2000. 71(11): 1104–10.

18. Convertino, V.A., Gender differences in autonomic functions associated with blood pressure regulation. American Journal of Physiology, 1998. 275(6 Pt 2): R1909–20.

19. Wirjosemito, S.A., J.E. Touhey, and W.T. Workman, Type II altitude decompression sickness (DCS): U.S. Air Force experience with 133 cases. Aviation, Space, and Environmental Medicine, 1989. 60(3): 256–62.

20. Waligora, J., D. Horrigan, and J. Gilbert, Incidences of symptoms and venous gas bubbles in male and female subjects after decompression [abstract]. Aviation, Space, and Environmental Medicine, 1989. 60: 511.

21. Webb, J.T., N. Kannan, and A.A. Pilmanis, Gender not a factor for altitude decompression sickness risk. Aviation, Space, and Environmental Medicine, 2003. 74(1): 2–10.

22. Bove, A.A., Fitness to dive, in *Bennett and Elliot's Physiology and Medicine of Diving*, A.O. Brubakk and T.S. Neuman, eds. 2003, Saunders, New York, pp. 700–17.

23. Dixon, G.A., R.W. Krutz, Jr., and J.R. Fischer, Decompression sickness and bubble formation in females exposed to a simulated 7.8 psia suit environment. Aviation, Space, and Environmental Medicine, 1988. 59(12): 1146–49.

24. Goldfarb, D.S., and F.L. Coe, Prevention of recurrent nephrolithiasis. American Family Physician, 1999. 60(8): 2269–76.

25. Geary, DC., Sex differences in modern society, in *Male, Female*. 1998, American Psychological Association, Washington, DC, pp. 307–31.

26. Geary, DC., Sexual selection in contemporary humans, in *Male, Female*. 1998, American Psychological Association, Washington, DC, pp. 121–58.

27. Kanas, N., and D. Manzey, Human interactions, in *Space Psychology and Psychiatry*. 2003, Kluwer Academic Publishers, Boston, MA, pp. 75–106.

28. Rosnet, E., et al., Mixed-gender groups: coping strategies and factors of psychological adaptation in a polar environment. Aviation, Space, and Environmental Medicine, 2004. 75(7 Suppl): C10–13.

29. Palinkas, L.A., et al., Incidence of psychiatric disorders after extended residence in Antarctica. International Journal of Circumpolar Health, 2004. 63(2): 157–68.

30. Walton, K.D., et al., Motor system development depends on experience, in *The Neurolab Spacelab Mission: Neuroscience Research in Space*, J.C. Buckey and J.L. Homick, eds. 2003, NASA, Houston, TX, pp. 95–104.

31. Raymond, J., et al., Development of the vestibular system in microgravity, in *The Neurolab Spacelab Mission: Neuroscience Research in Space*, J.C. Buckey and J.L. Homick, eds. 2003, NASA, Houston, TX, pp. 143–50.

32. Speroff, L., and P.D. Darney, Intrauterine contraception (the IUD), in *A Clinical Guide for Contraception*. 2001, Lippincott, Williams and Wilkins, New York, pp. 221–57.
33. Smith, T.W., *American Sexual Behavior: Trends, Socio-Demographic Differences, and Risk Behavior*. 2003, National Opinion Research Center, University of Chicago, Chicago.
34. Blum, D., Counterstrikes: Love, lust and rape, in *Sex on the Brain: The Biological Differences Between Men and Women*. 1997, Viking Penguin, New York, pp. 220–52.

11

Preflight Preparation and Postflight Recovery: Preparation and Rehabilitation

Introduction

When the Greek explorer Odysseus returned from his long-duration voyage, only his dog recognized him. Even though his loved ones had yearned for a reunion with him, when Odysseus finally arrived home, his reentry into his previous life was not smooth. Odysseus had to work hard to establish a relationship with his son and dispose of the suitors who were pursuing his wife. He literally had to fight his way back to his former position.

Odysseus's case, of course, differs from what an astronaut might experience. Although Odysseus had aged and his appearance had changed, he did not require physical rehabilitation. Also, he had not had the benefit of frequent communication with his family while he was away. But the story of Odysseus highlights an important, and sometimes overlooked, aspect of the preparation for and rehabilitation after spaceflight. Physical preparation and rehabilitation (e.g., exercise training) is only part of the challenge. An astronaut on a long-duration voyage may be leaving a spouse and family behind, and this family will adapt to the new circumstances. A family that is not fully supportive of the voyage at the outset may develop strong and long-lasting resentments. Even a supportive family will change over time as it learns to cope with the astronaut's absence. After the mission, the astronaut cannot expect to return to the life that existed before departure. Adjustment will be needed on all sides.

If the crew member has a spouse and/or children, the family needs to fully understand the commitment demanded by a long-duration mission. Also, the crew member needs to prepare physically and mentally before the mission and be ready for readjustment and rehabilitation challenges after returning. This chapter reviews preflight preparation and postflight rehabilitation and outlines strategies to address these issues.

Preparation and Rehabilitation

The need for preparation and rehabilitation depends on the mission. A relatively short stay on a space station differs substantially from a 2.5-year mission to Mars. Clearly, the Mars mission requires more preparation. The nature of the mission will also influence the rehabilitation needed afterward. If the countermeasure program was effective and the mission was completed successfully, the crew may be in good shape both mentally and physically. If, however, the mission overtaxed the crew or operational challenges interfered with the countermeasure program, then the crew members may need significant rehabilitation.

Physically, the main areas requiring rehabilitation are the musculoskeletal and neurovestibular systems (described in chapters 1, 4, and 6). Depending on the type and effectiveness of countermeasures used, some degree of muscle weakness and postural instability likely will be present. Too aggressive a rehabilitation program can lead to early postflight injury; too slow a program is inefficient for the crew members.

Mentally, the need for rehabilitation will depend on each crew member's individual situation. If the astronaut has a family, the long separation dictated by a Mars mission is likely to introduce several stresses into family life. During the mission, the family may have become more independent or may have established new preferences. Children will have grown and changed while the astronaut was away. In some cases, it is possible that astronauts will return to children who do not know them very well at all. Experience with wars and other stressful events shows that long separations may drive people apart. Divorce rates, for example, have increased after most wars [1]. Both physical and mental rehabilitation should be individualized. Nevertheless, guidelines can provide information on best practices and proven approaches.

Musculoskeletal Changes

Before a mission, a crew member needs to prepare for the physical challenges that training and the mission may present. Once in flight, extravehicular activities or excursions on a planetary surface require strength and endurance. After the mission, as outlined in chapters 1 and 4, the major musculoskeletal losses are likely to be in the postural muscles and the load-bearing portions of the skeleton. The crew member will need to work to get strength back.

The main problem after spaceflight is that the combination of postural muscle loss and vestibular instability leaves the astronaut more susceptible to muscle or joint injury. Athletes who try to work at peak performance levels after a long break from exercise can injure muscles and experience significant muscle soreness. Similarly, crew members who stress muscles excessively after landing can experience the same effects. Intense activation of atrophied muscle can lead to trauma and inflammation in the muscle. The damage can include sarcolemmal disruption, distortion of the myofibrils' contractile components, and cytoskeletal damage [2]. Studies on muscles from animals who flew in space showed edema, tears, and necrosis within the anti-gravity muscles after landing [3, 4]. Examination of muscles during spaceflight does not show these changes, demonstrating that the necrosis was due to reloading of the muscle after the spaceflight [3, 4].

Atrophy in the back muscles leave astronauts more susceptible to back injuries after flight [5]. Also, the instability in posture that occurs after spaceflight could lead to intense activation of some postural muscles if the astronaut gets off balance. These sudden, intense muscle activations, when they occur in muscles unaccustomed to regular use, can lead to injury.

Preflight preparation

Ideally, the crew member should start the mission confident, rested, and in good physical condition. A rushed or overly demanding training schedule can lead to a crew that is harried, tired, and deconditioned before entering orbit. A high level of physical fitness provides benefits before a mission. During training, fitness improves performance during strenuous activities. Maintaining strength and flexibility in joints also helps avoid preflight injury. Once in flight, an astronaut's reserves of strength or endurance could be called upon in an emergency or unexpected situation. Physical activity also has both mental and physical benefits, including a possible reduced risk of cancer. One disadvantage of extensive physical training is that detraining can occur quickly while in space. Training-induced increases in maximal oxygen uptake

Table 11-1. General training principles to maintain strength and endurance before spaceflight.

---

1. Aerobic activity above 50% of maximal, with peaks to maximal effort (wind sprints, interval training).
2. Circuit training on major muscle groups at 6–10 repeat maximum. Do not exercise same muscle group 2 days in a row.
3. Include sports such as racquetball, squash, and basketball to work on agility and balance.
4. Keep workouts to 1–1.5 hours to allow them to be scheduled effectively in the training flow.
5. Become familiar with strengthening exercises using body weight because these are useful for travel or busy periods (heel raises, toe raises, leg lifts, crunches, lunges, knee bends, push ups, pull ups).

The goal of the training program is to (1) improve maximal aerobic power and endurance, (2) increase muscle strength, and (3) maintain joint mobility, enhance metabolism of articular cartilage, and improve coordination.

will not be sustained in space unless a training program is pursued in space as well. Similarly, training-induced strength increases can disappear rapidly once a training program is stopped.

Strength can be maintained using circuit training with either free weights or isotonic weight machines. Aerobic exercise for 30 minutes 3 times a week at greater than 50% of maximal oxygen uptake is the minimum amount of exercise that can be done while still receiving a training effect [6]. Sports such as basketball, soccer, or racquet games help to sustain agility. In general, physical training should be incorporated into the preflight training flow to make sure that the crew members can sustain and even increase their levels of fitness before flight. General training principles to be used before flight are presented in table 11-1. A workout that can be kept to 1 or 1.5 hours will be easier to schedule and is more likely to be completed than a longer, more demanding session.

Postflight assessment

After a long-duration mission, an astronaut should first be assessed to establish the nature and degree of any musculoskeletal changes that have occurred. Because of its high precision and reproducibility, dual x-ray absorptiometry (DEXA) is used in clinical studies of osteoporosis to assess changes in bone density. [7]. This technique also has been used after spaceflight. Although this is the most sensitive method to assess bone density, it does increase the amount of radiation the crew members receive. Nevertheless, these postflight measurements are important because they provide objective data on the effectiveness of the countermeasure program, establish a postflight baseline, and add to the database on the skeletal changes that occur during spaceflight.

Magnetic resonance imaging offers a comprehensive assessment of muscle volume. With MRI, precise volume measurements can be made on muscles throughout the body [8]. Just as with the DEXA measurements, these measurements are critical after flight because they provide feedback on countermeasure effectiveness and allow for an individualized rehabilitation plan to be instituted. Simple anthropometric measurements can also provide important data in addition to the MRI data. Measures of

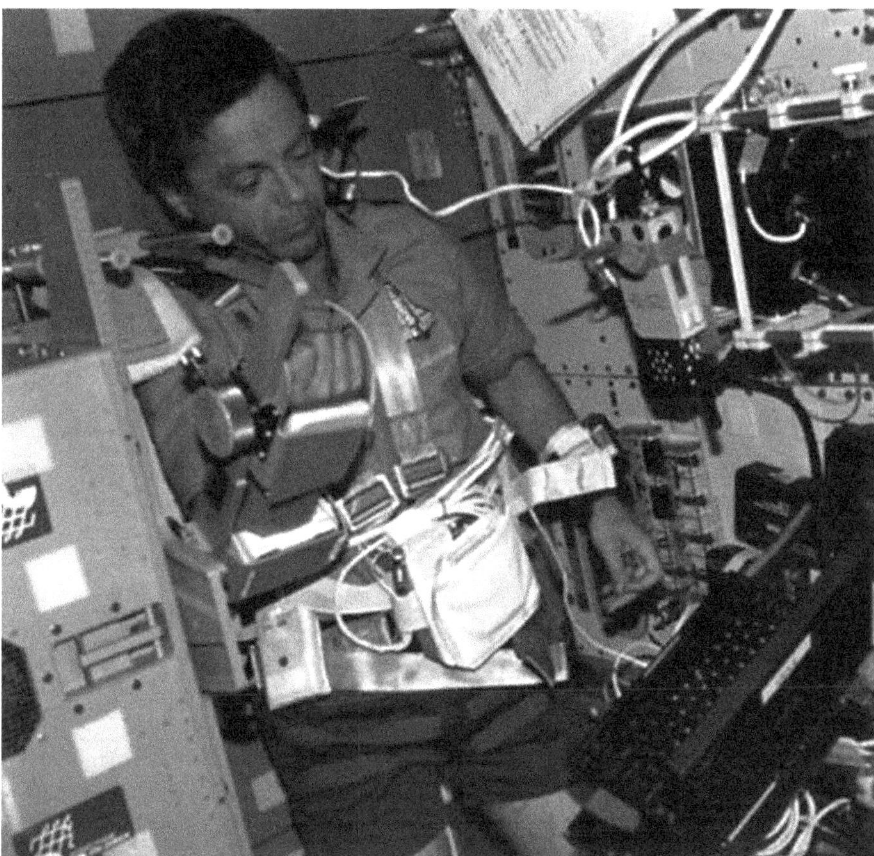

Figure 11-1. Picture of isokinetic dynamometer. Robert Thirsk works on the torque-velocity dynamometer on the STS-78 LMS mission. As he moves his right arm through a full range of motion, the resistance generated by the device varies to keep the velocity of movement constant. This approach provides a widely used method to estimate muscle strength. (Photo courtesy of NASA.)

calf circumference, leg volume, and arm circumferences can be used to follow the crew member's progress over time during the rehabilitation program [9, 10].

Although MRI imaging can provide information on muscle volume or muscle cross sectional area, it is not a direct measure of muscle strength. The most commonly used method to assess muscle function in rehabilitation settings is isokinetic dynamometry [11, 12]. As shown in figure 11-1, with isokinetic dynamometry the subject pushes or pulls against a lever arm connected to the dynamometer. While the subject produces a maximal effort, the machine works to keep the speed of the lever arm constant. This means that the resistance that the machine presents changes as the exercise is executed. For example, if muscle strength falls off as the exercise is being performed, then the resistance the machine presents also will go down to keep the velocity of the movement constant. This kind of muscle testing offers a safety advantage over test-

Table 11-2. Recovery times for various muscles after bed-rest periods of greater than 36 days.

| | | Days to recover | |
Region	Bed rest days	No countermeasure	Countermeasure
Leg volume	180		30–40
Calf circumference	62		>10
Heart size	70		30-60
Bone mineral, total body	119	~180 days	
Maximal oxygen uptake	45		~18 days
Muscular force (thigh, leg, trunk)	49	10–16 days	

Data extracted from Greenleaf and Quach [24]. The data shown are approximate times for the physiological parameters listed to return to baseline values after a bed-rest period of greater than 36 days. If the study had a countermeasure program implemented during the bed-rest period, the data are included in the "countermeasure" column. In general, the data show that muscular force is recovered rapidly, but it may take more time for leg volume and calf circumference to return to baseline. Bone mineral takes the longest time to recover.

ing with free weights or weight machines. If the muscles are weak after spaceflight, an astronaut might not be able to control a weight at all points during an exercise. The astronaut could struggle with the weight and injure joints, ligaments, or muscles while trying to regain control. With the isokinetic dynamometer, however, the machine will back off if the subject reaches a weak or painful part of the muscle activation.

Isokinetic dynamometry has some limitations, however. The maximal torque developed depends on the testing velocity the dynamometer is set to and the angle of the joint tested. Typical testing schemes use multiple velocities, and care is taken to ensure that the settings for each subject are individualized [11, 12].

Rehabilitation after disuse

Although there is a tendency to view rehabilitation after disuse as an activity requiring a variety of machines and interventions, the activities of daily life (walking, climbing stairs, etc.) can provide a potent stimulus to bones and muscles. Even without a structured strength training program, strength will increase rapidly as an individual begins to walk after a period of immobilization [13]. This suggests that while disuse atrophy can present a risk to crew members, overaggressive rehabilitation also can be a problem. Crew members who gradually increase their activity as tolerated after a spaceflight may do as well as those who follow a more rigorous regime.

The time course of recovery will vary from individual to individual depending on particular circumstances. Some general ranges on the amount of time that should be devoted to rehabilitation come from previous spaceflight experience and from studies on rehabilitation after bed rest. In the Russian program, formal rehabilitation can last for 30–40 days [14]. Initially the cosmonauts recover for 2–3 weeks at their cosmonaut training center, and then they are taken to a health sanatorium for further medical rehabilitation. Table 11-2 shows some of the recovery times for particular

Table 11-3. Muscle rehabilitation techniques.

Technique	Description	Application
Passive and active stretch	Movement of muscles through their normal range of motion	Helps extend the range of motion. Can be passive where traction is applied, or active where the individual moves the muscle.
Massage	Pressure on muscles	Believed to reduce muscle tightness and increase capillary circulation. Massage is also relaxing.
Pool exercises	Performance of exercises in water	Standing in a pool increases orthostatic tolerance and allows for upright exercise in someone who has orthostatic intolerance. Also minimizes the impact loading on bone, which may be useful in the early postflight period.
Swimming	Crawl, breaststroke, etc.	Method of endurance exercises that is low impact and less likely to lead to muscle strain/injury.
Resistance exercise	Free weights, isotonic devices	Best method to regain/improve muscle strength. Start with multiple reps at low load early after flight and increase load, decrease reps as rehabilitation proceeds.

body areas after bed rest studies. In general, muscle strength recovers quickly, but bone can take a long time to recover.

Rehabilitation techniques

There are several possible rehabilitation strategies that can be used to help recover from disuse. On landing day, crew members may have difficulty walking, so rehabilitation can consist of stretching, massage, and isometric contractions. To minimize injury, exercise can be done in a pool to allow for nonimpact running and low-impact plyometric exercises. Swimming is a good exercise immediately after spaceflight because it is also low impact. Table 11-3 lists a variety of different musculoskeletal rehabilitation techniques and their advantages and disadvantages. In general, the best approach is to individualize the program and advance each crew member as tolerated from less stressful, low-impact activities toward more demanding, high impact ones.

Neurovestibular Changes

The most frequent neurovestibular changes are motion sickness (both early in flight and immediately after flight) and postural instability after the mission. Motion sickness symptoms are typically short lived and disappear as the vestibular system adapts to the new environment. To date, there is no evidence of permanent effects on postural stability as a consequence of spaceflight. Postflight postural instability, however, can

Table 11-4. Postflight tests to assess neurovestibular function.

Test	Description	Comments
Romberg	Subject stands with arms folded across the chest, feet shoulder width apart, with eyes open and eyes closed. Standing on foam, which diminishes ankle movement and reduces proprioceptive input, can enhance the test. This makes it more specific for the vestibular system [25].	Results from the Romberg done on dense foam can be similar to the results from dynamic posturography. Rapid test that can be used to follow an astronaut during rehab to measure progress.
Tandem walk	Subject performs a heel-to-toe walk.	With impaired balance subject will be unstable. Easy to perform, can be used as a test to follow progress.
Standing on one leg	The time a subject can stand on one leg, with eyes closed, provides an overall measure of postural stability.	Also a simple test that can be used to follow progress.
Dynamic posturography	Subject stands on a platform that can move with the body's sway. This testing can help distinguish between vestibular, proprioceptive, and visual contributions to balance problems.	Requires the most equipment, but is standardized, repeatable and provides quantitative results [16].

interfere with rehabilitation. Instability also may lead to injury as a consequence of falls or as a consequence of vigorous muscular activation to avoid a fall.

Preflight preparation

Adaptation to stressful motion before spaceflight, in the hopes that this will reduce motion sickness symptoms in space, has been a regular component of cosmonaut training. In the final 2 weeks before launch, cosmonauts ride in a rotating chair daily while executing pitching head movements [15]. These motions usually provoke motion sickness, but over time the cosmonauts are able to tolerate more of this stressful motion. No definitive study exists that demonstrates that this program is effective, but it has remained part of the Russian program, indicating that it is perceived to be useful.

Postflight assessment

As mentioned in chapter 6, after spaceflight changes are seen in both vestibular function and proprioception. Several methods exist to quantify these changes (table 11-4). Simple functional tests, such as the Romberg, tandem walk, or one-legged standing tests, reveal changes in postural stability. These tests, however, are not quantitative and can be hard to standardize. These tests may be helpful in following a given crew member's progress over time. On landing day, however, a more quantitative test is needed because the goal of testing is not just to demonstrate the degree of change in a given crew member, but also to gather information on the overall effectiveness of the in-flight countermeasure program against neurovestibular changes.

Dynamic posturography (described in chapter 6) allows for visual, proprioceptive, and vestibular cues to be isolated in different test sequences so that the main contributions to postural instability can be identified. The test also provides quantitative measures of postural sway and other variables. Because dynamic posturography provides quantitative information on both the vestibular and proprioceptive systems, it is an attractive test for postflight evaluation [16]. The simpler clinical tests (Romberg, tandem walk) are useful for quick assessments of a crew member during the rehabilitation process. To establish a database on postflight neurovestibular changes and track the effectiveness of the countermeasure program, dynamic posturography is preferable.

Postflight neurovestibular rehabilitation

Crew members will show varying degrees of instability after flight. Also, as is the case for muscle recovery, being back in Earth's gravity and walking upright provides a powerful stimulus for readaptation. Experience in sports medicine has shown that certain exercises can help with retraining after injury. A proposed schedule for using these exercises is outlined in table 11-5. These exercises improve balance and coordination and may be particularly useful after spaceflight. Standing on one leg is one simple exercise. Other exercises include balancing on a rocker boards and hopping and exercising on a minitrampoline. Although these are called "proprioceptive" exercises, it is clear that they stimulate both the vestibular system and position sense.

Cardiovascular Changes

Crew members may become stressed in a demanding extravehicular activity or an emergency that may require sustained, intense work. The better their aerobic fitness, the better they will be able to confront these challenges. After spaceflight, they may have lost cardiac muscle mass and are likely to have some degree of orthostatic intolerance. Typically, a gradual resumption of normal Earth activities will reverse these problems. Aerobic fitness can be regained by regular aerobic training. No specific cardiovascular rehabilitation program is typically needed. Nevertheless, reliable tests of orthostatic tolerance and cardiac mass should be obtained after landing to allow for an ongoing assessment of the effectiveness of the overall countermeasure program. Orthostatic tolerance can be assessed with a stand test, tilt table test, or lower body negative pressure evaluation. Currently, the most precise method to determine cardiac mass is MRI [17].

In contrast to the musculoskeletal changes, an early assessment of maximal aerobic exercise capacity after spaceflight may not be essential. Maximal aerobic testing shortly after spaceflight may lead to muscle damage, and the data provided from the test would probably not alter the postflight countermeasure program. A maximal exercise test could be delayed for a few weeks after landing. At that point, if maximal

Table 11-5. Progression of proprioceptive exercises after injury.

Partial weight-bearing
1. Walking with support (crutches, cane) ensuring correct heel-toe movement
2. Seated with feet on rocker board, forward/backwards rocking for 2 minutes, first with both legs, then with one leg

Full weight-bearing
3. Multiaxial rocker
   - 2–3 minutes each way circling
   - Attempt to balance for 15 seconds, rest 10 seconds
   - Progressively increase complexity—arms out in front of body, with
     Arms crossed
     Eyes closed
     Knee bends
     Other leg swinging
     Bounce/catch ball
4. Balance on minitrampoline
   - Same progression as above
   - Hop and land
   - Hop and land with one-quarter turn and return
   - Progress to half turn, three-quarter turn and full turn
   - Rhythmical hopping, alternatively placing toe forwards and sideways
   - Rhythmical hopping across a line, forward/backwards, sideways
5. Jumping
   - Various patterns
6. Hopping without rebounder
   - Alternatively two hops on one leg, two hops on other leg
7. Skipping
   - On spot, both legs, forwards/backwards/sideways
   - Single leg, two hops on one leg, two hops on other leg
8. Advanced tasks
   - Walk/run across a steep hill each way
   - Run sideways up and down hill, each way
   - Walk along balance beam, then bounce a throw ball while walking
   - Sideways step-ups, gradually increasing height of step
9. Running drills
   - Straight
   - Backwards
   - Sideways
   - Circle (5 m diameter)
   - Cutting 90°
   - Zigzag through cones set at 45°

Table reproduced from Kinch [26], with permission of McGraw-Hill.

oxygen uptake is still decreased compared to a preflight baseline, then the rehabilitation program could be altered.

Psychosocial Aspects

The physical and schedule demands of preflight preparation and postflight rehabilitation can crowd out mental preparation and recovery. Family or relationship issues that are not addressed before the mission could lead to in-flight crises as a crew member strives to deal with a dissolving marriage or a troubled child. The feelings of guilt or anger that these situations engender could create psychological distress on the spacecraft. After the mission, inadequate attention to reintegration could potentially create significant tensions in families and relationships.

Preflight preparation

Before flight, psychosocial preparation involves both personal and, if the astronaut has a spouse, child, or other significant relationship, family preparation. On a personal level, the crew member needs to be confident and free from excessive worry or nervousness. One potential source of anxiety is concern about an inability to complete certain complex tasks successfully (e.g., an extravehicular activity) or of making an error at a critical time. To help reduce these concerns, several of the techniques from sports psychology can be used. Visualization, where the crew members can imagine working through some of the more demanding procedures they are likely to encounter, is an effective psychological technique [18]. Similarly, techniques such as progressive muscle relaxation and centering are also helpful. Progressive muscle relaxation involves contracting and relaxing specific muscle groups in turn. Centering involves controlled breathing. Both these techniques keep anxiety in check before demanding events [18].

To increase confidence in the ability to handle emergencies or complex systems, the training program should be carefully designed to escalate its demands in accord with the crew member's ability to deliver. Simulations need to be demanding and challenging but should allow the crew to develop mastery of the systems. Very demanding simulations can be used as a "wake-up call" if the crew is not meeting performance goals, but forcing the crew into extreme scenarios where they repeatedly fail can erode confidence among the crew and the team. This can be counterproductive.

A long-duration mission will present the crew with a challenging group-living situation. Crew members will need to function within a cohesive team, while also maintaining their individuality. They will need to be able to discuss problems openly, without creating long-lasting resentments. Each crew member will need to accept the weaknesses of other crew members but still be able to note when a crewmate's performance has declined to the point where it should be discussed. Crew members need training in group living, including instruction on conflict resolution, cultural differences, and teamwork. Drawing up a set of agreed-upon rules before the mission can help avoid conflicts later on. These rules could cover potentially explosive issues such as romance, criticism, courtesy, and privacy. Outdoor leadership and survival training, which forces the crew to work as team and to depend on each other, is one way to develop the bond that must exist among the crew [19–21].

If the crew member has a family, significant stresses can develop. As the mission approaches, the crew member will probably become the focus of attention within the family. The crew member will be drawn away for last-minute training at the same time that the approaching mission will likely be producing anxiety and concern at home. The impending separation could represent a significant stress. Studies of wartime separations show that divorce can often result. After World War II, the divorce rate spiked. Some attributed this to the large number of rapid romances culminating in wartime marriages just before the soldiers were deployed overseas. A study performed in one cohort of WWII veterans, however, showed that it was the older veterans, who had been married for some time before deployment, who had the highest divorce rate [1]. One possible conclusion from this study is that the individuals who married just before deployment knew that a time of separation would follow, while for those who had already been married for some time, the wartime separation may have represented a major change in expectations. It is possible that a similar dynamic could exist in some astronaut marriages. Most astronauts are accepted into the program after they have already begun a career in another area (e.g., military aviation, scientific research). It is possible that, in some marriages, the astronauts' spouses might perceive selection for a long-duration mission as something they had not bargained for.

The impeding separation can affect the family in various ways. Pearlman [22] describes several phases in the preparation for separation, including protest against accepting the separation, despair, and detachment. Isay [23] studied how separations affected the wives of submariners. The study showed that while most families were able to establish a method to accommodate the separations, in some cases there were serious consequences. In the relationships studied by Isay, some wives attempted suicide, and in some cases the only solution was for the sailor to leave the navy [23]. Clearly, the psychological support team needs to be acutely aware that upcoming long-duration mission will stress the family significantly, and an adequate support system should be in place for the family.

A counseling session with the family and crew member should be considered shortly after the crew member has been selected for the mission. This would provide the opportunity for the psychological support team to describe for both the family and the crew member the kinds of problems that could possibly develop as training progresses. The support team could help establish a common set of expectations for the flight. Also, the psychological support team should be aware that sometimes the preflight training period could be the most stressful. The family may be able to accept and adjust to the idea that the crew member will be away for a long time on a space mission. During the preflight training period, however, the crew member may be away for long periods of time training at various locations on Earth. This situation (the crew member is gone, but has not left on the mission yet) can create significant tensions for both the crew member and the family.

Postflight reintegration

After the mission, and particularly after a high-profile mission like a Mars journey, the crew member will again be the center of attention. The program will have an interest in the rehabilitation program and in ensuring that it goes well. The crew member will likely be in demand for public appearances, debriefs, and other activities. In this setting it could be easy for family and private concerns to be pushed aside. A definite period should be set aside to allow the crew member to reestablish ties with family and friends during rehabilitation. Public appearances and nonessential debriefs should be postponed until the crew member has completed rehabilitation and has begun reintegration into the family. The program would need to enforce this requirement strictly so that the crew member or the cremember's family, is not placed in the position of turning down requests from friends, co-workers, and organizations. A clear expectation should be established ahead of time that a certain period of time will be set aside for reintegration and that this time cannot be scheduled for other concerns.

Recommendations Based on Current Knowledge

Although the nature and amount of preflight preparation and postflight rehabilitation needed will vary from crew member to crew member, some basic guidelines can be formulated. In general, preflight preparation is geared toward providing fit, confident, well-trained astronauts who have the full support of their family and friends. Postflight rehabilitation should be designed to allow the astronauts to regain strength, balance, and functional capability while getting integrated back into their previous lives. Some pre- and postflight guidelines are given below:

Preflight

1. Incorporate physical training, both for strength and endurance, into the scheduled training flow for the crew member. If the training is not scheduled and is instead left up to the crew member, this creates a situation where the crew member will need to choose between physical training and other issues in their personal life (like family time) as the mission approaches. This increases the likelihood that the crew member will start to lose strength and endurance in the time crunch that occurs as the mission approaches.

2. If the crew member has a family or other significant relationship, a counseling session should be held shortly after mission selection to go through the likely challenges that the mission will present. This session should offer the opportunity for expectations to be clearly stated and for reservations about the mission to be discussed. For long-term or high-profile missions, on-going counseling should be offered for the whole family. Often it is not necessary to identify these as formal counseling sessions, but as get-togethers to discuss how mission preparation is going. Some issues to be considered before flight are financial stresses, domestic responsibilities, loss of intimacy, parenting, and mutual support among the crew families.

3. Crew members should receive training in conflict resolution and cultural differences. They should know how to recognize depression. They should also have training in how to be good leaders, good followers, and considerate group members. Outdoor survival courses often provide this kind of training.

4. Instruction in sports psychology (e.g., visualization, relaxation) could be helpful in preparing for certain technical tasks.

Postflight

1. Accurate assessment of muscle, bone, and neurovestibular changes is essential. This serves as a rehabilitation baseline for the individual crew member and adds to the database of physiological changes after spaceflight. Dual-x-ray absorptiometry is an accurate method to measure bone changes. MRI provides a comprehensive assessment of muscles, including cardiac muscle, throughout the body. Isokinetic dynamometry provides the most widely accepted estimates

of muscle strength. Posturography provides an objective measure of balance. These tests should be done as soon as possible after landing. Although testing can present a physical challenge to the returning crew and may interfere with the reunion with family and friends, adequate time should be devoted to the testing. The postlanding measurements will be critical for assessing the countermeasure program and will be essential for the continuing success of the program.

2. An individualized musculoskeletal rehabilitation program should be created for each crew member based on the results of the postflight assessments. Pool exercises can be used to minimize impact loads in crew members with significant bone losses. Exercising in the pool also combats orthostatic intolerance. Stretching and swimming are particular useful in the few days after landing.

3. Protected time should be set aside so that crew members can reestablish relationships with family and friends. For a Mars mission, this time period should be significant, perhaps several months long. Although postflight testing is important, a clear end time for the testing must be set and strictly adhered to.

References

1. Pavalko, E.K., and G.H. Elder, World War II and divorce: A life-course perspective. American Journal of Sociology, 1990. 95(5): 1213–34.
2. Kasper, C.E., L.A. Talbot, and J.M. Gaines, Skeletal muscle damage and recovery. AACN Clinical Issues, 2002. 13(2): 237–47.
3. Riley, D.A., Review of primary spaceflight-induced and secondary reloading-induced changes in slow antigravity muscles of rats. Advances in Space Research, 1998. 21(8–9): 1073–75.
4. Fitts, R.H., D.R. Riley, and J.J. Widrick, Functional and structural adaptations of skeletal muscle to microgravity. Journal of Experimental Biology, 2001. 204(Pt 18): 3201–8.
5. Greenisen, M.C., et al., Functional performance evaluation, in *Extended Duration Orbiter Medical Project Final Report*, C.F. Sawin, G.R. Taylor, and W.L. Smith, eds. 1999, NASA, Houston, TX, pp. 3.1–3.24.
6. Astrand, PP., et al., Physical training, in *Textbook of Work Physiology*, P.P. Astrand, et al., eds. 2003, Human Kinetics, Champaign, IL, pp. 313–68.
7. Fuleihan, G.E., et al., Reproducibility of DXA absorptiometry: a model for bone loss estimates. Journal of Bone and Mineral Research, 1995. 10(7): 1004–14.
8. LeBlanc, A., et al., Muscle volume, MRI relaxation times (T2), and body composition after spaceflight. Journal of Applied Physiology, 2000. 89(6): 2158–64.
9. Thornton, W.E., G.W. Hoffler, and J.A. Rummel, Anthropometric changes and fluid shifts, in *Biomedical Results from Skylab*, R.S. Johnston and L.F. Dietlein, eds. 1977, NASA, Washington, DC, pp. 330–38.
10. Thornton, W.E., and J.A. Rummel, Muscular deconditioning and its prevention in space flight, in *Biomedical Results from Skylab*, R.S. Johnston and L.F. Dietlein, eds. 1977, NASA, Washington, DC, pp. 191–97.
11. Osternig, L.R., Isokinetic dynamometry: implications for muscle testing and rehabilitation. Exercise and Sport Sciences Reviews, 1986. 14: 45–80.
12. Baltzopoulos, V., and D.A. Brodie, Isokinetic dynamometry. Applications and limitations. Sports Medicine, 1989. 8(2): 101–16.
13. Herbert, R., Human strength adaptations—implications for therapy, in *Key Issues in Musculoskeletal Physiotherapy*, J. Crosbie and J. McConnell, eds. 1993, Butterworth-Heinemann, Boston, pp. 142–71.
14. Stupin I., V., V.V. Bogomolov, and T.D. Vasil'eva, Sanatorium-health resort stage of medical rehabilitation of astronauts after long-term space flight [in Russian]. Kosmicheskaia biologiia i aviakosmicheskaia meditsina, 1991. 25(2): 18–21.

15. Clement, G., et al., Effects of cosmonaut vestibular training on vestibular function prior to spaceflight. European Journal of Applied Physiology, 2001. 85(6): 539–45.
16. Furman, J.M., and S.P. Cass, Vestibular laboratory resting, in *Vestibular Disorders: A Case-Study Approach*, J.M. Furman and S.P. Cass, eds. 2003, Oxford University Press, New York, pp. 30–40.
17. Perhonen, M.A., et al., Cardiac atrophy after bed rest and spaceflight. Journal of Applied Physiology, 2001. 91(2): 645–53.
18. Blundell, N., Maximizing performance: Psychology, in *Clinical Sports Medicine*, P. Brukner and K. Khan, eds. 2002, McGraw-Hill, New York, pp. 639–48.
19. Holland, A.W., Psychology of spaceflight, in *Human Spaceflight: Mission Analysis and Design*, W.J. Larson and L.K. Pranke, eds. 2000, McGraw Hill, New York, pp. 155–91.
20. Kanas, N., Psychological, psychiatric, and interpersonal aspects of long-duration space missions. Journal of Spacecraft and Rockets, 1990. 27(5): 457–63.
21. Kanas, N., and D. Manzey, Psychological countermeasures, in *Space Psychology and Psychiatry*. 2003, Kluwer Academic Publishers, Boston, MA, pp. 131–72.
22. Pearlman, C.A., Jr., Separation reactions of married women. American Journal of Psychiatry, 1970. 126(7): 946–56.
23. Isay, R.A., The submariners' wives syndrome. Psychiatric Quarterly, 1968. 42(4): 647–52.
24. Greenleaf, J., and D.T. Quach, Recovery after prolonged bed-rest deconditioning. 2003, NASA-Ames Research Center, Moffett Field, CA.
25. Furman, J.M., and S.P. Cass, Physical examination of the dizzy patient, in *Vestibular Disorders: A Case-Study Approach*, J.M. Furman and S.P. Cass, eds. 2003, Oxford University Press, New York, pp. 21–29.
26. Kinch, M., Principles of rehabilitation, in *Clinical Sports Medicine*, P. Brukner and K. Khan, eds. 2002, McGraw-Hill, New York, pp. 160–85.

12

Long-Duration Flight Medical Planning: Medical Care on the Way to the Moon and Mars

Introduction

In 1999, in the middle of the Antarctic winter, the sole physician at the South Pole Station felt a mass in her breast. She could not be evacuated because a plane had never landed at the South Pole in midwinter. So, using materials at hand, the team biopsied the mass, stained the cells, and transmitted pictures to specialists in the United States. The images showed breast cancer. Chemotherapeutic drugs and other supplies were airdropped to the station. When weather improved sufficiently to allow an evacuation, the doctor returned home to definitive medical care. Fortunately, the cancer had been controlled, and she made a full recovery.

This event, described in Jerri Nielsen's autobiographical book *Ice Bound* [1], illustrates some of the medical problems that could be faced on a Mars mission. Providing a doctor for an expedition is only a partial solution to medical problems because the doctor can become sick. Also, crew members may develop conditions that no one onboard will be prepared to handle. If, for example, breast cancer is diagnosed on a Mars mission, neither evacuation nor resupply will be possible. The crew could attempt to remove the cancer using onboard equipment, but success could not be guaranteed. Ultimately, Mars mission planners must accept that not every medical contingency can be anticipated and addressed. The challenge in medical planning for long-duration space missions is providing the greatest capability within the missions' strict limits on mass and volume.

Fortunately, experience has shown that, with proper screening, it is often possible to avoid medical crises during space expeditions. In this chapter, the medical risks of long-duration spaceflight are presented, along with methods to keep them under control.

Medical Risks on Long-Duration Space Missions

The range of possible medical and surgical conditions that could occur in a specific population over a given period of time is vast. Crews in Antarctica, submarines, and undersea habitats have experienced intracerebral hemorrhage, stroke, myocardial infarction, appendicitis, bone fracture, cancer, and psychiatric illness—among many other conditions [2, 3]. Although the range of conditions is wide, the overall rate of serious medical or surgical illness is low, particularly with proper crew screening and close attention to hygiene.

Even in the days before antibiotics and advanced surgical techniques, challenging expeditions often succeeded without high mortality rates. Lewis and Clark lost only one person on their 2.5-year trek (1804–1806) through the Louisiana Purchase (the cause of the death is thought to be appendicitis). All the members of Sir Ernest Shackleton's 2-year expedition (1914–1916) made it back alive, despite being completely isolated in the ice off the coast of Antarctica. Although some stories of sea voyages from the nineteenth century describe the dirty, confined, and unhealthy steerage compartments, when hygiene was a priority, mortality on long sea trips could be very low. In *Life and Death in the Age of Sail* [4], Haines notes that by including a shipboard doctor and by paying close attention to hygiene, mortality on sailing ships traveling to Australia from the United Kingdom was comparable to the mortality in the villages the immigrants had left behind. The regimen onboard included daily scrubbing, disinfecting, and deodorizing of the steerage quarters, supervision of personal cleanliness, inspection of dietary rations, and quarantine of the sick. More than half of the ships bound for South Australia had less than six deaths per voyage. Most of these, unfortunately, were children who were particularly susceptible to infectious disease. If these ships had carried only young, healthy adults, the overall mortality would have been quite low.

These historical examples illustrate that, in the past, the greatest mortality on expeditions resulted from infectious disease. When appropriate sanitary measures were taken, mortality and morbidity were reduced substantially. Nevertheless, medical risks remain. Billica et al. [5] reviewed data from submarines, Antarctic expeditions, military aviation and U.S. and Russian spaceflight experience. They calculated the rate of significant illness or injury in these populations. A significant illness or injury was defined as an event that would require either an emergency room visit or a hospital admission. In these populations, the incidence of significant medical events was approximately 0.06 events/person-year [5]. If these data were extrapolated to a 2.5-year Mars mission involving 6 crew members, the expected rate of a significant medical event would be 0.9 person/mission (i.e., 1 significant medical event could be expected).

Medical Problems Caused by the Environment and Physiological Adaptation

Examining analog populations to estimate the rate of medical events on a Mars mission is useful, but such an analysis does not account for the unique problems produced by the spacecraft environment and the physiological adaptation to weightlessness. The isolation and confinement of the spacecraft can lead to psychosocial problems (chapter 2), and isolated and confined communities have been shown to develop altered immune responses. The loss of calcium from bone can increase the saturation of calcium in the urine and increase the risk for kidney stones (chapter 1). Extravehicular activities carry the risk of decompression sickness (chapter 5).

Clinical psychosocial problems

Even though the crew members will be well screened and adaptable, situations can develop where factors such as sleep deprivation, isolation, interpersonal conflict, and operational breakdowns combine to overwhelm psychological defenses. In submarines, anxiety reactions and depression are the top two psychiatric diagnoses [2]. These diagnoses are also frequent among researchers enduring long Antarctic winters [3]. Any medical care planned for long-duration spaceflight should include the capability to diagnose and treat depression and anxiety.

Immune responses

Numerous factors such as sleep deprivation, psychological stress, and poor nutrition can reduce immune responses. Individuals wintering over in Antarctica have altered immune responses, including a reduced proliferative response of T-cells to stimulation and reactivation of latent virus infections [6]. Studies of immune function performed after spaceflight have shown reductions in natural killer (NK) cell function, reduced proliferation of peripheral blood mononuclear cells in response to stimulation, and reactivation of latent viruses [7, 8]. Studies done in flight have shown a reduction in delayed type hypersensitivity in some crew members.

The immunological changes seen both during and after spaceflight are difficult to put into context. The changes may result in part from weightlessness or may be a response to the stresses of a particular flight. In general, findings from spaceflights are not correlated with other mission events. It is not known, therefore, if the laboratory data from missions where crews lost significant weight, slept poorly, or had major psychological stress show greater changes in immunological parameters compared with crew data from missions deemed to be less stressful. Data collected on landing day may reflect the increases in stress hormones that occur on that day in addition to any changes due to spaceflight. Also, baseline data are limited. Crews may experience significant stress during training, but ground-based normative data on immunological parameters during these events are not available for comparison to flight results.

Because of these uncertainties, it is difficult to determine whether crew members must accept a higher risk of suppressed immunity during spaceflight or whether, with proper attention to sleep, stress, and psychological well-being, immune function can be maintained at preflight levels. Because it is known that a high level of psychological stress can reduce immune responses [9] (although some confinement studies show very minimal changes in immunological parameters [10]), an important component of the in-flight medical program will be close attention to the social and work environment on the station.

Kidney stone risk

On the ground, immobilization and bed rest both increase urinary calcium excretion and renal stone risk [11]. In space, because urinary calcium excretion also increases markedly, renal stone risk would be expected to increase as well. Symptoms consistent with a kidney stone have occurred in the Russian space program, although it is not known if the stone formed in space or if it was already in the kidney at the time of launch [12]. Studies by Whitson et al. [11] showed an increase in the urinary supersaturation of calcium oxalate and a decrease in urinary citrate in space. These data suggest that renal stone risk is increased in space.

The mainstays of kidney stone prevention are to increase fluid intake, increase urinary citrate, or decrease urinary calcium. Typically, a fluid intake of 2 l a day will prevent kidney stones. Potassium citrate is an excellent source of citrate (which alkalizes the urine) and has been studied for use in space. Calcium excretion can be reduced pharmacologically. Both bisphosphonates (e.g., alendronate) and thiazide diuretics (e.g., hydrochlorothiazide) reduce urinary calcium excretion. Alendronate has been shown to reduce hypercalciuria and stone-forming propensity during bed rest [13], and hydrochlorothiazide is an established therapy to reduce stone risk because of its effects on renal calcium excretion [14].

Successful kidney stone prevention programs also incorporate frequent monitoring to establish that urinary supersaturation of calcium is controlled [15, 16]. When urinary supersaturation values are kept within normal limits, stone formation is rare. These data suggest that the capability to monitor urinary parameters (at a minimum urinary calcium) would be important to establish that the kidney stone prevention program is working.

If, however, a stone should form, the most likely outcome is it would be passed without incident. Rarely, however, stones can obstruct a ureter. If this were to occur on a spaceflight, it could obstruct the urinary flow from one kidney and create a medical emergency. Ultrasound can be used to locate stones within the kidney and ureter. Intense ultrasound (lithotripsy) could be used to break up obstructing stones and allow them to pass.

Decompression sickness

The risk of decompression sickness among astronauts depends on their tasks during the mission and the spacecraft's atmospheric pressure. If the spacecraft is maintained at sea-level pressure and atmospheric composition (such as on the Shuttle and the International Space Station), decompression sickness is a risk for every spacewalk. The crew members must reduce the nitrogen load in their bodies to transition safely from the sea level atmosphere in the spacecraft into the low pressure within their suit. If instead the spacecraft is kept at a low atmospheric pressure similar to the extravehicular activity (EVA) suits (which was the case in the Apollo command module and lunar module), then moving between the spacecraft and suit does not present a risk of decompression sickness.

If the spacecraft and mission design requires moving between different pressure environments, then decompression sickness is a potential risk. If decompression sickness develops, the bubbles can disrupt the endothelium in blood vessels, leading to thrombosis and local ischemia. These bubbles can mechanically block blood vessels. In the worst case, a bubble in the brain could cause a stroke and incapacitate the crew member. The main treatments for decompression sickness are intravenous fluids, anti-inflammatory drugs, and recompression with hyperbaric oxygen [17]. Perfluorocarbon emulsions also may be a treatment for decompression sickness. Perfluorocarbons have a high solubility for both oxygen and nitrogen, and they have surfactant properties. Studies in animal models have shown that perfluorocarbons can improve outcome in severe decompression sickness [18, 19]. Perfluorocarbons are also the main component of most proposed synthetic blood products.

On Earth, the definitive treatment for decompression sickness is recompression therapy in a hyperbaric chamber. Providing hyperbaric capability on a spacecraft involves a trade-off between risk and mission resources. The Russian *Salyut* and *Mir* stations did not have hyperbaric capability, although the EVA suits worked at a higher pressure. The proposed Space Station *Freedom* design incorporated a hyperbaric chamber, but this was not retained in the design for the International Space Station [20]. For planetary surface exploration, where EVA may be a daily event, a hyperbaric capability could be important if moving between the habitat and the suit involved a significant pressure change.

Hearing loss

Temporary and in some cases permanent hearing threshold shifts have been noted after long-duration spaceflights [21]. These shifts have been presumed to be noise induced, but other factors, such as toxins, carbon dioxide levels, or fluid shifts also may play a role. The best approach to this problem is adherence to noise standards and the use of hearing protection when necessary. Just as in any ground-based hearing conservation program, the ability to test hearing in space is important. New hearing testing technologies (such as otoacoustic emissions) may be appropriate for this application [21].

Dental Problems

Dental problems are rarely life threatening, but they could make a crew member's life miserable. On the *Soyuz 28* flight in 1978, Yuri Romanenko was reported to have taken pain medication for the last 2 weeks of his flight because of a severe toothache [22]. In other isolated settings, dental problems can require dental intervention or even transfer. In 100 British Polaris submarine patrols, there were 30 fillings and 7 extractions performed [23]. Rarely, dental problems have been a cause for a transfer at sea in the American Polaris submarine program [2]. Extractions and fillings have also been required in Antarctica [24]. Risk estimations done for the Skylab program in the early 1970s placed the chance of a dental problem that would significantly impair a crew member's ability to work at 0.92% for a 3-man, 28-day mission [25].

The approach to dental problems in space includes two critical components: prevention and treatment. Prevention of dental caries includes the use of fluoride rinses, appropriate diet, brushing, and flossing. The Skylab program included recommendations to brush twice a day and floss once a day [25]. Since the Skylab era, an additional element is a better understanding of the role of the role of bacteria (particularly *Streptococcus mutans*) in the formation of dental caries. Possible treatments now include the use of antibacterial (chlorhexidine) rinses [26]. When preventive dentistry principles (such as fluoride gels, antibacterial rinses when appropriate, flossing, and brushing) are actively applied to high-risk adults, the risk of dental caries can be reduced dramatically [27]. If dental caries do occur, one possible approach is atraumatic restorative treatment, in which cavities are filled with an adhesive restorative material that seals the cavity [26].

Despite aggressive preventive care, dental problems may still arise. The Skylab dental kit included tools for anesthesia and tooth removal. On the Shuttle, the capability to place a temporary filling is also provided. The International Space Station dental kit includes the tools needed to treat toothaches, exposed pulp, and dislodged teeth. In addition, tools and training are provided for dental anesthesia, placing a temporary filling, and extracting a tooth. For a Mars mission or an extended stay on the surface of the Moon the ability to maintain proficiency in these areas will become important.

Medical Illness

Even if the crews did not launch into space, they would still face a yearly risk of developing a medical illness. For the population at large, the three top causes of death in the 35- to 44-year-old age group are cancers, accidents, and heart disease [28]. Because the astronaut corps has to meet medical selection standards, the corps is not a representative sample from the population, so these data might not be directly applicable. Other data suggest, however, that although astronauts as a group are healthier than the general population, the major causes of death are the same. At NASA, the Longitudinal Study of Astronaut Health was established to follow astronauts over their life span and compare their morbidity and mortality to a comparison group. In the years 1959–1991, there were 20 deaths among the astronaut population, 16 from accidents, 2 from coronary disease, 1 from cancer, and 1 from unknown causes. Compared to the general population, the death rate from accidents is higher, and the death rate from cardiovascular disease and cancer is lower in the astronaut corps. Nevertheless, accidents, cardiovascular disease, and cancer remain the top three causes of death among astronauts [29, 30].

Another important medical consideration for an isolated group is infectious disease. Before antibiotics were available, infectious disease was the leading cause of death on sailing voyages and among groups living in close quarters. Improved sanitation and antibiotics have drastically reduced the hazards from infectious agents, but infection can still be a problem when crews are stressed, supplies are low, or sanitation is lax.

Infectious disease

In the early days of the space program, infectious diseases developed early on in several missions. These events were thought to be due to exposure to infectious agents during the preflight period. Because the crew members spent so much time together, people outside of the crew were the most likely sources of infection. A program was instituted to limit contact 1–2 weeks before flight between the crew and those who might have an infectious disease. This dramatically reduced the incidence of infectious disease during spaceflights [31].

Once a crew is isolated on a spacecraft, they are, in effect, quarantined. The pathogens that can cause infections among the crew are onboard at the outset of the flight. As a result, epidemics of influenza or other respiratory viruses that are common on Earth have not been a concern on long-duration spaceflights. The main kinds of infections that could occur would be due to breaks in skin or mucosal barriers or a reduction in immunity. There is also the possibility for the microbial population to change over time in response to environmental pressure on the spacecraft (just as populations of microbes can develop resistance to antibiotics when they are stressed). This will require monitoring the spacecraft microbial environment to track whether any changes are occurring that might be hazardous [32].

Cardiovascular disease

Coronary artery disease is a leading cause of mortality for individuals in the age range of most astronauts. Also, despite improvements in screening and the identification of cardiac risk factors, often the first manifestation of coronary artery disease is sudden death or myocardial infarction. In a population of U.S. Air Force aviators, the 5-year annual cardiac event rate by age group was 0.0054% (30–34 years),

0.018% (35–39 years), 0.038% (40–44 years), 0.14% (45–49 years), and 0.13% (50–54 years) [33]. Of the events that did occur, 21% were sudden death and 61% were myocardial infarction. In other words, despite screening, very often the first manifestation of coronary artery disease in this population was a significant cardiac event. Although this group of aviators is not screened as aggressively as astronauts, they must pass annual exams, and they are aware that their livelihoods depend on meeting medical standards. For these reasons, these data are probably roughly applicable to the astronaut corps and show that the risk of sudden death or heart attack still exists in space despite screening.

Risk factors for coronary artery disease include elevated low density lipoprotein (LDL) cholesterol levels, diabetes, hypertension (blood pressure > 140/90), smoking, and a family history of premature coronary disease (men < 45 years, women < 55 years) [34]. High-density lipoprotein (HDL) cholesterol levels < 40 mg/dl are also a risk factor. The risk for coronary artery disease is influenced by other factors such as an atherogenic diet, high homocysteine levels, and impaired fasting glucose. Prothrombotic and proinflammatory factors, such as C-reactive protein, fibrinogen, activated factor VII, plasminogen activator inhibitor-1, tissue plasminogen activator, von Willebrand factor, factor V Leiden, protein C, and antithrombin III also can influence risk. The most important biochemical factor, however, is the LDL cholesterol level. In the absence of other risk factors, current guidelines recommend a LDL cholesterol level <160 mg/dl, and drug therapy is recommended when LDL levels reach 190 mg/dl [34].

The current screening program for astronauts excludes those with diabetes, hypertension, or familial hypercholesterolemia. Additionally, astronauts have regular exercise stress testing. No screening program is perfect, however, and coronary events can occur in individuals who have had negative exercise stress tests [35]. It is possible that the risk of cardiovascular disease in the astronaut population could be reduced further through more aggressive screening. Several approaches exist, including the use of electron-beam computed tomography to detect calcium in coronary arteries [36], measurement of carotid intimal thickening with ultrasound [34], genetic screening [37], and measurement of homocysteine [38].

Cardiovascular disease prevention strategies in those with acceptable cholesterol levels include frequent physical activity, smoking avoidance, cholesterol monitoring, and an appropriate diet. These interventions are used in the astronaut corps. Other possible interventions include daily aspirin use and a change in diet to include items that lower cholesterol (such as plant stanols/sterols and fiber). Plant stanol esters help block absorption of cholesterol from the digestive tract, which in turn helps lower LDL cholesterol blood levels without reducing HDL cholesterol levels. Approximately 2–3 g/day of plant stanols/sterols are effective [39–41]. Fiber (such as from psyllium seed or other sources), 10–25 g/day, can also lower LDL cholesterol. It is not known how much further benefit can be obtained by aspirin, plant stanols/sterols, or fiber in an already low-risk person.

If cholesterol levels are elevated, various drugs are available. The 3-hydroxy-3-methylglutaryl coenzyme A reductase inhibitors (statins; e.g., lovastatin, pravastatin, atorvastatin, simvastatin) are very effective at bringing cholesterol levels into the normal range. Although typically well tolerated, statins may rarely elevate liver enzymes or cause a myopathy. One added benefit of statins in the context of spaceflight is that statins may help bone in addition to lowering serum cholesterol. Animal studies show that statins can increase bone formation [42], and epidemiological studies show that patients on statins have a lower risk of fracture [43–46]. Other drugs used to lower cholesterol include bile acid sequestrants (e.g., cholestyramine), nicotinic acid, and fibric acid derivatives (e.g., gemfibrozil, clofibrate).

If cholesterol were controlled aggressively in the astronaut population to levels below what are deemed acceptable for the general population, it is possible that the risk of cardiovascular disease would also continue to fall. Exercise and nutritional interventions make sense and are low risk, but the use of drugs to reduce cholesterol beyond established guidelines is controversial. Even though statin drugs are well tolerated, they are not free of side effects, so any benefit from the lowered cholesterol would have to be balanced against the risk of drug treatment. No data at present support drug treatment to lower risk in already low-risk individuals [47]. Also, what LDL cholesterol target level should be set for the astronaut population to maintain a low risk profile is not known. The National Cholesterol Education Program guidelines, designed for the general population, provide the best available guidance.

Despite adequate preventive measures, it is still possible that a cardiac event could occur in flight. On Earth, the standard therapy for a myocardial infarction is aspirin, oxygen, and thrombolytics. Morphine is also used. The ability to perform early defibrillation markedly improves the outcome if there is loss of consciousness due to an acute cardiac ischemic event (also called "sudden death"). In some cases, the patient survives the heart attack but develops congestive heart failure and/or dangerous arrhythmias. If that should happen on a spaceflight, it would likely overtax the medical kit because it may not be possible to bring the range and amounts of medications needed to sustain an individual with a severe, chronic cardiac condition.

Cancer

In the general population, cancer is a major cause of death in the astronaut age group. Cancers of the colon, prostate, and breast are particularly common [48]. Lung cancer, while an issue for the public at large, is rare in the nonsmoking population, and most astronauts are nonsmokers. In pilots, a similar population to astronauts, skin cancer is also a risk. Epidemiological evidence suggests that the risk of melanoma and other skin cancers is elevated in pilots [49–51]. As this may be due to radiation exposure at altitude, it is possible that astronauts on long space missions also will run an increased risk for melanoma.

Treatment of some cancers is straightforward. For example, many skin cancers can be excised easily or treated with topical agents. In general, however, once cancer is established, treatment usually involves either surgery, radiation, chemotherapy, or some combination of the three. Providing definitive care for the variety of potential cancers that could arise on a long-duration mission would require a considerable amount of volume and mass. This would most likely not be practical for a long-duration mission. For the bulk of cancers, the best approach is screening and prevention.

It has long been known that cancer runs in families, and a strong family history of cancer could lead to an individual not being selected for the astronaut corps. The development of genetic testing has allowed for more precise genetic diagnoses [52, 53]. Mutations in the *BRCA1* and *BRCA2* genes are associated with a higher risk of

Table 12-1. Tumor-suppressor genes associated with an inherited predisposition to cancer.

Gene	Diseases	Functions of gene
BRCA1, *BRCA2*	Breast cancer, ovarian cancer, pancreatic cancer (possible)	Both participate in the DNA repair process
ATM	Homozygous: ataxia-telangiectasia and greatly increased cancer risk; heterozygous: probable, but small, increase in breast cancer risk	Phosphorylates *BRCA1* in response to DNA damage
hMSH2, *hMLH1*, *PMS1*, *PMS2*	Hereditary nonpolyposis colorectal cancer, endometrial cancer, ovarian cancer, urinary tract cancer	DNA mismatch repair genes
APC	Familial adenomatous polyposis	Loss of function in *APC* tumor suppressor gene
CDKN2A	Melanoma	Produces cell-cycle control protein p16, which inhibits excessive cell proliferation

Mutations in these genes damage the tumor suppression function and can increase the susceptibility to cancer. In general, the genes listed above either have a role in blocking uncontrolled cellular proliferation or in repairing DNA damage. Some mutations have a several variants, and the significance of the variants may not be known. Also, while some mutations have a strong link to future cancer (*BRCA1* and *BRCA2*), for others the association is weak (*ATM* heterozygotes).

developing breast and ovarian neoplasms. Up to 10% of breast cancer cases may be associated with an inherited predisposition for the disease. Hereditary nonpolyposis colon cancer is associated with mutations on the *HMSH2*, *hMLH1*, *PMS1*, and *PMS2* genes. Mutations in the *CDKN2A* gene can lead to an increased risk of melanoma. Table 12-1 summarizes some of the mutations associated with an increased risk of cancer. In clinical use, genetic tests are usually run because of a family history of cancer or because a close family member has had a positive test. One issue for astronaut selection is the ethics of whether to use genetic testing as a screening device, even without a positive family history [26].

Other, nongenetic risk factors for cancer also can be assessed. Infection with human papilloma virus increases the risk of cervical cancer [54]. Smoking increases the risk of lung, tongue, and laryngeal cancer. Excessive radiation exposure increases cancer risk. A history of extensive ultraviolet light exposure increases the risk of skin cancer. Chronic gastroesophageal reflux is associated with a higher rate of esophageal cancer [55]. These environmental and occupational exposures can increase cancer risk and may need to be taken into account during screening.

Certain procedures can be performed to screen for cancer or detect it early. Colonoscopy is often recommended at age 50 to screen for colon cancer. During colonoscopy, polyps can be removed that might go on to form more significant cancers. The digital rectal exam and prostate specific antigen blood test can provide information on the development of prostate cancer. Breast self-exams and mammography can be used to detect breast cancer at an early stage. Pap smears can detect cervical cancer before it becomes clinically significant [56]. These various tests allow crews to embark on a long space trip with the knowledge that, as best can be ascertained, they are cancer free.

Although established cancer screening methods (such as pap smears and colonoscopy) are available for some organ systems, for other organs no established screening technologies exist. Advanced medical imaging such as total-body MRI can assess the entire body and find lesions as small as approximately 2 mm [57]. This technology has not been used extensively as a screening tool, so few data exists to guide its use. This technology could, however, provide data on organs such as the brain, thyroid, pancreas, and ovaries.

Although screening can be used to minimize the risk of launching an astronaut with cancer, once the crew members are on a mission, the risk of developing cancer should be kept as low as possible. The main risk is radiation (chapter 3), which can be reduced through proper shielding. Antioxidants may help reduce tissue radiation damage. Crew members also can attempt to reduce cancer risk through diet. Research shows that diets low in fat and high in fiber, fruits, vegetables, and whole-grain products are associated with reduced risks for many cancers (such as colon, prostate, and breast) [58, 59]. Some drugs may help reduce cancer risk (such as aspirin to prevent colon cancer), but the use of drugs also carries the risk of side effects (such as an increased risk of bleeding complications with aspirin). Because most cancers take time to reach a significant size, if crew members were screened close to launch for a long (2–3 year) mission, it is unlikely that they would develop a significant cancer during the mission.

If a cancer should arise, treatment options are limited. Some drugs (such as estrogen blockers for breast cancer or androgen blockers for prostate cancer) can slow the growth of tumors. The likelihood that these drugs would be used is low, so providing space for them onboard would be hard to justify. Cancer treatment that requires radiation, specialized surgery, or complex chemotherapeutic regimens would most likely be too difficult for a Mars crew to provide. Tumors that can be felt and have clear margins could potentially be excised using the onboard surgical tools.

Surgical Illness

Minor surgery, such as suturing of lacerations, removing skin lesions, or draining superficial abscesses, can be performed in a variety of demanding locations on Earth and should be possible in space as well. Surgery, technically demanding dissections, and perfusion fixation were all done in space on research animals as part of the Neurolab mission in 1998 [60]. These experiments demonstrated that problems with the use of surgical instruments, management of bleeding, and control of fluids could be solved in weightlessness.

The main issue for long-duration spaceflight is how much capability should be provided to perform major (as opposed to minor) surgery. In the general population, the three most common major surgical operations are hernia repair, appendectomy, and cholescystectomy [61]. (It is likely that a hernia would be repaired before spaceflight.) In the submarine program, the most common general surgical condition has been appendicitis [2]. A death occurred in Antarctica due to appendicitis [26]. These experiences indicate that the most common type of major surgery is abdominal surgery. Surgery for trauma is also a possibility for a Mars mission. Fractures, dislocations, and traumatic amputations have all occurred on submarines. In the submarine experience reported by Tansey et al. [2], however, there was never a need for thoracic surgery.

Appendicitis

One approach to appendicitis is prevention. In the Australian Antarctic program, physicians who are wintering over in Antarctica must have a prophylactic appendectomy [3]. For travel to low Earth orbit, prophylactic appendectomy is not currently required because evacuation to the ground is possible. Even if immediate evacuation were not possible, appendicitis could be treated successfully with antibiotics for a short time until an appendectomy could be done at a later time [26]. For a long-duration Mars mission, however, prophylactic appendectomy may make sense. Treatment of appendicitis on a Mars mission would use up antibiotics and medical supplies that otherwise might be needed to treat medical conditions that are not preventable. Also, a previous appendectomy would remove appendicitis from the differential diagnosis of abdominal pain, which would simplify the diagnosis of other abdominal conditions requiring surgery. The need to maintain proficiency in treating this condition also would be removed. Unfortunately, no hard data exist that can be used to balance any benefits of prophylactic appendectomy on a Mars mission against the risk of the preflight surgery.

Cholecystitis

Although cholecystectomy is a common operation in the general population, it has not been a significant issue in submarines or at Antarctica [2, 3, 26]. Part of this may be due to the fact that cholecystitis is more common in females (female to male ratio 3:1 up to the age of 50), and to date most submarine crews and Antarctica participants have been male. Cholecystitis without gallstones (acalculous cholecystitis) is rare (4–8% of all cases), so the majority of acute cholecystitis occurs in individuals with gallstones. One preventive measure that could be taken for space crews is to screen for gallstones and remove any existing stones before flight.

Another strategy is to prevent the initial formation of gallstones. Some epidemiological evidence suggests that physical activity reduces the risk of gallstone formation and that this may be due to a reduction in biliary cholesterol levels. It is possible that interventions taken to reduce the risk of cardiovascular disease (physical exercise, diet to reduce total cholesterol levels) may serve to reduce gallstone formation as well.

Trauma

Trauma treatment in space will require the ability to establish airways, stop bleeding, set bones, and provide both general and regional anesthesia. In addition, for a long-duration mission, a system will be needed to maintain the proficiency of the crew to treat potential trauma. The range of possible traumatic events that can occur is vast, but experience in analog settings has shown that such events will occur rarely. As a consequence, refresher training and simulations are essential for maintaining a reasonable level of skill within a long-duration crew.

Supplies will be needed for wound closure, wound care, casting, splinting, eye emergencies, and minor surgery. The capability would be similar to what exists on a submarine. Ultrasound would be useful as a multipurpose diagnostic and treatment tool; ultrasound can be used to image a variety of body sites [62] as well as to stop intra-abdominal bleeding [63].

Miscellaneous Medical Events

Because a spacecraft has a variety of potential toxins onboard, the possibility of toxic exposure or poisoning exists. Crew members will need access to a reliable database on all the potentially hazardous substances onboard and what actions to take if they come in contact with the substance. Allergy symptoms could become an issue if a crew member develops an allergy to one of the many compounds onboard. Anaphylactic reactions are possible if substances are injected for experiments or medical reasons.

Prevention Strategies

Although technological advances have made minimally-invasive surgery possible and increased the variety of treatments available for many illnesses, the mainstay of medical care for a Mars mission is prevention. A significant medical or surgical problem drains resources from the primary mission and can put the flight at risk. The keys to prevention include both appropriate preflight selection to reduce the likelihood of illness and careful monitoring during the mission to detect problems before they become unmanageable.

Selection

Screening the general population for disease (or propensity to disease) presents several challenges. If the prevalence of the disease in the population is low, even a screening test with a high sensitivity and specificity can produce more false positive than true positive results. This can potentially lead to further costly or risky medical procedures to follow up on the positive tests. The cost of screening has to be balanced against the possible benefit that can be gained. While a person with a true positive result would benefit from the screening program, those with false positive results would not benefit and may even be harmed. In the case of genetic testing, a positive genetic test might trigger employment and insurance discrimination, even if the significance of the finding is not completely clear.

For the crew of a Mars or long-duration Moon mission, the screening issues are different. Long-duration spaceflight is a new and pioneering effort, so the medical preparations and testing can be considered part of an ongoing research project. Detailed knowledge about the crew member's physiologic and genetic makeup may turn out to be important at some point in the mission or may provide information that can help future crews. This perspective changes the approach to collecting medical data—including data for medical screening. Although in an Earth-based setting doctors would not collect screening data until research shows its utility, for a Mars mission, screening data might be collected until experience proves it is not needed.

This approach was used early in the Apollo program. At the beginning of the space program, crews received extensive preflight examinations. Crew members were disqualified from flight for conditions that later in the program were not considered significant. At the time, however, the consequences of mission failure were high, and medical findings could not be ignored unless solid data existed to show they were unimportant. Once more spaceflight experience was gained, it was apparent that some people had been unnecessarily disqualified from spaceflight. What this reflected was a bias toward excessive screening to minimize the possibility of inflight medical problems. For a demanding exploration flight (such as the first flight to Mars), it is likely that the same bias will exist.

Currently, astronauts have a careful review of their medical history and a comprehensive medical exam to screen for potentially significant medical problems. Candidates with chronic medical problems (diabetes, cardiovascular disease, cancer) are not selected into the astronaut corps [64]. Also, those with a family history of a cancer syndrome (e.g., hereditary polyposis) are not be selected. For a Mars mission, where the possibility of an evacuation does not exist, more stringent criteria could be considered.

Cancer-risk screening

Current prevention guidelines for the general population call for routine Pap smears, mammography, digital rectal exams, and colonoscopy [56]. Prostate specific antigen (PSA) tests are also frequently done. The recommended schedules of these tests are typically designed to administer the test at those ages that maximize the tests' utility. For example, testing for colon cancer is not begun until age 50 because by that time the prevalence of the disease in the population is high enough that a positive test has a reasonable chance of being a true positive (i.e., the test has a reasonable positive predictive value). When screening for a long-duration spaceflight, however, the most important consideration is not necessarily maximizing the positive predictive value of a screening test. Instead, the important factor is high sensitivity and minimizing the rate of false negatives. In other words, because launching an astronaut with an undetected cancer would be unacceptable, testing is be done earlier and more frequently than in the general population.

New technologies offer more comprehensive screening. As mentioned earlier, one new screening technique is total-body MRI [57]. With total-body MRI, lesions as small as a few millimeters can be detected. The main difficulty with this approach is that in an asymptomatic, normal population, the testing is more likely to detect insignificant or incidental findings (e.g., cysts, adenomas, anatomic variants) than serious pathology. This in turn can lead to a variety of medical procedures (each with its own risk) to track down abnormal findings and establish they are not significant. Nonetheless, this kind of screening could (1) detect potentially significant abnormalities that would otherwise go unnoticed, (2) provide a database about the crew member that could be useful if medical problems should arise on a mission, and (3) replace other less sensitive screening tests (abdominal ultrasound, sinus x-rays, chest x-rays) that are currently done on astronauts. One way to use this technology is to perform a baseline study at initial selection and then have a follow-up study at an assigned time in the future. If there had been a questionable finding that changed in size over time, this might be a strong indication that the finding on the MRI represented significant pathology. The data from the MRI scans also could be used to

Table 12-2. Genetic monitoring tests to determine if an exposure to a toxin or environmental hazard has produced DNA damage.

Gene/test	Function	Comments
Sister chromatid exchange	Indication of DNA damage	Sister chromatid exchanges are examined in cells, relationship between sister chromatid exchanges and a defined health effect is not established
Micronuclei (detected by fluorescence in situ hybridization)	Micronuclei indicate previous chromosomal aberrations	General marker of exposure to toxic agents
Single-cell gel electrophoresis assay (COMET assay)	Detects DNA strand breaks	May indicate a significant toxic exposure
DNA repair assay (challenge assay)	Measures the ability of DNA to repair after exposure to an insult (e.g., radiation)	Cells previously exposed to hazardous agents or patients with faulty DNA repair mechanisms may be comprised in their ability to repair DNA in response to an insult
Host-cell reactivation assay	Measures DNA repair capability	Lymphocytes are transfected with damaged plasmid DNA; the host cell repairs the DNA to produce the gene product

Theoretically, cells from individuals with compromised DNA repair mechanisms would show increased damage on a challenge assay. The use of this kind of assay for screening, however, has not been fully studied. If a crew member had a clearly abnormal result on a test of DNA repair capability, this would present a tough selection decision. The crew member might be experiencing high radiation levels for 2–3 years as part of a Mars mission.

create anatomic models of the crew members that might be useful for diagnosis or training while in flight.

Another possible method to prevent cancer on a spaceflight is to use genetic markers to select potential crew members. Currently, a family history of an inherited cancer syndrome disqualifies a candidate for selection. This is a crude form of genetic screening. The advent of genetic testing makes it possible to identify gene mutations that correlate with an increase risk of cancer. These mutations are typically associated with tumor-suppressor genes, which are genes that produce proteins involved in regulating the cell cycle or in DNA repair. Table 12-1 lists some of the gene mutations that have been associated with an increased susceptibility to cancer. Some of these genes, such as the *BRCA2* gene, are associated with significant increases in cancer risk. Other gene mutations are either rare or of unknown significance. While genetic screening offers the potential to prevent cancer and to avoid exposing susceptible individuals to high cancer-risk environments (such as the radiation environment in interplanetary space), it also can present several problems. There can be several different types of mutations for a given gene, and the significance of each mutation may not be known. Although the links of some gene mutations to the development of cancer is strong, in other cases the links may be weak or tenuous.

Another possible cancer screening technique is to use functional testing of DNA repair mechanisms. The susceptibility to some cancers has been linked to a reduced ability to repair DNA. Bondy et al. [65], for example, demonstrated that patients with gliomas have a lower ability to repair DNA strand breaks. Similarly, Landi et al. [66] showed that DNA repair capacity could modify the risk of developing melanoma. Individuals with low tanning ability and a low DNA repair capability were much more likely to develop cutaneous melanoma compared to those with high tanning ability and high DNA repair capability. Table 12-2 lists some of the tests that can determine DNA repair capability. Just as is the case with genetic mutations, the tests are much easier to perform than they are to interpret. In many cases the significance of a particular test result cannot be known with precision.

Cardiovascular risk determination

High cholesterol remains a strong marker of cardiovascular disease risk. The National Cholesterol Education program stratifies cardiovascular risk into three groups based on the likelihood of a significant cardiac event over 10 years: >20%, 10–20%, <10% [34]. Astronauts should have their cholesterol checked and risk factors assessed, with the goal of staying in the low-risk category. A new approach to screening is the use of a coronary calcium score. Calcium in the coronary arteries is strongly correlated with subsequent cardiac events and the development of clinically important athero-sclerotic plague. The absence of coronary calcium, however, has a strong negative predictive value [67]. If an astronaut has an acceptable cholesterol level, no significant risk factors (hypertension, smoking, family history of coronary artery disease) and no coronary calcium, the risk of a cardiovascular event in that individual is quite low. The most commonly used test for coronary calcium is electron-beam computed tomography, which exposes the individual to a significant dose of radiation. Other approaches to assessing coronary arteries are multislice computed tomography and intimal thickening determined by ultrasound.

Kidney stone risk

Currently, astronauts have ultrasound scans of their kidneys to assess for occult stones. In addition, an assessment of kidney stone risk can be obtained from a 24-hour urine collection. In this collection, calcium, oxalate, citrate, sodium, potassium, uric acid, and other metabolites are measured. From this a stone-risk profile can be developed that indicates whether the supersaturation of the various salts is high enough to allow the development of kidney stones. The major urinary risk factors for stone formation include hypercalciuria, hyperoxaluria, hypocitraturia, and hyperuri-cosuria [16].

Patent foramen ovale/right-to-left shunts

Astronauts who transition between different atmospheric pressure environments run the risk of decompression sickness. If a crew member has a right-to-left shunt, the hazard is increased. The bubbles of decompression sickness could traverse the shunt and lodge in the central nervous system, producing neurological decompression sickness. The decision to screen for right-to-left shunts is complex because the incidence of decompression sickness is low, and neurological decompression sickness can occur even without a right-to-left shunt. An improved screening technique involving injecting an intravenous contrast solution while monitoring middle cerebral artery flow may provide a better functional assessment of a shunt than just ultrasound imaging of the heart. Although right-to-left shunts often can be repaired, the repair surgery exposes the astronaut to some medical risk. For a Mars mission medical planners will want a detailed study of every crew member's physiology before the flight. This may tilt the balance toward screening for right-to-left shunts.

Monitoring

Selection and screening are designed to minimize medical events by choosing the qualified individuals least likely to have a medical problem. Despite these precautions, however, medical events can still occur. Careful monitoring of the crew can determine when a medical risk is developing or increasing. For example, crew members who exercise infrequently or who do not diligently follow the countermeasure program may experience increases in urinary calcium, thereby elevating their kidney stone risk. Similarly, chronic conflict and sleep loss onboard may impair immune function. Astronauts who are not wearing hearing protection according to guidelines or who are particularly sensitive to noise-induced hearing loss may start to lose hearing. Astronauts will need the same kind of preventive services they would have had if they stayed on the ground (e.g., cholesterol testing).

Urinary composition monitoring

Urine contains several important biochemical markers that could be useful in tracking crew members. As discussed in chapter 1, urinary calcium increases markedly in weightlessness. This increase in urinary calcium is a marker for bone loss and indicates an increased risk for kidney stone formation. If urinary calcium levels could be tracked over time, then the crew members would have feedback on how well they were doing with their countermeasure program. Urine contains other more specific markers as well. An increased N-telopeptide level in the urine reflects bone breakdown. A variety of other hormones and cytokines in the urine could provide useful information on health.

The need for a variety of reagents and lab supplies for processing and analyzing urine limits the number of assays that can be done on a spacecraft. New, compact mass spectrometers that can measure a variety of biochemical compounds may offer the possibility of simpler and more frequent urinary monitoring.

Immune function monitoring

Several studies have shown that spaceflight can affect immune function, and thus a formal monitoring program would be worthwhile. Data on white cell populations, cytokine levels, and viral reactivation could be collected through blood samples. These data could help determine whether changes in immune function are an expected consequence of spaceflight or whether there is a correlation between changes in immune function and other mission events. Also, a straightforward total white blood cell count would be useful for diagnosing infections.

Hearing testing

Hearing testing is a component of most occupational health programs for workers exposed to noise. The best approach to minimizing hearing loss is strict adherence to acoustic standards. (These standards are specified for each piece of hardware aboard a spacecraft.) Experience, however, has shown that sometimes other factors within the space program can lead to these restrictions being waived. As a result, noise levels may increase and hearing testing may become necessary. Evaluating hearing in a noisy environment presents problems because threshold-based hearing testing requires a quiet environment, which can be hard to achieve on a spacecraft. One possible approach to testing in space is the use of otoacoustic emissions. This method of testing provides an objective, short, and non-threshold–based hearing assessment [21].

Blood chemistry monitoring

On Earth, a variety of blood chemistry tests are done on otherwise healthy people. Cholesterol testing is done routinely as part of physical exams and could potentially be useful for monitoring in space. Liver enzyme tests are done after a toxin exposure and to help diagnose conditions such as cholecystitis. Tests of kidney function, such as blood urea nitrogen and creatinine, are useful when there is a concern about urinary obstruction or kidney damage due to a medication reaction. Serum electrolyte levels, such as sodium and potassium, can be important if there is a cardiac arrhythmia or severe dehydration. These tests also may be useful in space. Devices such as the i-STAT portable clinical blood analyzer [68] and the Reflotron blood analyzer [69] have flown in space as part of the American and Russian space programs, respectively. They are used for routine health monitoring and to help with medical diagnosis.

Intervention Strategies

Although screening and monitoring can keep medical events at a low rate, there will still be a need for medical care. As the submarine, Antarctic, and space experience shows, significant medical and surgical events can occur even in a well-screened population. Medications will be needed to deal with infections (among other medical problems), and instruments will be needed if surgical care is required.

Medical Capabilities

Medical kits have been assembled for the Skylab, Salyut, Space Shuttle, Mir, and International Space Station programs. The composition of a typical medical kit is outlined in table 12-3. Kits have included a variety of analgesics, antibiotics, anxiolytics, antidepressants, and muscle relaxants. These medications are designed to pro-

Table 12-3. Classes of drugs that may be needed in a medical kit.

Class of medication	Examples	Rationale
Allergy treatments	Epinephrine injector; Benadryl, oral; prednisone, oral; albuterol inhaler; dexamethasone, IV	Toxins or other compounds in cabin atmosphere may lead to allergic reaction; dust on Moon or Mars may irritate lungs
Antibiotics	Ciprofloxacin, broad-spectrum oral; amoxicillin with clavulanate, oral; trimethoprim/Sulfamethoxaszole, oral; cephalexin, oral; ceftriaxone, broad-spectrum IV	Skin, urinary, and upper respiratory infections may occur; Antibiotics may be part of treatment for serious abdomenal conditions
Analgesics	Acetaminophen, oral; ibuprofen, oral; acetaminophen with hydrocodone, oral; Fentanyl, patch; morphine, IV; lidocaine, SC (local anesthetic)	Trauma and other illnesses may require pain control; lidocaine and other local anesthetics are useful for trauma and lacerations; morphine could be used during a heart attack and after trauma.
Antiemetic	Promethazine, IV or PR; prochlorperazine, PR	Vomiting may be part of an onboard illness
Stimulants	Dextroamphetamine, oral; methylphenidate, oral	Life-threatening fatigue in emergency situations
Cardiovascular drugs	Epinephrine, IV; atropine, IV; beta-blocker; calcium-channel blocker; tissue plasminogen activator (possible)	Drugs for cardiac resuscitation should be available and for the treatment of common arrhythmias, thrombolytics are the definitive treatment for myocardial infarction
Ophthalmic preparations	Proparacaine drops; fluorescein; dexamethasone/neomycin/polymyxin drops	Corneal abrasions may need to be diagnosed and treated. Eye infections may occur
Otic preparations	Neomycin/polymyxin/hydrocortisone drops; Cerumenex; acetic acid/hydrocortisone drops	Ear canal irritation can lead to infection; ear wax may accumulate over time and require removal
Laxatives/Antidiarrheals	Senna, laxative; loperaminde, antidiarrheal	Lack of bulk in diet may lead to constipation, diarrhea may occur
Hypnotics	Zoleplon; zolpidem	Sleep medication may be needed in stressful situations or when circadian rhythms shift

Drugs for motion sickness, anxiety, depression, bone loss, and radiation protection are discussed in other chapters. In addition to the usual over-the-counter remedies (e.g., ibuprofen, bismuth subsalicylate), various prescription drugs should be available for a variety of conditions. The amount of a drug to bring on the mission depends on the mission length and crew size.

vide a wide range of medical capabilities within a small volume. For a Mars or other long-duration exploration mission, one unique problem facing medical planners is extending the shelf-life of the drugs onboard. For example, propofol, an excellent anesthetic agent, is provided as a lipid emulsion with a shelf life of approximately 24 months and has limited tolerance of temperature extremes. The amount of a given drug to be included in the kit is also an issue. If a crew member develops a chronic disease, such as an allergy, that requires medication on a regular schedule, this might overtax the contents of the kit. Sufficient quantities of supplements that might be used for the entire mission (e.g., vitamins, antioxidants) would need to be onboard.

Ultrasound

One technology that may be very useful for medical diagnosis in space is ultrasound. Several ultrasound units have flown in space, and the International Space Station has a multipurpose ultrasound onboard. Ultrasound has a variety of uses both for monitoring and diagnosis. Cardiac ultrasound could provide a measure of cardiac mass, which may be important in following the cardiovascular adaptation to weightlessness (chapter 7). Abdominal ultrasound could be used to detect kidney stones, urinary obstruction, and gallstones. In addition, ultrasound may be useful in the diagnosis of appendicitis, an intra-abdominal abscess, or fluid collection. Thoracic ultrasound may help determine if a pneumo- or hemothorax is present [70–72]. Ultrasound can also be used to image the vascular system, eyes, testes, and muscles. Technological advances have reduced the size of ultrasound machines to the point where handheld units are now available.

Emergency/Surgical Care Capabilities

Trauma on a spacecraft or on the Moon or Mars could lead to bleeding, loss of consciousness, and perhaps pneumo- or hemothorax. Abdominal surgery may be needed. The ability to provide advanced trauma life support and advanced cardiac life support may be critical. Table 12-4 lists some of the devices that are needed to provide a minimal emergency care capability. These items (with the exception of the defibrillator) are recommended for high-risk wilderness outings when individuals with special medical training are available [73].

Advances in the technology for minimally invasive surgery make the provision of advanced surgical care a possibility, even when the space available for stowage is small. Laparoscopic surgical techniques allow for appendectomies and other abdominal surgery to be done without a large abdominal incision. This capability, however, comes at a significant training cost. Performing minimally invasive surgery successfully requires training and proficiency that needs to be maintained over time. Providing this training requires time and resources.

Airway control

In the event of a serious emergency or if anesthesia is administered for surgery, the crew will need to maintain positive control over the airway. In a hospital setting, the most frequent choice is endotracheal intubation, which offers the best airway control and protects the trachea from aspiration. This is also the most technically demanding method and may not be essential. Different airways were studied by Keller et al. [74] in a neutral buoyancy tank using manikins. This study showed that endotracheal intubation had the highest failure rate, compared to the cuffed oropharyngeal airway, the standard laryngeal mask airway, and the intubating laryngeal mask airway. The

Table 12-4. Devices whose use requires medical training that may be useful on a space mission.

Problem	Devices	Comments
Airway management	Laryngeal mask airway, cricothyroidotomy kit; oxygen	Laryngeal mask provides airway control in case of resuscitation or surgery; cricothyroidotomy may be necessary in emergencies or with trauma
Hemo/Pneumo thorax	Chest tube kit	Trauma may lead to fluid or air in the chest
Eye injuries	Ophthalmoscope	With blue filter and fluorescein strips, corneal abrasions can be diagnosed
Secretions/fluid accumulations	Suction device	Suction will help with surgery, chest tubes, resuscitation
Surgery	Needles; Forceps; Scissors Clamps; Tool sterilization; Wound adhesive	Minor surgery, such as draining abscesses or closing wounds, will require a basic surgical kit; the tools will need to be sterilized
Vital signs monitoring	Blood pressure (manual cuff or automatic device); Heart rate (may be incorporated in pulse oximeter); Pulse oximeter (useful for surgical monitoring)	Vital signs will be needed during resuscitation or surgery
Cardiac resuscitation	Defibrillator with transcutaneous pacer	Essential for advanced cardiac life support
Urinary retention	Foley catheter	Drugs onboard, such as the antihistamines, anthicholenergics, or adrenergic agents, may lead to urinary retention; a catheter may be needed in cases of trauma or for surgery

laryngeal mask airway may offer the best choice for space applications because of its ease of use [62].

Anesthesia

Inhalational anesthetics can present a problem in the spacecraft because the volatile gases could be expelled into the cabin atmosphere and would have to be scrubbed out by the atmospheric control system [62]. Several good intravenous anesthetics exist that could be readily adapted to spaceflight. Propofol, for example, is a short-acting anesthetic used frequently in day surgery and other applications. For longer lasting anesthesia, intravenous pentobarbital is also an option. Short-acting intravenous narcotics (Fentanyl) can help with pain control. Anesthetic infusion pumps, such as might be used to administer intravenous anesthesia, have already been used in space as part of science experiments on the Neurolab mission. Injection anesthesia has been used successfully on rodents in space as part of the *Spacelab Life*

Sciences-2 and *Neurolab* missions [60]. The insertion of intravenous lines has been part of several spaceflights, particularly the *Spacelab Life Sciences-1*, *Spacelab Life Sciences-2*, *Deutsche-2*, and *Neurolab* missions. With proper training, it is likely that a crew could provide adequate anesthesia using intravenous anesthetics and a laryngeal mask airway.

Advanced trauma life support/ advanced cardiac life support capability

Considerable advanced trauma and cardiac life support capability already exist on most space stations. A defibrillator is onboard the International Space Station. Medical kits include airways, intravenous lines, chest tubes, and so on. The feasibility of performing several advanced trauma life support procedures (including artificial ventilation, intravenous infusion, laceration closure, tracheotomy, Foley catheter drainage, chest tube insertion, and peritoneal lavage) has been demonstrated on animals in weightlessness aboard the KC-135 aircraft [75]. Rapid intravenous infusions have been done in space. The *Deutsche-2* Spacelab mission included an experiment where 2 l of saline were rapidly administered [76]. Although this was done to study the physiology of fluid balance in weightlessness, the experiment also demonstrated that it is possible to infuse large amounts of fluid in space, such as might be needed in case of an episode of major trauma.

One unresolved issue in trauma life support during spaceflight is the provision of blood products in case of a major hemorrhage. For a long-duration mission, if blood products are considered important to have onboard, they will need to have a long shelf life. Artificial blood (such as a fluorocarbon emulsion) might be a potentially useful approach to providing an oxygen-carrying fluid with a long shelf life.

Laparoscopic surgery

A laparoscope is a fiber optic scope that can be passed through a small incision in the skin (typically the abdominal wall) to view internal anatomy. Tools such as scissors, staplers and forceps can be passed through the scope and manipulated by the operator. In laparoscopic abdominal procedures, the abdomen is usually filled with carbon dioxide gas to create a pneumoperitoneum that improves visibility [77]. Laparoscopic surgical techniques have been demonstrated in short periods of weightlessness on the KC-135 aircraft [78]. Abdominal laparoscopic surgery was found to be feasible, although thorascopy was difficult. Also, the round shape of the abdominal cavity during 0-*G* exposure increased the anterior to posterior diameter compared to 1-*G*, which improved laparoscopic visualization and manipulation. Laparoscopic surgical capability on a spacecraft might allow for definitive treatment in cases of abdominal trauma, cholecystitis, or appendicitis (among other conditions). Still, it is an open question whether the investment in training and equipment needed for this technique would be worthwhile. The cost of providing laparoscopic surgery capability would have to be balanced against the likelihood that problems requiring surgical intervention would occur in weightlessness. For long-term stays on a planetary surface, a laparoscopic capability may make sense. Other options include using focused ultrasound to stop bleeding that would otherwise require laparoscopy [63].

Training

If the selection and prevention strategies work, the demand for medical and surgical care on a long-duration mission should be low. This means, however, that if a medical or surgical situation should arise, no one on the crew would be likely to have had any recent medical and surgical experience to draw upon. On Earth medical practitioners evaluate patients daily and so maintain proficiency and currency with medical problems. On a space mission, however, the crew members will spend most of their time involved in mission tasks. For crew members to be ready to deal with medical and surgical emergencies, they will need to use simulations, workbooks, and texts to maintain proficiency.

One possible approach to keeping the medical officers on the spacecraft well-trained is the development of "virtual mentor." This would be a computer-based system that could present the crew members with a variety of medical or surgical problems and have them work through the issues on the simulator. Simulators that allow physicians to practice laparoscopic surgical skills have been built, and some studies show that the use of the simulators by surgeons reduces errors and improves performance [79].

Recommendations Based on Current Knowledge

The best approach to medical and surgical events on a spaceflight is to prevent them from occurring in the first place. The overhead in resources and training for providing complex medical care is high. To keep medical risks low while providing the best capability to deal with the medical problems that do arise, the following recommendations can be made:

1. Standard field hygiene measures should be carefully followed. The spacecraft should be kept clean and free of areas where excessive microbial buildup can occur. Food, water, and air should be kept free of contamination. A program of microbial monitoring on the spacecraft should be in place.

2. Crews need to be trained in how to recognize and prevent common psychological problems. Ongoing refresher training should be provided onboard.

3. Immunological parameters should be monitored in a way that allows the findings to be interpreted within the context of a particular flight. The significance of immunological changes that have been noted in space remains unclear, and the best solution to this problem is careful monitoring.

4. Urine should be monitored for kidney-stone risk parameters (urinary calcium would be perhaps the single most important measure). If the urinary supersaturation of calcium compounds is kept within an acceptable range, the likelihood of forming a kidney stone is quite low. Potassium citrate could be taken routinely to minimize the risk of stone formation.

5. A hyperbaric chamber should be provided if the rate of EVA is high (i.e., on a planetary surface) and the crews will be moving frequently between different pressure environments.

6. Routine hearing testing should be performed to detect any early changes in hearing.

7. An aggressive preventive dentistry program should be followed, which could include fluoride rinses, chlorhexidine mouthwash, frequent brushing, and frequent flossing. The capability to provide dental anesthesia, pull teeth, and place fillings should be provided. A simulator to provide refresher training in these techniques would also be important. Training in the use of atraumatic restorative treatment could be considered.

8. For long voyages beyond Earth orbit, crews should be screening aggressively for cardiovascular disease and cancer risks. Measures of coronary calcium may offer a reliable way to establish the future risk of cardiovascular disease. If the risk of cardiovascular disease can be lowered sufficiently, it might be possible to minimize the cardiovascular treatment capability and save space onboard. In addition to standard cancer screening techniques, total-body MRI may offer a way both to screen for cancer and to provide a detailed anatomic model of a crew member. This baseline MRI may be useful if medical issues occur in flight.

9. A genomic database may be needed for individuals who fly on exploration-class (long-duration Moon or Mars) missions in space. Genetic testing has significant problems. The results of the testing will be hard to assess. Individuals may be removed from flight status for questionable genetic testing findings that may subsequently be found to be insignificant. Ultimately, however, the value of genetic testing for screening and medical care cannot be established unless the data are collected. The crew members would need to know at the time they enter they program that they are part of an ongoing research program on the medical aspects of long-duration spaceflight and that this program includes genetic testing. This program carries along with it the risk of medical disqualification based on imperfect data. The goal would be to find, over time, the genetic tests that have validity for screening and medical care.

10. Functional testing of DNA repair capability should be considered for those embarking on long-duration interplanetary flights. One major drawback of this approach is that it is not clear where to draw the line between a "normal" and "abnormal" result. This would require a consensus or "best practices" approach until more reliable data could be obtained.

11. Crew members should follow an aggressive cardiovascular and cancer prevention program. This would include a low-cholesterol, high-fiber diet and regular exercise. The onboard food system should be designed to provide adequate intake of essential vitamins and minerals. At present, there is not enough evidence to support taking medications, such as aspirin, for preventing colon cancer or cardiovascular events in a low-risk population such as the astronaut corps.

12. The medical kit should provide an adequate mix of medications to treat common medical conditions. Methods to provide the longest possible shelf life will be needed. The ability to measure blood parameters should be provided.

13. For exploration-class missions, the risk of surgical illness could possibly be reduced by a prophylactic appendectomy and the preflight removal of any existing gallstones. This approach could be considered if mission constraints make the provision of advanced surgical care impossible. Laparoscopic surgery capability could be considered if space is available and the ability to maintain proficiency exists.

14. The equipment to provide for emergency care (airways, defibrillator, intravenous fluids, surgical instruments, anesthetics) would be needed for cases of

trauma or illness. A training system also would be essential to provide refresher training throughout the flight so that proficiency in dealing with emergencies could be maintained.

References

1. Nielsen, J., *Ice Bound.* 2001, Hyperion, New York.
2. Tansey, W.A., J.M. Wilson, and K.E. Schaefer, Analysis of health data from 10 years of Polaris submarine patrols. Undersea Biomedical Research, 1979. 6 (Suppl): S217–46.
3. Lugg, D.J., Antarctic medicine. Journal of the American Medical Association, 2000. 283(16): 2082–84.
4. Haines, R., *Life and Death in the Age of Sail.* 2003, University of New South Wales Press, Sydney
5. Billica, R.D., et al., Perception of the medical risk of spaceflight. Aviation, Space, and Environmental Medicine, 1996. 67(5): 467–73.
6. Lugg, D., and M. Shepanek, Space analogue studies in Antarctica. Acta Astronautica, 1999. 44(7–12): 693–99.
7. Sonnenfeld, G., Extreme environments and the immune system: effects of spaceflight on immune responses. Journal of Allergy and Clinical Immunology, 2001. 107(1): 19–20.
8. Borchers, A.T., C.L. Keen, and M.E. Gershwin, Microgravity and immune responsiveness: implications for space travel. Nutrition, 2002. 18(10): 889–98.
9. Sonnenfeld, G., Space flight, microgravity, stress, and immune responses. Advances in Space Research, 1999. 23(12): 1945–53.
10. Chouker, A., et al., Effects of confinement (110 and 240 days) on neuroendocrine stress response and changes of immune cells in men. Journal of Applied Physiology, 2002. 92(4): 1619–27.
11. Whitson, P.A., R.A. Pietrzyk, and C.Y. Pak, Renal stone risk assessment during Space Shuttle flights. Journal of Urology, 1997. 158: 2305–10.
12. Lebedev, V., *Diary of a Cosmonaut: 211 Days in Space*, ed. D. Puckett and C.W. Harrison. 1988, PhytoResource Research, College Station, TX.
13. Ruml, L.A., et al., Prevention of hypercalciuria and stone-forming propensity during prolonged bedrest by alendronate. Journal of Bone and Mineral Research, 1995. 10: 655–62.
14. Yendt, E.R., and M. Cohanim, Prevention of calcium stones with thiazides. Kidney International, 1978. 13: 397–409.
15. Lingeman, J., et al., Medical reduction of stone risk in a network of treatment centers compared to a research clinic. Journal of Urology, 1998. 160(5): 1629–34.
16. Goldfarb, D.S., and F.L. Coe, Prevention of recurrent nephrolithiasis. American Family Physician, 1999. 60(8): 2269–76.
17. Moon, R.E. and D.F. Gorman, Treatment of the decompression disorders, in *Bennett and Elliot's Physiology and Medicine of Diving*, A.O. Brubakk and T.S. Neuman, eds. 2003, Saunders, New York, pp. 600–50.
18. Dromsky, D.M., B.D. Spiess, and A. Fahlman, Treatment of decompression sickness in swine with intravenous perfluorocarbon emulsion. Aviation, Space, and Environmental Medicine, 2004. 75(4): 301–5.
19. Spiess, B.D., et al., Treatment of decompression sickness with a perfluorocarbon emulsion (FC-43). Undersea Biomedical Research, 1988. 15(1): 31–37.
20. Powell, M.R., et al., Extravehicular activities, in *Space Physiology and Medicine*, A.E. Nicogossian, C.L. Huntoon, and S.L. Pool, eds. 1994, Williams and Wilkins, Baltimore, MD, pp. 128–40.
21. Buckey, J.C., et al., Hearing loss in space. Aviation, Space, and Environmental Medicine, 2001. 72(12): 1121–24.
22. Newkirk, D., *Almanac of Soviet Manned Space Flight.* 1990, Gulf Publishing, Houston, TX.

23. Glover, S.D., and E.W. Taylor, Surgical problems presenting at sea during 100 British Polaris submarine patrols. Journal of the Royal Naval Medical Service, 1981. 67(2): 65–69.
24. Lisney, S.J., Dental problems in Antarctica. British Dental Journal, 1976. 141(3): 91–92.
25. Brown, L.R., et al., Skylab Oral Health Studies, in *Biomedical Results from Skylab*, R.S. Johnston and L.F. Dietlein, eds. 1977, NASA, Washington, DC, pp. 35–44.
26. Institute of Medicine, Managing risks to astronaut health, in *Safe Passage: Astronaut Care for Exploration Missions*, J.R. Ball and C.H. Evans, eds. 2001, National Academy Press, Washington, DC, pp. 75–116.
27. Rask, P.I., et al., Effect of preventive measures in 50–60-year-olds with a high risk of dental caries. Scandinavian Journal of Dental Research, 1988. 96(6): 500–4.
28. NCHS, Death: leading causes by age, race and sex. 2000, National Center for Health Statistics, Hyattsville, MD.
29. Peterson, L.E., et al., Longitudinal study of astronaut health: mortality in the years 1959–1991. Radiation Research, 1993. 133(2): 257–64.
30. Hamm, P.B., et al., Risk of cancer mortality among the Longitudinal Study of Astronaut Health (LSAH) participants. Aviation, Space, and Environmental Medicine, 1998. 69(2): 142–44.
31. Billica, R., S.L. Pool, and A.E. Nicogossian, Crew Health-Care Programs, in *Space Physiology and Medicine*, A.E. Nicogossian, C.L. Huntoon, and S.L. Pool, eds. 1994, Williams and Wilkins, Baltimore, MD, pp. 402–423.
32. Pierson, D.L., Microbiology, in *Space Physiology and Medicine*, A.E. Nicogossian, C.L. Huntoon, and S.L. Pool, eds. 1994, Williams and Wilkins, Baltimore, MD, pp. 157–66.
33. Osswald, S., et al., Review of cardiac events in USAF aviators. Aviation, Space, and Environmental Medicine, 1996. 67(11): 1023–27.
34. NCEP, Third report of the National Cholesterol Education Program (NCEP) on detection, evaluation and treatment of high blood cholesterol in adults. 2002, National Heart Lung and Blood Institute, Bethesda, MD.
35. MacIntyre, N.R., et al., Eight-year follow-up of exercise electrocardiograms in healthy, middle-aged aviators. Aviation, Space, and Environmental Medicine, 1981. 52(4): 256–59.
36. Redberg, R.F., and L.J. Shaw, A review of electron beam computed tomography: implications for coronary artery disease screening. Preventive Cardiology, 2002. 5(2): 71–8.
37. Hauser, E.R., et al., Design of the Genetics of Early Onset Cardiovascular Disease (GENECARD) study. American Heart Journal, 2003. 145(4): 602–13.
38. Gauthier, G.M., J.G. Keevil, and P.E. McBride, The association of homocysteine and coronary artery disease. Clinical Cardiology, 2003. 26(12): 563–68.
39. Bays, H., and E.A. Stein, Pharmacotherapy for dyslipidaemia—current therapies and future agents. Expert Opinion on Pharmacotherapy, 2003. 4(11): 1901–38.
40. de Jong, A., J. Plat, and R.P. Mensink, Metabolic effects of plant sterols and stanols. Journal of Nutritional Biochemistry, 2003. 14(7): 362–9.
41. Katan, M.B., et al., Efficacy and safety of plant stanols and sterols in the management of blood cholesterol levels. Mayo Clinic Proceedings, 2003. 78(8): 965–78.
42. Mundy, G., et al., Stimulation of bone formation in vitro and in rodents by statins. Science, 1999. 286(5446): 1946–49.
43. Chan, K.A., et al., Inhibitors of hydroxymethylglutaryl-coenzyme A reductase and risk of fracture among older women. Lancet, 2000. 355(9222): 2185–88.
44. Edwards, C.J., D.J. Hart, and T.D. Spector, Oral statins and increased bone-mineral density in postmenopausal women [letter]. Lancet, 2000. 355(9222): 2218–19.
45. Meier, C.R., et al., HMG-CoA reductase inhibitors and the risk of fractures. Journal of the American Medical Association, 2000. 283(24): 3205–10.
46. Wang, P.P.S., et al., HMG-CoA reductase inhibitors and the risk of hip fractures in elderly patients. Journal of the American Medical Association, 2000. 283(24): 3211–16.
47. Jackson, P.R., et al., Statins for primary prevention: at what coronary risk is safety assured? British Journal of Clinical Pharmacology, 2001. 52(4): 439–46.

48. Longo, D.L., Approach to the patient with cancer, in *Harrison's Principles of Internal Medicine*, A.S. Fauci, et al., eds. 1998, McGraw-Hill, New York, pp. 492–98.
49. Blettner, M., et al., Mortality from cancer and other causes among male airline cockpit crew in Europe. International Journal of Cancer, 2003. 106(6): 946–52.
50. Pukkala, E., et al., Cancer incidence among 10,211 airline pilots: a Nordic study. Aviation, Space, and Environmental Medicine, 2003. 74(7): 699–706.
51. Rafnsson, V., J. Hrafnkelsson, and H. Tulinius, Incidence of cancer among commercial airline pilots. Occupational and Environmental Medicine, 2000. 57(3): 175–79.
52. Keku, T.O., T. Rakhra-Burris, and R. Millikan, Gene testing: what the health professional needs to know. Journal of Nutrition, 2003. 133(11 Suppl 1): 3754S-57S.
53. McKelvey, K.D., Jr., and J.P. Evans, Cancer genetics in primary care. Journal of Nutrition, 2003. 133(11 Suppl 1): 3767S-72S.
54. Schiffman, M.H., et al., Epidemiologic evidence showing that human papillomavirus infection causes most cervical intraepithelial neoplasia. Journal of the National Cancer Institute, 1993. 85(12): 958–64.
55. Lagergren, J., et al., Symptomatic gastroesophageal reflux as a risk factor for esophageal adenocarcinoma. New England Journal of Medicine, 1999. 340(11): 825–31.
56. U.S. Preventive Services Task Force, Guide to clinical preventive services. 2004, Agency for Healthcare Research and Quality, Rockville, MD.
57. Kavanagh, E., C. Smith, and S. Eustace, Whole-body turbo STIR MR imaging: controversies and avenues for development. European Radiology, 2003. 13(9): 2196–205.
58. Key, T.J., et al., Diet, nutrition and the prevention of cancer. Public Health Nutrition, 2004. 7(1A): 187–200.
59. Thomson, C.A., et al., Nutrition and diet in the development of gastrointestinal cancer. Current Oncology Reports, 2003. 5(3): 192–202.
60. Buckey, J.C., D.R. Williams, and D.A. Riley, Surgery and recovery in space, in *The Neurolab Spacelab Mission: Neuroscience Research in Space*, J.C. Buckey and J.L. Homick, eds. 2003, NASA, Houston, TX, pp. 275–78.
61. Rutkow, I.M., Surgical operations in the United States. Then (1983) and now (1994). Archives of Surgery, 1997. 132(9): 983–90.
62. Institute of Medicine, Emergency and continuing care, in *Safe Passage: Astronaut Care for Exploration Missions*, J.R. Ball and C.H. Evans, eds. 2001, National Academy Press, Washington, DC, pp. 117–35.
63. Noble, M.L., et al., Spleen hemostasis using high-intensity ultrasound: survival and healing. Journal of Trauma, 2002. 53(6): 1115–20.
64. Pool, S.L., et al., Medical evaluations for astronaut selection and longitudinal studies, in *Space Physiology and Medicine*, A.E. Nicogossian, C.L. Huntoon, and S.L. Pool, eds. 1994, Williams and Wilkins, Baltimore, MD, pp. 375–93.
65. Bondy, M.L., et al., Gamma-radiation sensitivity and risk of glioma. Journal of the National Cancer Institute, 2001. 93(20): 1553–57.
66. Landi, M.T., et al., DNA repair, dysplastic nevi, and sunlight sensitivity in the development of cutaneous malignant melanoma. Journal of the National Cancer Institute, 2002. 94(2): 94–101.
67. Schmermund, A., S. Mohlenkamp, and R. Erbel, Coronary artery calcium and its relationship to coronary artery disease. Cardiology Clinics, 2003. 21(4): 521–34.
68. Smith, S.M., et al., Assessment of a portable clinical blood analyzer during space flight. Clinical Chemistry, 1997. 43: 1056–65.
69. Markin, A., et al., The dynamics of blood biochemical parameters in cosmonauts during long-term space flights. Acta Astronautica, 1998. 42(1–8): 247–53.
70. Dulchavsky, S.A., et al., Prospective evaluation of thoracic ultrasound in the detection of pneumothorax. Journal of Trauma, 2001. 50(2): 201–5.
71. Hamilton, D.R., et al., Sonographic detection of pneumothorax and hemothorax in microgravity. Aviation, Space, and Environmental Medicine, 2004. 75(3): 272–77.
72. Sargsyan, A.E., et al., Ultrasound evaluation of the magnitude of pneumothorax: a new concept. American Surgery, 2001. 67(3): 232–35 [discussion 235–36].

73. Zell, S.C., and P.H. Goodman, Wilderness Preparation, equipment, and medical supplies, in *Wilderness Medicine*, P.S. Auerbach, ed. 2001, Mosby, Philadelphia, pp. 1662–85.
74. Keller, C., et al., Airway management during spaceflight: a comparison of four airway devices in simulated microgravity. Anesthesiology, 2000. 92(5): 1237–41.
75. Campbell, M.R., et al., Performance of advanced trauma life support procedures in microgravity. Aviation, Space, and Environmental Medicine, 2002. 73(9): 907–12.
76. Arbeille, P., et al., Effect of microgravity on renal and femoral flows during LBNP & intravenous saline load. Journal of Gravitational Physiology, 1996. 3(2): 91–2.
77. Eubanks, S., and P.R. Schauer, Laparoscopic surgery, in *Textbook of Surgery. The Physiological Basis of Modern Surgical Practice*, D.C. Sabiston and H.K. Lyerly, eds. 1997, W.B. Saunders, Philadelphia, pp. 791–807.
78. Campbell, M.R., et al., Endoscopic surgery in weightlessness: the investigation of basic principles for surgery in space. Surgical Endoscopy, 2001. 15(12): 1413–18.
79. Grantcharov, T.P., et al., Randomized clinical trial of virtual reality simulation for laparoscopic skills training. British Journal of Surgery, 2004. 91(2): 146–50.

Index

Printed in the USA/Agawam, MA
January 31, 2020

749276.005